CHAOS AND COMPLEXITY

CHAOS AND COMPLEXITY RESEARCH COMPENDIUM, VOLUME 1

Chaos and Complexity

Series Editors: Franco F. Orsucci and Nicoletta Sala

This new series presents leading-edge research on artificial life, cellular automata, chaos theory, cognition, complexity theory, synchronization, fractals, genetic algorithms, information systems, metaphors, neural networks, non-linear dynamics, parallel computation and synergetics. The unifying feature of this research is the tie to chaos and complexity.

Chaos and Complexity Research Compendium, Volume 1
2011. ISBN: 978-1-60456-787-8

Chaos and Complexity Research Compendium, Volume 2
2011. ISBN: 978-1-60456-750-2

CHAOS AND COMPLEXITY

CHAOS AND COMPLEXITY RESEARCH COMPENDIUM, VOLUME 1

FRANCO F. ORSUCCI
AND
NICOLETTA SALA
EDITORS

Nova Science Publishers, Inc.
New York

For permission to use material from this book please contact us:
Telephone 631-231-7269; Fax 631-231-8175
Web Site: http://www.novapublishers.com

LIBRARY OF CONGRESS CATALOGING-IN-PUBLICATION DATA

ISSN: 2158-8066
ISBN: 978-1-60456-787-8

Published by Nova Science Publishers, Inc. † New York

CONTENTS

PREFACE

This new book presents leading-edge research on artificial life, cellular automata, chaos theory, cognition, complexity theory, synchronization, fractals, genetic algorithms, information systems, metaphors, neural networks, non-linear dynamics, parallel computation and synergetics. The unifying feature of this research is the tie to chaos and complexity.

In: Chaos and Complexity Research Compendium
Editors: F.F. Orsucci and N. Sala, pp. 1-7
ISBN: 978-1-60456-787-8

Chapter 1

EDITORIAL

Franco F. Orsucci
University College, London

For his course is not round; nor can the Sunne
Perfit a Circle or maintaine his way
One inche direct; but where he rose to day
He comes no more, but with a cousening line,
Steales by that point, and so is Serpentine.

John Donne, An Anatomie of the World, 1611

A State of the Art

The ancient English of these verses brings a remarkable insight we ought to the poet John Donne. This poem highlights how the consciousness of complexity has been present for very long times in human cultures, even long time before these verses. It is quite recently, however, that it has become suitable of a scientific approach.

John von Neumann, circa 1950, affirmed:

> "All stable processes, we shall predict. All unstable processes, we shall control." (cited in Dubè, 2000).

But, in his 1985 Giord Lectures, Freeman Dyson (1988) expressed his quite different opinion:

> "A chaotic motion is generally neither predictable nor controllable. It is unpredictable because a small disturbance will produce exponentially growing perturbation of the motion. It is uncontrollable because small disturbances lead only to other chaotic motions and not to any stable and predictable alternative."

Was Von Neumann's a mistake to imagine that every unstable motion could be nudged into a stable motion by ***small pushes and pulls applied at the right places*****?** If chaos is one of the possible marriages between order and disorder, habit and life, how much is it possible ***taming complex systems***?

The enterprise is still at the beginning, and the science of complexity, at this stage of development, still resembles a sea of ignorance with some small islands where results are known and applicable. The important thing is that now we accept that this sea exists and we can explore it.

This is well represented in the sketch below that we owe to Francisco Varela (1991) and Thomas Schreiber (Schreiber, 1999).

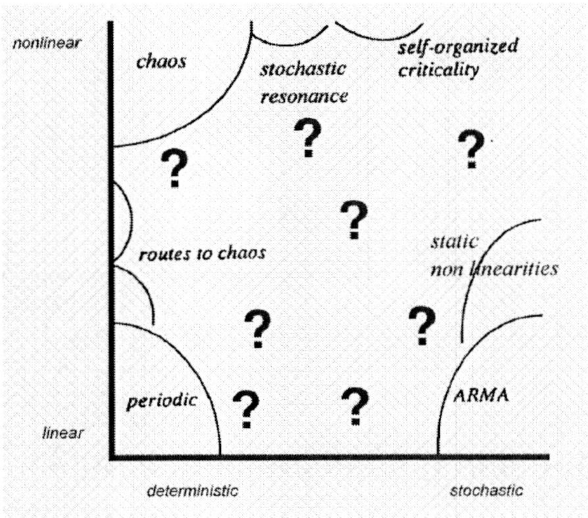

A state of the art in complexity theory (Varela, 1991; Schreiber, 1999)

Complexity in Metaphor

These small islands of knowledge called ***chaos, SOC or stochastic resonance*** are like candles in the darkness: we can finally have intuitions of the elephant's whole shape.

There is a Sufi tale called ***The Elephant in the Dark*****:**

> Some Hindus had brought an elephant for exhibition and placed it in a dark house. Crowds of people were going into that dark place to see the unknown beast. Finding that ocular inspection was impossible, each visitor felt it with his palm in the darkness.
>
> The palm of one fell on the trunk.
> 'This creature is like a water-spout,' he said.
> The hand of another lighted on the elephant's ear.
> To him the beast was evidently like a fan.
> Another rubbed against its leg.
> 'I found the elephant's shape is like a pillar', he said.
> Another laid his hand on its back.
> 'Certainly this elephant was like a throne', he said.

"The sensual eye is just like the palm of the hand. The palm has not the means of covering the whole of the best. The eye of the Sea is one thing and the foam another. Let the foam go, and gaze with the eye of the Sea. Day and night foam-flecks are flung from the sea: amazing! You behold the foam but not the Sea.

We are like boats dashing together; our eyes are darkened, yet we are in clear water" (Rumi, 1995).

16th century Chinese painting about the story.

The enterprise is crucial but not new. Classical knowledge was called ***Philosophia Naturalis***, ***Naturwissenschaften*** or ***Natural Philosophy***, until it had been fragmented and oversimplified in many sub-disciplines. Natural Philosophy, instead, had been rooted on the integration of different ways to approach a reality recognized as complex and multi-ordered.

The Sufi tale is clear: when the object is so large and complex, your perspectives can be partial and misleading. Sometimes the attempts to handle complexity have produced just the abuse of Occam's tools, with some risks for the epistemological survival of the object.

Dynamical Systems Theory

The gradual emergence of a set of formal and methodological tools called Dynamical Systems Theory, or Complexity Theory could finally make the scene different. This discipline has been also called Nonlinear Science as a marker of the shift in scientific paradigms (Kuhn, 1996). The name by exclusion might seem surprising for someone, as Stanislaw Ulam said: "Calling a science 'nonlinear' is like calling zoology 'the study of non-human animals' ". Anyway, the shift in scientific paradigms has been strong: a real scientific revolution.

Almost every system is complex, dynamical and nonlinear by nature. It is the limit of our approaches or our deliberate choice that makes us see them as linear. Yet the distinction has been necessary in the natural sciences, which became so accustomed to linear systems, because they were more treatable: linearization could "tame" their "wild" complexities.

The discovery of the possible scientific study of dynamical behaviors unrestricted by linearity is one of the greatest scientific revolutions of all times. It is becoming even a revolution in our everyday perception of reality, as trees and lightning become scientific objects under the name of fractals, just as cubes and cones have been for centuries.

It is clear that Maxwell and Boltzmann, the founders of statistical physics, were acutely aware of the property of sensitivity to initial conditions and its consequences. Not before Poincaré (1892) however, could ascertain the existence of this property in a system with few degrees of freedom, namely the reduced 3-body problem. In the continuing history of nonlinear dynamical systems, the first evidence of physical chaos is associated with the name of Edward Lorenz (1963; 1994) whose discovery of the first strange attractor in a simplified meteorological model containing only 3 state variables has led to a remarkable explosion in the study of chaos and its properties.

More recent years have seen the definition of a new frontier in complexity studies: the theory and application of control and synchronization. This Journal is proud to include in its Board many of the founders of this new wave of nonlinear studies.

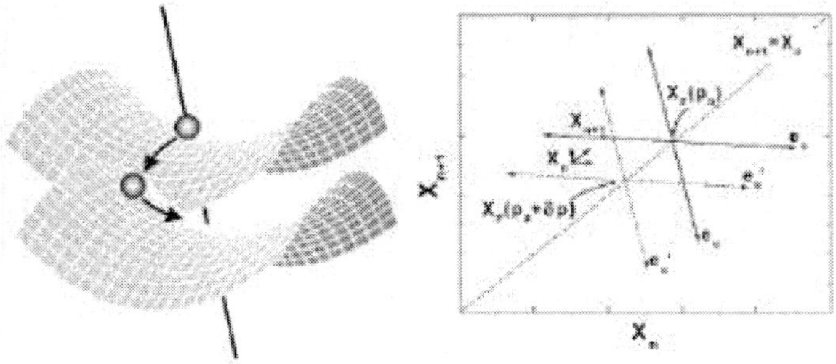

Local geometry of control: left 2D saddle dynamics and right linearization of the stable and unstable manifolds (Dubè, 2000)

Variation and Selection

Chaos theory becomes also a crucial way to understand some deep implications of Darwin's research on biological laws. If chaos is a source of optimal variation, targeting desirable states within chaotic attractors is a preliminary phase of selection and co-evolution. One of the major problems in the above process is that one can switch on the control only when the system is sufficiently close to the desired behavior. This is warranted by the ergodicity of chaos regardless of the initial condition chosen for the chaotic evolution, but it may happen that the small neighborhood of a given attractor point (target) may be visited only infrequently, because of the locally small probability function.

This is just one of the many questions that are still open. David Ruelle (1994), almost ten years ago, summarized some methodological *caveats:* "Suppose that you have concocted a mathematical model in biology or economics; you put this model on your computer and you discover a Feigenbaum period-doubling cascade (…) is this result interesting?" He answered that probably it hasn't a lot of interest: you should care about the relation between your model and the real empirical situations. Real systems are not directly equivalent to computer models: "Computer study of a model is an important method of investigation, but the results can be only as good as the model".

However, the greatest challenge will remain for some time the application to complex biological systems: in particular to mind and brain dynamics (Freeman, 1999; Guastello, 1995; Orsucci, 2003).

The perspective of unifying the techniques of deterministic chaos control with a statistical description as a possible therapeutic strategy against dynamical diseases is the challenge for next years.

The following is a photo about the meeting between a man and some cetaceans, in a quasi topological and dynamical presentation of a coevolving interaction: a smart way to deal with big animals. It confirms that sometimes artists can find some metaphorical knowledge that scientists are trying to conquer in more formal ways (Verhulst, 1994).

Cetaceans and man play synchronized underwater (Colbert, 2002)

Our Mission

The *International Journal of Dynamical Systems Research: Chaos & Complexity Letters* is born to collect and disseminate complexity science related information to anybody interested in the topic. We know that nowadays there are several other journals in this area but the idea of this new journal was welcomed by many and important scientist, as our Scientific Board illustrates. This new Journal is born to:

(1) Speed up the evolutionary development of complexity science;
(2) Extend its interactions crossing over disciplines, levels of knowledge and geography to find new research and new applications.

We will have both a paper and a digital version (on cd and the web). The digital version will allow the exploitation of all the multimedia opportunities and the allocation space offered by this format. Scientific papers, for example, will have the opportunity to publish movies (in various formats) of plots, experiments and any other knowledge material. We will also publish a special section of raw data, available to the scientific community in order to compare different empirical approaches. Finally we will also publish GNU software free for the scientific community public testing. In any case digital media are offering a lot of opportunities that are not yet completely exploited. For example, the possibility of making interactive scientific publications is another perspective to be explored.

The structure of CCL is specifically designed to add value to the trans-disciplinary approach while, at the same time, differentiating the epistemology of different contributions. You will find modeling, simulations, data analysis and even a metaphors section, but clearly differentiated.

We will try to follow and stimulate research on the edge of new frontiers by also stimulating new focuses by special issues devoted to the new frontiers in theory and applications. We are just now planning new special issues on new mathematics and the arts, noise and synchronization, and new challenges in the neurosciences.

In this enterprise we will be sustained by the memory and example of two great companions that in different ways shared our project during its prehistory: Ilya Prigogine and Francisco Varela.

Their trajectories in life and research design some contours of the new science to come.

Franco F. Orsucci,
Editor in Chief
franco.orsucci@ixtu.org

Rome and London, October 2003

References

Colbert D (2002) *Ashes and Snow*, Venice: Biennale Monographs and Catalogues.

Dubè JL; Desprès P (2000) The Control of Dynamical Systems - Recovering Order from Chaos, in Itikawa,Y (Ed.) *The Physics of Electronic and Atomic Collisions*, Woodbury, N.Y.: AIP Conference Proceedings.

Dyson F (1988), *Infinite in All Directions*, New York: Harper and Row Publishers.

Freeman WJ (1999). *How Brains Make up their Minds*, London: Weidenfeld & Nicolson.

Gardner H (1985) *The mind's new science, a history of the cognitive revolution*. New York: Basic Books.

Guastello SJ (1995) *Chaos, catastrophe, and human affairs: applications of nonlinear dynamics to work, organizations, and social evolution,* Mahwah NJ: Lawrence Erlbaum Associates.

Kuhn TS (1996) *The Structure of Scientific Revolutions*. Chicago, IL: University of Chicago Press.

Lorenz EN (1963) *J. of Atmos. Sci.* **20**, 130 .

Lorenz EN (1994) *The Essence of Chaos (The Jessie and John Danz Lecture Series)*, University of Washington Press.

Orsucci F (2003) *Changing Mind: Transitions in Natural and Artificial Environments,* Singapore: World.Scientific.

Poincaré H (1892) *Les Methodes Nouvelles de la Mecanique Celeste*, Paris: Gauthier-Villars et fils 13, 1 .

Ruelle D, (1994) Where can one hope to profitably apply the ideas of chaos?, *Physics Today* **47**, 7, 24-30.

Rumi JJ-D & Barks C (1995) *The Essential Rumi*. San Francisco, CA: Harper.

Schreiber T (1999) Interdisciplinary application of nonlinear time series methods, *Physics Reports* **308**, 1-64.

Varela FJ, Thompson E & Rosch E (1991) *The Embodied Mind: Cognitive Science and Human Experience*, Cambridge, Mass: MIT Press.

Verhulst F (1994) Metaphors for psychoanalysis, *Nonlinear Science Today* **4** (1):1-6.

In: Chaos and Complexity Research Compendium
Editors: F.F. Orsucci and N. Sala, pp. 9-11
ISBN: 978-1-60456-787-8

Chapter 2

MEMORIAL: ILYA PRIGOGINE AND HIS LAST WORKS

Gonzalo Ordonez

During his long and extremely fruitful career, Prof. Ilya Prigogine worked on many different subjects, from the theory of molecular solutions, to the theory of vehicular traffic and the big bang. He was one of the pioneers in the field of non equilibrium thermodynamics, especially with his work on dissipative structures, which showed that the increase of entropy, usually associated with increasing disorder, could also lead to self-organization and complexity in open systems.

He often spoke of unification between man and nature, connected through time in its creative role. He was inspired by Bergson, who said (I. Prigogine, Autobiographie, Florilège des Sciences en Belgique II, 1980):

> "The more deeply we study the nature of time, the better we understand that duration means invention, creation of forms, continuous elaboration of the absolutely new."

The study of time was a recurring theme in his scientific career. He kept working on this theme, and other problems in physics derived from it, throughout his last years. I had the great privilege of working with Prof. Prigogine during this period.

When talking about physics, he had as much enthusiasm as a freshly graduated student. He was always looking into the future, coming up with new problems to work on. This was quite consistent with his philosophical views on time. "The future is open," "we are only at the beginning" were common phrases he used.

One of the subjects that most interested Prof. Prigogine in his last years was the study of entropy and its connection to dynamics. Traditionally, this connection has been made through the introduction of supplementary assumptions or approximations. But many questions remain surrounding entropy: how to define entropy for dense systems, as well as for systems far from equilibrium? Prof. Prigogine believed that to answer these questions one should look more closely at the transition from the dynamical description, given by classical or quantum dynamics, to the thermodynamical description. This transition occurs at the limits of dynamics, when the solutions of equations of motion become so irregular that they cannot be written in a compact way. As described by Poincarè, they become "non-integrable." Prof. Prigogine and collaborators showed, however, that one can introduce new representations that yield thermodynamic behavior, without approximations. Quoting from his autobiography:

> "… I was prompted by a feeling of dissatisfaction, as the relation with thermodynamics was not established by our work in statistical mechanics, nor by any other method … the question of the nature of dynamical systems to which thermodynamics applies was still without answer."

> "…If irreversibility does not result from supplementary approximations, it can only be formulated in a theory of transformations which expresses in "explicit" terms what the usual formulation of dynamics does "hide". In this perspective, the kinetic equation of Boltzmann corresponds to a formulation of dynamics in a new representation … In conclusion: dynamics and thermodynamics become two complementary descriptions of nature, bound by a new theory of non-unitary transformations."

On this subject, together with collaborators he was able to make much progress during the last years. We could precisely define, at least in simple cases, a "microscopic entropy," derived from a non-unitary transformation. This entropy could measure the "age" of a system out of equilibrium.

Related to this, Prof. Prigogine thought he could find a new view on the "twin paradox."

This is a well know paradox in the theory of relativity, used to show that acceleration can lead to slower aging, due to relativistic time dilation. This fact prompted him to study the effects of acceleration on age, age defined through the microscopic entropy mentioned above. I am sure he had a deeper question in mind: what is the relation between the "geometric" time of relativity, and the "thermodynamic" time, connected with increasing entropy? In this context we considered first a non-relativistic situation, showing that acceleration can indeed lead to "rejuvenation." (here due to different causes than in the twin paradox). We started to write a paper on this. I prepared a draft, following many suggestions from Prof. Prigogine. The next step was for him to have a look on it, but unfortunately he became ill and passed away.

Here I want to say that Prof. Prigogine kept interest in his work until the very end. He even wanted to work on this paper while he was in the hospital. This was not uncharacteristic of him. In a previous, less serious occasion, he had to stay in the hospital for a few days. Students and colleagues would visit him to discuss physics. At some point we thought that we should bring a blackboard to his hospital room!

The editors of "Chaos and Complexity Letters" have kindly accepted to publish this paper on acceleration and entropy after Prof. Prigogine's death. As I mentioned, this paper really shows work in progress. I hope that it will draw attention to some of the subjects that Prof. Prigogine was working on, and which I think are worth pursuing further.

I believe Prof. Prigogine has made a deep, lasting contribution to our understanding of nature, giving us a glimpse on the mechanisms of self-organization, a key element in our understanding of life. For the people who knew him, he left as well unforgettable memories, marked by his great, warm human nature and contagious enthusiasm. As a tribute to him we can continue his work, guided by his aim of unification between different disciplines, unification between man and nature. As he said, time is not an illusion; the future is widely open for us to shape.

Gonzalo Ordonez
Austin, Texas, June 30, 2003.

In: Chaos and Complexity Research Compendium
Editors: F.F. Orsucci and N. Sala, pp. 13-25
ISBN 978-1-60456-787-8

Chapter 3

Acceleration and Entropy: A Macroscopic Analogue of the Twin Paradox

I. Prigogine and G. Ordonez
Center for Studies in Statistical Mechanics and Complex Systems,
The University of Texas at Austin, Austin, TX 78712 USA
and
International Solvay Institutes for Physics and Chemistry,
CP231, 1050 Brussels, Belgium

Abstract

The twin-paradox described in relativity theory shows that acceleration leads to slower aging. Motivated by this, we consider the effects of acceleration on entropy. We consider a macroscopic, non-relativistic analogue of the twin effect on a 2-D weakly coupled gas. We introduce a dynamical entropy ($\mathcal{H}$ function), which measures the "age" of the system. In previous papers we have considered the effect of rejuvenation by velocity inversion of every particle. Here we generalize our results by studying how the $\mathcal{H}$-function changes as a result of rotation of the velocities by a given angle. The rotation is the result of some acceleration. Therefore acceleration leads to entropy flow. The degree of "rejuvenation" of the system depends on the angle. As a special case, a rotation by $\pi/2$ approximately resets the $\mathcal{H}$-function to its initial value. In thermodynamic terms, there is a compensation between entropy production and entropy flow.

1. Introduction

The twin paradox is explained by acceleration on the moving twin. It has been verified by the time delay of unstable particles, related to relativistic field theory. Still, it is not clear why this would have an effect on chemical reactions or living material. To study this one would require a relativistic formulation of nonequilibrium physics. Therefore it is not without interest to consider a non-relativisitic effect of acceleration on aging. It is generally accepted that the age of such systems is related to entropy. We define non-equilibrium entropy through an $\mathcal{H}$-function, analogous to Boltzmann's $\mathcal{H}$-function, but incorporating correlations [2]. When we start from a non-equilibrium state, the time evolution leads to

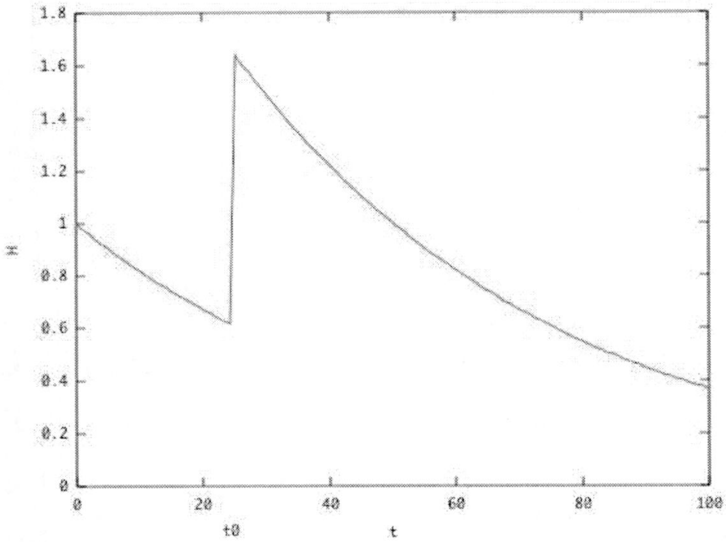

Figure 1. Schematic plot of the Lyapounov function $\mathcal{H}(t)$, showing the effects of velocity inversion at a given time t_0. Velocity inversion creates correlations and causes $\mathcal{H}$ to jump up.

approach to equilibrium through relaxation processes. If we then invert velocities we have a highly complex behavior corresponding to a time-reversed evolution. This corresponds to an injection of "negative entropy," leading to a rejuvenation of the system. Post-collisional correlations are turned into pre-collisional correlations, and this leads to a "jump" or discontinuity in the $\mathcal{H}$-function. After the jump the $\mathcal{H}$-function continues to decrease, due to the decay of the pre-collisional correlations (see Fig. 1). This is the answer we give to the Lochsmidt paradox [2].

In thermodynamics the change of entropy is

$$dS = d_i S + d_e S \tag{1}$$

where $d_i S$ is the entropy production and $d_e S$ is the entropy flow. The inversion of velocities corresponds to a flow of entropy. In a 2-D system velocity inversion corresponds to a rotation of the velocity of every particle by an angle $\phi = \pi$. The question we want to study is the generalization of this problem to arbitrary angles ϕ for velocity rotations. We will study a 2-D, weakly interacting classical gas. We will estimate what is the effect of finite velocity rotations, by an angle ϕ, on the $\mathcal{H}$-function. As we will see rejuvenation will now depend on the angle. So, already in a non-relativistic setting, acceleration has an effect on age. This is of course a highly idealized situation because we need a force which would turn all the velocities by the same angle ϕ. But we could consider systems for which we can turn the velocities in a finite region. Then the effect would disappear after a time depending on the extension of the region.

2. Weakly Coupled Gas

We consider a classical 2-D gas with Hamiltonian

$$H = \sum_{n=1}^{N} \frac{\mathbf{p}_n^2}{2m} + \sum_{n<j}^{N} \lambda V(|\mathbf{r}_n - \mathbf{r}_j|) \tag{2}$$

We write the potential energy as

$$V(r) = \left(\frac{2\pi}{L}\right)^2 \sum_{\mathbf{k}} V_k \exp(i\mathbf{k}\cdot\mathbf{r}) \tag{3}$$

where $k = |\mathbf{k}|$ and L is the size of the system. We are interested in the thermodynamic limit $N \to \infty$, $L \to \infty$, keeping a finite concentration

$$c = N/L^2 > 0 \tag{4}$$

$$\lambda \ll 1 \tag{5}$$

Hereafter we will use units where $m = 1$ and we will work with the velocities $\mathbf{v}_n = \mathbf{p}_n/m = \mathbf{p}_n$ rather than with the momenta. Ensembles $\rho(\mathbf{r}_1, \ldots \mathbf{r}_N, \mathbf{v}_1, \ldots \mathbf{v}_N, t)$ or $\rho(t)$ in short, evolve according to the Liouville equation,

$$i\frac{\partial}{\partial t}\rho(t) = L_H\rho(t), \qquad \rho(t) = \exp(-iL_H t)\rho(0) \tag{6}$$

where

$$L_H = -i\sum_{n=1}^{N}\left(\frac{\partial H}{\partial \mathbf{v}_n}\cdot\frac{\partial}{\partial \mathbf{r}_n} - \frac{\partial H}{\partial \mathbf{r}_n}\cdot\frac{\partial}{\partial \mathbf{v}_n}\right) \tag{7}$$

with

$$\begin{aligned}\frac{\partial}{\partial \mathbf{v}} &= \left(\frac{\partial}{\partial v_x}, \frac{\partial}{\partial v_y}\right)\\ \frac{\partial}{\partial \mathbf{r}} &= \left(\frac{\partial}{\partial r_x}, \frac{\partial}{\partial r_y}\right)\end{aligned} \tag{8}$$

We decompose the Liouvillian as $L_H = L_0 + \lambda L_V$ where

$$L_0 = -i\sum_{n}^{N} \mathbf{v}_n \cdot \frac{\partial}{\partial \mathbf{r}_n} \tag{9}$$

$$L_V = \sum_{n<j}^{N}\sum_{\mathbf{k}} V_k e^{i\mathbf{k}\cdot(\mathbf{r}_n - \mathbf{r}_j)} i\mathbf{k}\cdot\left(\frac{\partial}{\partial \mathbf{v}_n} - \frac{\partial}{\partial \mathbf{v}_j}\right) \tag{10}$$

We consider initial ensembles that depend only on the velocities

$$\rho(0) = P^{(0)}\rho(0) \tag{11}$$

where $P^{(0)}$ is the projector into homogenous distributions:

$$P^{(0)}\rho = \frac{1}{L^{3N}} \int d^2r_1 \dots d^2r_N\, \rho \tag{12}$$

We denote the complement projector as

$$Q^{(0)} = 1 - P^{(0)} \tag{13}$$

The unperturbed Liouville operator satisfies the property

$$L_0 P^{(0)} = P^{(0)} L_0 = 0 \tag{14}$$

because functions in the $P^{(0)}$ subspace are independent of the positions of the particles.

3. The Λ Transformation

In the following sections we will introduce an "entropy" operator to define the "age" of our system. This operator is constructed starting with a transformation Λ we have introduced previously. In this section we will give an overview of the Λ transformation. More details can be found in Refs. [2, 4]. We define inner products between functions of phase-space variables and ensembles as

$$\langle\langle f|\rho\rangle\rangle = \int d^2r_1 \dots d^2v_N f^*(\mathbf{r}_1 \dots \mathbf{v}_N)\rho(\mathbf{r}_1 \dots \mathbf{v}_N) \tag{15}$$

With this inner product we define Hermitian conjugation as usual

$$\langle\langle f|O|\rho\rangle\rangle = \langle\langle \rho|O^\dagger|f\rangle\rangle^* \tag{16}$$

Canonical transformations U are unitary:

$$U^\dagger = U^{-1} \tag{17}$$

For integrable systems in the sense of Poincaré, we can construct by perturbation series or otherwise a canonical transformation U that eliminates the interactions. We define new phase space variables

$$\begin{aligned} \tilde{\mathbf{r}}_n &= U^\dagger \mathbf{r}_n \\ \tilde{\mathbf{v}}_n &= U^\dagger \mathbf{v}_n \end{aligned} \tag{18}$$

such that the Hamiltonian is only a function of the new velocities $H(\mathbf{r}_1 \dots \mathbf{v}_N) = \tilde{H}(\tilde{\mathbf{v}}_1 \dots \tilde{\mathbf{v}}_N)$. In terms of the new variables, the equations of motion are enormously simplified. For integrable systems the operator $\tilde{L}_0 = U L_H U^{-1}$ gives the Liouville operator of free particles, so we have (see Eq. (14))

$$P^{(0)}\tilde{L}_0 = \tilde{L}_0 P^{(0)} = 0 \tag{19}$$

One can construct U using standard perturbation theory. For example, acting on the homogenous subspace $P^{(0)}$ we have

$$P^{(0)}U = P^{(0)} + P^{(0)}\lambda L_V \frac{1}{-L_0} Q^{(0)} + O(\lambda^2) \tag{20}$$

$$U^{-1}P^{(0)} = P^{(0)} + Q^{(0)} \frac{1}{-L_0} \lambda L_V P^{(0)} + O(\lambda^2) \tag{21}$$

However, the situations where this can be done are exceptional. Most systems are non-integrable, due to divergences (vanishing denominators) in the perturbation expansion of U, caused by resonances. Denominators appear as

$$\frac{1}{\omega} \tag{22}$$

where, e.g., $\omega = \mathbf{k} \cdot \mathbf{v}$. We have shown in Refs. [2, 4] that one can remove the divergences through regularization of denominators

$$\frac{1}{\omega} \Rightarrow \frac{1}{\omega \pm i\epsilon} = \mathcal{P}\frac{1}{\omega} \mp \pi i \delta(\omega) \tag{23}$$

The denominators are interpreted as distributions (generalized functions). This leads to a non-unitary transformation Λ, which replaces U. We no more obtain a description in terms of free particles. Instead, we obtain a "kinetic" description with broken time-symmetry. The sign of $\pm i\epsilon$ is chosen depending on the types of transitions (from lower to higher correlations or vice versa), which corresponds to a "dynamics of correlations" [1, 8, 4]. In terms of the new variables, we obtain now probabilistic, irreversible equations. It is remarkable that Λ is invertible. We can go back and forth between the dynamic description and the kinetic description. Instead of the operator $\tilde{L}_0$ we introduce the operator

$$\tilde{\theta} = \Lambda L_H \Lambda^{-1} \tag{24}$$

The Λ transformation satisfies the block-diagonal property

$$P^{(0)}\tilde{\theta} = \tilde{\theta}P^{(0)} \equiv \tilde{\theta}^{(0)} \tag{25}$$

which replaces Eq. (19). $\tilde{\theta}$ is now a collision operator, as used in kinetic theory. In the perturbation expansion we have now

$$P^{(0)}\Lambda = P^{(0)} + P^{(0)}\lambda L_V \frac{1}{i\epsilon - L_0} Q^{(0)} + O(\lambda^2) \tag{26}$$

$$\Lambda^{-1}P^{(0)} = P^{(0)} + Q^{(0)} \frac{1}{i\epsilon - L_0} \lambda L_V P^{(0)} + O(\lambda^2) \tag{27}$$

which are the extensions of Eqs. (20), (21). Note that the sign of ϵ is the same in both expressions. Due to this, Λ is no more unitary:

$$\Lambda^{-1} \neq \Lambda^{\dagger} \tag{28}$$

Instead, it is "star-unitary" [2, 4]

$$\Lambda^{-1} = \Lambda^{\star} \tag{29}$$

One of the most interesting features of the Λ transformation is that it allows us to introduce an "entropy" operator or, more precisely speaking, an $\mathcal{H}$-function.

4. $\mathcal{H}$-function

In our earlier work [2, 7] we have shown that if there exists a dynamical entropy it must be an operator. For systems that present Poincaré resonances, one can introduce the entropy operator

$$M = \Lambda^\dagger \Lambda \tag{30}$$

As shown in Ref. [2], the average of M is a positive, monotonic function of time. In other words, it is a Lyapiunov function. In this sense we can associate M with a generalized entropy. If Λ were replaced by the unitary transformation U, we would find that $M = U^\dagger U = 1$. So for integrable systems our entropy remains constant, just like Gibbs' entropy. The operator M depends on all the particles of the system. One can introduce reduced operators depending only on a limited number of particles [8]. We will consider the reduced operator

$$M_1 = \Lambda^\dagger |f_1\rangle\rangle\langle\langle f_1|\Lambda \tag{31}$$

where

$$\langle\langle f_1|\rho\rangle\rangle = \int d^2r_1 \ldots d^2r_n \int d^2v_1 \ldots d^2v_n\, f_1(\mathbf{v}_1)^* \rho(\mathbf{r}_1, \cdots \mathbf{r}_N, \mathbf{v}_1, \cdots \mathbf{v}_N) \tag{32}$$

is the reduced, one-particle velocity distribution function. We define the $\mathcal{H}$-function as

$$\mathcal{H}(t) = \langle\langle \rho(t)|M_1|\rho(t)\rangle\rangle \tag{33}$$

As we will show below, this is a Lyapounov function of the system [8].[1] For simplicity we will consider functions f_1 that depend only on the magnitude of the velocity

$$f_1(\mathbf{v}) = f_1(v) \tag{34}$$

where $v = |\mathbf{v}|$. We write the $\mathcal{H}$-function as

$$\mathcal{H}(t) = \mathcal{A}^2(t) \tag{35}$$

where

$$\mathcal{A}(t) = \langle\langle f_1|\Lambda|\rho(t)\rangle\rangle \tag{36}$$

The $\mathcal{H}$-function is a Lyapounov function because it is positive and it is non-increasing for all t. Indeed we have (see Eq. (24))

$$\mathcal{A}(t) = \langle\langle f_1|\Lambda e^{-iL_H t}|\rho(0)\rangle\rangle = \langle\langle f_1|e^{-i\tilde{\theta}t}|\tilde{\rho}(0)\rangle\rangle \tag{37}$$

where

$$|\tilde{\rho}(0)\rangle\rangle = \Lambda|\rho(0)\rangle\rangle \tag{38}$$

[1] A very simple example of our $\mathcal{H}$-function has been given in Refs. [6, 3] for the quantum Friedrichs model, which consists of a discrete state ("atom") coupled to a continuum of field modes ("photons").

Noting that $\langle\langle f_1| = \langle\langle f_1|P^{(0)}$ we obtain (see Eq. (25))

$$\mathcal{A}(t) = \langle\langle f_1|e^{-i\tilde{\theta}^{(0)}t}|\tilde{\rho}(0)\rangle\rangle \tag{39}$$

As shown in Refs. [2, 9] the operator $\tilde{\theta}^{(0)}$ is purely imaginary,

$$[\tilde{\theta}^{(0)}]^{\dagger} = -\tilde{\theta}^{(0)} \tag{40}$$

and as shown below it has the property

$$i\tilde{\theta}^{(0)} \geq 0 \tag{41}$$

which breaks time-symmetry in Eq. (39). This shows that $\mathcal{A}(t)$ is non-increasing. This function may be interpreted as a renormalized velocity distribution, corresponding to the velocity distribution of dressed quasiparticles. The evolution of quasiparticles is strictly Markovian and irreversible. In contrast, the time evolution of the original particles (bare particles) is not Markvian, because of the existence of dressing processes. The Λ transformation thus allows us to separate dressing processes from irreversible processes. Age corresponds to the evolution of dressed particles. To show Eq. (41), we note that due to the similitude relation in Eq. (24), the operators $\tilde{\theta}$ and L_H share the same eigenvalues [8]. The eigenvalues of L_H are given by the singularities of the resolvent operator

$$R(z) \equiv \frac{1}{z - L_H} \tag{42}$$

It is well-known [1] that, depending on the analytic continuation of the resolvent (from the upper to the lower half-plane of z or viceversa) all the complex singularities are either on the lower or on the upper half plane. We choose the analytic continuation from the upper to the lower half-planes, since we are interested in extracting contributions to the time evolution operator $\exp(-iL_H t)$ that decay for $t > 0$. In this branch of the resolvent, all the eigenvalues of L_H (and hence of $\tilde{\theta}$) are thus either real or on the lower half-plane. This proves Eq. (41). Our $\mathcal{H}$-function extracts the exponential decay processes during the approach to equilibrium of the system for $t > 0$. As discussed in [2], if we perform a velocity inversion (equivalent to a time inversion) we have the change $\tilde{\theta} \Rightarrow -\tilde{\theta}$ in the exponential in Eq. (39). Indeed, introducing the velocity inversion operator I we have

$$IL_H = -IL_H \tag{43}$$

After the inversion the $\mathcal{A}$-function changes as

$$\mathcal{A}(t) \Rightarrow \mathcal{A}_I(t) = \langle\langle f_1|\Lambda I|\rho(t)\rangle\rangle \tag{44}$$

Due to the property (43), instead of exponential decay we have now exponential growth and $\mathcal{A}$ jumps (see Fig. 1). The jump in the $\mathcal{A}$-function after velocity inversion is related to the injection “negative entropy” that creates anomalous correlations between the particles [2]. We obtain a rejuvenation of the system. The more time we wait, the larger the negative entropy needed to perform the inversion. Indeed, $\mathcal{A}_I(t)$ grows exponentially with t. In a 2-D space, velocity inversions are equivalent to velocity rotations by an angle π. Next we will study the effects of velocity rotations for arbitrary angles ϕ.

5. Effects of Velocity Rotations

We introduce a velocity rotation operator R_ϕ acting on a velocity $\mathbf{v} = (v_x, v_y)$ as

$$R_\phi \mathbf{v} = (v_x \cos\phi + v_y \sin\phi, -v_x \sin\phi + v_y \cos\phi) \tag{45}$$

and on functions of the velocities

$$R_\phi F(\mathbf{v}_1, \mathbf{v}_2, \ldots \mathbf{v}_N) = F(R_\phi \mathbf{v}_1, R_\phi \mathbf{v}_2, \ldots R_\phi \mathbf{v}_N) \tag{46}$$

We will assume that the initial ensemble depends only on the magnitudes of all the velocities

$$\rho(\mathbf{r}_1, \ldots \mathbf{r}_N, \mathbf{v}_1, \ldots \mathbf{v}_N, 0) = \rho(\mathbf{r}_1, \ldots \mathbf{r}_N, |\mathbf{v}_1|, \ldots |\mathbf{v}_N|, 0) \tag{47}$$

This means that

$$R_\phi \rho(0) = \rho(0) \tag{48}$$

for any angle ϕ. After a velocity rotation the function $\mathcal{A}(t)$ in Eq. (36) changes as

$$\mathcal{A}(t) \Rightarrow \mathcal{A}_\phi(t) = \langle\langle f_1 | \Lambda R_\phi | \rho(t) \rangle\rangle \tag{49}$$

or

$$\mathcal{A}_\phi(t) = \langle\langle f_1 | \Lambda R_\phi e^{-iL_H t} | \rho(0) \rangle\rangle \tag{50}$$

Defining

$$L_H(\phi) \equiv R_\phi L_H R_\phi^{-1} \tag{51}$$

we have

$$\mathcal{A}_\phi(t) = \langle\langle f_1 | \Lambda e^{-iL_H(\phi)t} R_\phi | \rho(0) \rangle\rangle \tag{52}$$

Using Eq. (48) we have then

$$\mathcal{A}_\phi(t) = \langle\langle f_1 | \Lambda e^{-iL_H(\phi)t} | \rho(0) \rangle\rangle \tag{53}$$

To calculate the "rotated" Liouvillian $L_H(\phi)$ we use the definition (7). Noting that

$$R_\phi \frac{\partial}{\partial \mathbf{v}} f(\mathbf{v}) = \frac{\partial}{\partial \mathbf{v}_\phi} f(\mathbf{v}_\phi) = \frac{\partial}{\partial \mathbf{v}_\phi} R_\phi f(\mathbf{v}) \tag{54}$$

where $\mathbf{v}_\phi = R_\phi \mathbf{v}$, and that H is independent of the orientation of the velocites, we obtain

$$R_\phi L_H \rho = -i \sum_{n=1}^{N} \left(\frac{\partial H}{\partial \mathbf{v}_{\phi n}} \cdot \frac{\partial}{\partial \mathbf{r}_n} - \frac{\partial H}{\partial \mathbf{r}_n} \cdot \frac{\partial}{\partial \mathbf{v}_{\phi n}} \right) R_\phi \rho \tag{55}$$

Since this is true for any ρ we get

$$L_H(\phi) = -i \sum_{n=1}^{N} \left(\frac{\partial H}{\partial \mathbf{v}_{\phi n}} \cdot \frac{\partial}{\partial \mathbf{r}_n} - \frac{\partial H}{\partial \mathbf{r}_n} \cdot \frac{\partial}{\partial \mathbf{v}_{\phi n}} \right) \tag{56}$$

For the velocity derivatives we have (see Eq. (45))

$$\begin{aligned}\frac{\partial}{\partial v_{\phi x}} &= \cos\phi\frac{\partial}{\partial v_x} + \sin\phi\frac{\partial}{\partial v_y}\\ \frac{\partial}{\partial v_{\phi y}} &= -\sin\phi\frac{\partial}{\partial v_x} + \cos\phi\frac{\partial}{\partial v_y}\end{aligned} \tag{57}$$

Substituting this in Eq. (56) and separating the cosine and sine terms we obtain

$$\begin{aligned}L_H(\phi) = &- i\cos\phi\sum_{n=1}^{N}\left(\frac{\partial H}{\partial v_{nx}}\frac{\partial}{\partial r_{nx}} + \frac{\partial H}{\partial v_{ny}}\frac{\partial}{\partial r_{ny}} - \frac{\partial H}{\partial r_{nx}}\frac{\partial}{\partial v_{nx}} - \frac{\partial H}{\partial r_{ny}}\frac{\partial}{\partial v_{ny}}\right)\\ &- i\sin\phi\sum_{n=1}^{N}\left(\frac{\partial H}{\partial v_{ny}}\frac{\partial}{\partial r_{nx}} - \frac{\partial H}{\partial v_{nx}}\frac{\partial}{\partial r_{ny}} - \frac{\partial H}{\partial r_{nx}}\frac{\partial}{\partial v_{ny}} + \frac{\partial H}{\partial r_{ny}}\frac{\partial}{\partial v_{nx}}\right)\end{aligned} \tag{58}$$

or

$$L_H(\phi) = \cos(\phi)L_H + \sin(\phi)L_H^{\perp} \tag{59}$$

where $L_H^{\perp}$ is the coefficient of $\sin\phi$ in Eq. (58). Note that for $\pm\pi$ rotations (velocity inversion) we have

$$L_H(\pm\pi) = -L_H \tag{60}$$

which is equivalent to Eq. (43). For $\pm\pi/2$ rotations we have

$$L_H(\pi/2) = -L_H(-\pi/2) \tag{61}$$

Now we come back to the amplitude of the $\mathcal{H}$-function (see Eq. (53))

$$\mathcal{A}_\phi(t) = \langle\langle f_1|\Lambda\exp\left[-i(L_H\cos\phi + L_H^{\perp}\sin\phi)t\right]|\rho(0)\rangle\rangle \tag{62}$$

Defining

$$\tilde{\theta}^{\perp} = \Lambda L_H^{\perp}\Lambda^{-1} \tag{63}$$

and using also the definitions in Eqs. (24), (38) we have

$$\mathcal{A}_\phi(t) = \langle\langle f_1|\exp\left[-i(\tilde{\theta}\cos\phi + \tilde{\theta}^{\perp}\sin\phi)t\right]|\tilde{\rho}(0)\rangle\rangle \tag{64}$$

With no rotation ($\phi = 0$) we have only the operator $\tilde{\theta}$ in the exponential, This operator breaks time-symmetry, as discussed in Sec. 4., giving exponential decay in the positive t direction. If we perform a $\phi = \pi$ rotation, corresponding to a velocity inversion we have $\mathcal{A}_\pi(t) = \mathcal{A}_I(t)$. As discussed in Sec. 4., $\mathcal{A}$ jumps to a higher value, the jump increasing exponentially with t. Now, if we perform a $\phi = \pi/2$ rotation we expect that the change of $\mathcal{A}$ will be the same as if we perform a $\phi = -\pi/2$ rotation, because nothing in the Hamiltonian H or the Λ transformation makes a distinction between the sense of the rotations. This implies that in contrast to $\tilde{\theta}$, the operator $\tilde{\theta}^{\perp}$ cannot be a dissipative operator breaking time-symmetry. If it were, then $\mathcal{A}_{\pi/2}(t)$ would contain terms growing exponentially in the future t direction while $\mathcal{A}_{-\pi/2}(t)$ would contain terms decaying exponentially, or viceversa (see Eq. (61)). We would have an asymmetry between clockwise and counter-clockwise velocity rotations. So, the function $\mathcal{A}_{\pm\pi/2}(t)$ should only contain time invariant components plus oscillating components. The spectrum of frequencies of the oscillating components is continuous, so by the Riemann-Lebesgue theorem, the oscillating components added together will give a contribution decreasing as an inverse power of t after a time scale t_c.

$$\mathcal{A}_{\pi/2}(t) \approx \mathcal{A}_{\pi/2}(0), \quad \text{for}\, t \gg t_c \tag{65}$$

For weak coupling one can estimate that $t_c \sim 1/(\lambda c^{1/2})$. Eq. (65) means that for $t \gg t_c$ we can neglect the contributions coming from the operator $\tilde{\theta}^{\perp}$. Thus we have

$$\mathcal{A}_{\phi}(t) \approx \langle\langle f_1 | \exp\left[-i(\tilde{\theta}\cos\phi)t\right] |\tilde{\rho}(0)\rangle\rangle \tag{66}$$

The $\mathcal{H}$-function is then even with respect to the angle of rotation ϕ. The $\mathcal{H}$-function jump increases with ϕ from $\phi = 0$ (no jump) to $\phi = \pi$ (maximum jump). For $\phi = \pi/2$ we have

$$\mathcal{H}_{\pi/2}(t) \approx \mathcal{H}(0) \tag{67}$$

In other words, a $\pi/2$ rotation resets the $\mathcal{H}$-function to its initial value. We have a complete rejuvenation. For $0 < \phi < \pi/2$ we have partial rejuventations, that is, the system becomes"younger" (more ordered) but not as young as it was at $t = 0$. For $\pi/2 < \phi < \pi$ we have an "over-rejuvenation:" the system becomes younger than it was at $t = 0$ due to the presence of anomalous correlations (see Fig. 2).

6. Comparison with the Twin Effect

Now we come to our analogy with the twin effect. Suppose we can "sit" on one of the particles (particle 1).

After a collision with another particle (particle 2), we see particle 2 move away from particle 1 with some velocity g. Let us say that the distance between the particles at the moment of the rotation was r. After a velocity rotation, the distance between the particles will change at the rate

$$\dot{r}(t) = g\frac{r\cos\phi + gt}{\sqrt{(r\cos\phi + gt)^2 + r^2\sin^2\phi}} \tag{68}$$

The distance between the particles will continue to increase if $\phi \leq \pi/2$, but it will decrease if $\phi > \pi/2$. The minimum distance between the particles will be

$$r_{min} = r\sin\phi < r, \quad \text{for}\, \pi/2 < \phi \leq \pi \tag{69}$$

As we have seen, it is precisely for this range of angles that we have an "over-rejuvenation" of the system, that is, after the rotation the system will become younger than

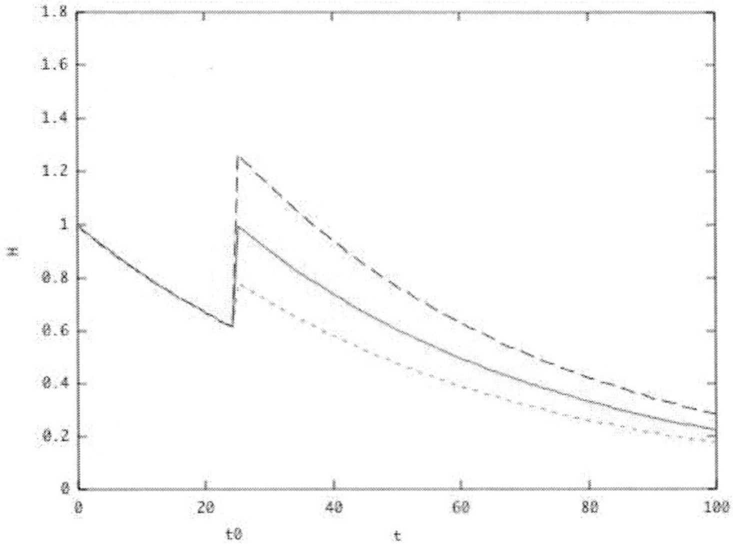

Figure 2. Schematic plot of the Lyapounov function $\mathcal{H}(t)$, showing the effects of velocity rotation at a given timet_0. Velocity rotations create correlations and cause $\mathcal{H}$ to jump up. The dashed line corresponds to a rotation by an angle $\pi/2 < \phi < \pi$ giving "over-rejuvenation", while the solid line corresponds to a $\pi/2$ rotation reseting $\mathcal{H}(t)$ to its initial value and the dotted line corresponds to a $0 < \phi < \pi/2$ rotation giving a partial rejuvenation.

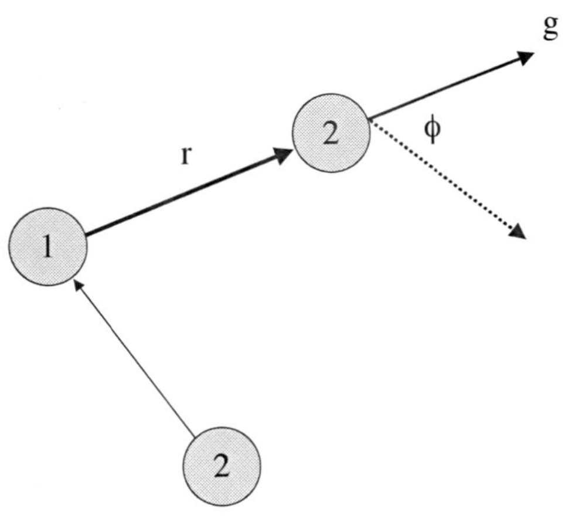

Figure 3. After colliding with particle 1, particle 2 moves away from particle 1 with a speed g. When they are separated a distance r, we perform a velocity rotation by an angle ϕ. This process, applied to all collisions inside the gas, leads to a "rejuvenation" of the system, due to the creation of new correlations among the particles. In the twin effect, an acceleration on the moving twin (analogous to particle 2) slows down his aging, as compared to the twin at rest. This effect is due to relativistic time dilation.

it was at $t = 0$. The closer the particles come together as a result of the rotation, the more the system is rejuvenated. For angles $0 \leq \phi \leq \pi/2$ the rate at which the particles move away is decreased, and we have a partial rejuvenation (note that $|\dot{r}(t)| < g$ for any angle ϕ). All this is reminiscent of the twin effect, replacing particles 1 and 2 by the twins. In the twin effect, the travelling twin becomes younger than the twin at rest, as a consequence of acceleration, which is necessary to bring the twins together. The larger the acceleration, the larger the differences in ages. A small acceleration, not strong enough to bring the twins together would still lead to a small effect, slowing slightly the aging of the moving twin. Of course, this is a relativistic effect due to dilation of time intervals. So there are similarities with the twin effect, but there are also differences. In contrast to the twin effect, the situation we have considered is non-relativistic, and the rejuvenation is a collective effect involving all the particles of the system (not only particles 1 and 2). It is due to the injection of correlations among the particles, which turns post-collisional correlations into pre-collisional correlations.

7. Concluding Remarks

External forces acting on a system, leading to acceleration, can have an effect on the entropy of the system. This effect has a relativistic component, as in the twin paradox, as well as non-relativistic components, an example of which we have discussed in this paper. There is also a distinction between global forces, leading to an overall acceleration (again as in the twin paradox) and local forces, depending on the state of each particle, as we considered here.

Acknowledgments

We thank Dr. E. Karpov and Dr. T. Petrosky for helpful comments and suggestions. We acknowledge the International Solvay Institutes for Physics and Chemistry, the Engineering Research Program of the Office of Basic Energy Sciences at the U.S. Department of Energy, Grant No DE-FG03-94ER14465, the Robert A. Welch Foundation Grant F-0365, and the European Commission Project HPHA-CT-2001-40002 for supporting this work.

References

[1] I. Prigogine, *Non Equilibrium Statistical Mechanics* (Wiley Interscience, 1962).

[2] I. Prigogine, C. George, F. Henin, L. Rosenfeld, *Chemica Scripta* **4**, 5 (1973).

[3] T. Petrosky, I. Prigogine and S. Tasaki, *Physica A* **173**, 175 (1991).

[4] G. Ordonez, T. Petrosky and I. Prigogine, *Phys. Rev. A* **63**, 052106 (2001).

[5] T. Petrosky, G. Ordonez and I. Prigogine, *Phys. Rev. A* **64**, 062101 (2001).

[6] M. de Haan, C. George, and F. Mayne, *Physica A* **92**, 584 (1978).

[7] I. Prigogine, *From being to becoming* (Freeman, New York, 1980).

[8] T. Petrosky and I. Prigogine, *Adv. Chem. Phys.* **99**, 1 (1997).

[9] C. George, *Physica* **39**, 251 (1968).

In: Chaos and Complexity Research Compendium
Editors: F. Orsucci and N. Sala, pp. 27-46
ISBN: 978-1-60456-787-8

Chapter 4

WILLIAM JAMES ON CONSCIOUSNESS, REVISITED

Walter J. Freeman*

Department of Molecular & Cell Biology, LSA 142
University of California at Berkeley CA 94720-3200

Abstract

According to the behavioral theory of pragmatism described most effectively by William James in collaboration with Charles Peirce and John Dewey, knowledge about the world is gained through intentional action followed by learning. In terms of neurodynamics, when the intending of an act comes to awareness through reafference, it is perceived as a cause. When the consequences of an act come to awareness through exteroception and proprioception, they are perceived as effects. These become the cause of a new act. Cycles of such states of awareness comprise consciousness, which can grow in complexity to include self-awareness. Intentional acts do not require awareness, whereas voluntary acts require self-awareness. Awareness of the action-perception cycle provides the cognitive metaphor of linear causality as agency. Humans apply this metaphor to objects and events in the world in order to predict and control them, and to assign social responsibility. Thus linear causality is the bedrock of social contracts and technology.

Complex material systems with distributed nonlinear feedback, such as brains and the activities of their neural and behavioral substrates, cannot be explained by linear causality. They can be said to operate by circular causality without agency. The nature of self-control is described by breaking the circle into a forward limb, the intentional self, and a feedback limb, awareness of the self and its actions. The two limbs are realized through hierarchically stratified kinds of neural activity. Actions are governed by the microscopic neural activity of cortical and subcortical components in the brain that is self-organized into mesoscopic wave packets. The wave packets form by state transitions that resemble phase transitions between vapor and liquid. The cloud of action potentials driven by a stimulus condenses into an ordered state that gives the category of the stimulus. Awareness supervenes as a macroscopic ordering state that defers action until the self-organizing mesoscopic process has reached closure in reflective prediction. Agency, which is removed from the causal hierarchy by the appeal to circularity, re-appears as a metaphor by which objects and events in the world are anthropomorphized

* E-mail address: wfreeman@socrates.berkeley.edu. TEL 510-642-4220 FAX 510-643-6791

and assigned the human property of causation, so that they can be assimilated as subject to the possibility of observer control.

Key words: causality, consciousness, intentionality, nonlinear dynamics, reafference

1. Introduction

Within a single generation of the publication by Charles Darwin of "The Origin of Species", the basic concepts of evolution had been grasped and put to use by William James. He wrote: "*A priori* analysis of both brain and conscious action shows us that if the latter were efficacious it would, by its selective emphasis, make amends for the indeterminacy of the former; whilst the study *à posteriori* of the distribution of consciousness shows it to be exactly such as we might expect in an organ added for the sake of steering a nervous system grown too complex to regulate itself" (James 1879, p. 18). This is classic James: elegant, urbane, a bit fey, and precisely on target, as far as he went. But, did he go far enough? In my view, he did not. There are several shortcomings. Firstly, he proposed the addition of a new part to the brain for the addition of consciousness. We have no evidence that consciousness resides in or operates from any newly added part of the human brain, including those for language. Secondly, his definition finessed the questions whether language is necessary for consciousness, and, if not, whether animals evolved the necessary brain part, and therefore consciousness, early in phylogenetic evolution. Thirdly, he gave no indication of how consciousness might execute its steering function. Some kind of control is modeled by cybernetics, a term that Norbert Wiener coined from the Greek word for "steersman", but even to the present there is no widely accepted explanation of the nature and role of consciousness. Fourthly and more generally, he assigned a causal role to consciousness, even though he allowed (James 1890): "The word 'cause' is ... an altar to an unknown god."

What he did do was to raise and answer the question whether consciousness had a biological basis that could be selected for in the race for the survival of the fittest. He disposed of alternative views that consciousness was an epiphenomenal appendage, which was produced by the brain but which had no role in behavior, and that consciousness was an endowment from God by which humans might come to know the Almighty. He stated clearly that consciousness is known through experience of the activities of one's own body, and by observation of the bodies of others. He laid the foundation for the biological study of the properties of consciousness and of its roles in the genesis and regulation of behaviors. These properties are fair targets for experimental analyses and modeling, unlike the questions whether it arises from the soul (Eccles, 1994), or from panpsychic properties of matter (Whitehead, 1938; Penrose, 1994; Chalmers, 1996), or as a necessary but unexplained and inexplicable accompaniment of brain operations (Searle, 1992; Dennett, 1991; Crick, 1994). In the way James phrased his conception, the pertinent questions are — however it arises and is experienced — how and in what senses does consciousness cause the functions of brains and bodies, and how do brain and bodily functions cause it? How do actions cause perceptions? How do perceptions cause awareness? How do states of awareness cause actions? How can the action potentials of neurons cause consciousness, and how can consciousness shape the patterns of neural firing?

2. The Typology of Causality

Analysis of causality is an essential step to understand consciousness, because the forms that answers to these questions take, and even whether answers can exist, depend on the choice among meanings that are assigned to "cause": (a) to make, move and modulate (an agency in linear causality); (b) to explain rationalize and assign credit or blame (comprehension in circular causality without agency); or (c) to flow in parallel as a meaningful experience, by-product, or epiphenomenon (noncausal interrelations as in predictors, statistical "risk factors" or Leibnizian monads). The troublesome and problematic nature of "causes" is reflected in the variety of synonyms that have been proposed over the centuries: "dispositions" by Thomas Aquinas (1272); "tendencies" by John Stuart Mill (1843); "anomalous monism" by David Davidson (1980); "propensities" by Karl Popper (1982); and "capacities" by Nancy Cartwright (1989).

The prior question I raise here is, why is it that we seek for an explanation of consciousness, by which it is both an effect of neural activity and a cause of behavior? In other words, what are the properties of the neural mechanisms of human thought that lead us to phrase questions in just this way? Obviously there are many answers available to us, but there is no agreement on what the basis might be for finding human satisfaction in answers that invoke causality. My aim here is to show why we humans are addicted to causality.

Linear causality is exemplified in stimulus-response determinism. A stimulus (S) initiates a chain of events including activation of receptors, transmission by serial synapses to cortex, integration with memory, selection of a motor pattern descending transmission to motor neurons, and activation of muscles. At one or more nodes along the chain awareness occurs, and meaning and emotion are attached to the response (R). Temporal sequencing is crucial; no effect can precede or occur simultaneously with its cause. At some instant each effect becomes a cause. This step is inherently problematic, because awareness cannot be defined at points in time. The demonstration of causal invariance must be based on repetition of trials, in which universal time is segmented. The time line for each observation is re-initiated at zero in observer time, and S-R pairs are collected. Some form of generalization is used over the pairs, and various forms of abstraction are used to control and exclude extraneous factors and correlations in the attempt to define true agencies.

Noncausal relations are described by statistical models, differential equations, phase portraits, and so on, in which time may be implicit and/or reversible. Once the constructions are completed by the calculation of risk factors and degrees of certainty from distributions of observed events and objects, the assignment of causation is optional. In describing brain functions by these techniques, consciousness is treated as irrelevant, epiphenomenal, or unscientific and of little further interest (Dennett 1991; Crick 1994).

Circular causality defies simple summary. My approach in this essay is to explain it at three levels of brain function: the macroscopic level of brain body and mind in relation to behavior; the mesoscopic level of neuron populations within the brain; and the microscopic level at which individual neurons act in concert to create populations. These concepts can be applied to animal consciousness, on the premise that the structures and activities of brains and bodies are comparable to those of humans over a broad variety of animals. The hypothesis is that the elementary properties of consciousness have emerged and are manifested in even the simplest of extant vertebrates. Structural and functional complexity of mind, brain and body increased with the evolution of brains into higher mammals. There were quantum leaps in complexity upon the addition of new parts subserving language and other social functions, but the dynamics of intentionality in brains was and remains couched in neural operations that construct goal-oriented behavior, for which language is neither necessary nor sufficient. The brains of invertebrates are not hereby consigned to mindless machines, because cuttlefish and bees appear to have the capacity for play, but their brains are sufficiently different topologically from those of vertebrates as to make inclusion too difficult for present purposes.

3. Level 3 - Macroscopic: The Circular Causality of Intentionality

An elementary process requiring the dynamic interaction between brain, body and world in all animals is an act of observation. This is not a passive receipt of information from the world, as expressed linear causality. It is the culmination of purposive action by which an animal directs its sense organs toward a selected aspect of the world and abstracts, generalizes, and learns from the resulting sensory stimuli. This principle was the starting point for Charles Peirce and William James in the development of pragmatism (Menand 2001; James 1893). Each such act requires a prior state of readiness that expresses the existence in the actor of a goal, a preparation for motor action by positioning the sense organs and selectively sensitizing the sensory cortices. Before stimulus arrival their excitability has already been shaped by the past experience that is relevant to the goal and the expectancy of stimuli. A concept that can serve as a principle by which to assemble and interrelate these multiple facets is intentionality. Aquinas (1272) introduced this concept in his program to Christianize Aristotelian doctrine. He conceived it on the basis of the fundamental integrity of the soul, mind and body of the individual, and the power of the individual to know God by taking action into the world ("stretching forth") and suffering the consequences. Descartes and Kant deliberately carried out revolutions against Thomist doctrine. They replaced intentionality with the concept of representationalism, in which forms and information come from the world and are transformed into images that are interpreted according to the laws of logic and reason. Brentano (1889) resurrected intentionality but only to denote the relation between mental representations and the objects and events being represented, thus reinforcing Descartes' subject-object dichotomy.

The properties of Thomist intentionality are (a) its intent, directedness toward some future state or goal; (b) its unity; and (c) its wholeness in the integration of a life-long remembrance of experiences (Freeman 1995). (a) Intent comprises the endogenous initiation, construction and direction of behavior into the world, combined with changing the self by learning in accordance with the perceived consequences of the behavior. Intent originates within brains. Humans and other animals select their own goals, plan their own tactics, and choose when to begin modify, and stop sequences of action. Humans at least are subjectively aware of themselves acting. This facet is commonly given the meaning of purpose and motivation by psychologists, because, unlike lawyers, they usually do not distinguish between intent and motive. Intent is the potential for a forthcoming action, whereas motive is the reason for the action to be taken. Intentions are biological; motives are mental.

(b) Unity appears in the combining of input from all sensory modalities into Gestalts (multisensory perceptions) in the coordination of all parts of the body, both musculoskeletal and autonomic, into adaptive, flexible, yet focused movements, and in the full weight of all past experience in the directing of each action. Subjectively, unity may appear in the awareness of self. Unity and intent find expression in modern analytic philosophy as "aboutness", meaning the way in which beliefs and thoughts symbolized by mental representations refer to objects and events in the world, whether real or imaginary (Searle 1992). The distinction between inner image and outer object invokes the dichotomy between subject and object that was not part of the originating, non-representationalist Thomist view.

(c) Wholeness is revealed by the orderly changes in the self and its behavior that constitute the development and maturation of the self through learning, within the constraints of its genes and its material, social and cultural environments. Subjectively, wholeness is revealed in the striving for the fulfillment of the potential of the self through its lifetime of change. Its root meaning is "tending", the Aristotelian view that biology is destiny. Its biological basis is seen in the process of healing of the brain and body from damage and disruption. The concept appears in the description by a 14th century surgeon LaFranchi of Milan of two forms of healing, by first intention with a clean scar, and by second intention with suppuration.

Intentionality cannot be explained by linear causality, because, under that concept, actions must be attributed solely to environmental (Skinner, 1969) and genetic determinants (Herrnstein and Murray, 1994), leaving no leeway for self-determination. Acausal theories (Hull, 1943; Grossberg, 1982) describe statistical and mathematical regularities of behavior without reference to intentionality. Circular causality explains intentionality in terms of "action-perception cycles" (Merleau-Ponty, 1945) and affordances (Gibson 1979), in which each perception concomitantly is an outcome of a preceding action and a condition for a following action. Dewey (1914) phrased the same idea in different words; an organism does not react to a stimulus but acts into it and incorporates it. That which is perceived already exists in the perceiver, because it is posited by the action of search and is actualized in the fulfillment of expectation. The unity of the cycle is reflected in the impossibility of defining a moving instant of 'now' in subjective time, as an object is conceived under linear causality. The Cartesian distinction between subject and object does not appear, because they are joined by assimilation in a seamless flow.

4. Level 2 - Mesoscopic: The Circular Causality of Reafference

Brain scientists have known for over a century that the necessary and sufficient part of the vertebrate brain to sustain minimal intentional action as a component of intentionality, is the ventral forebrain including those parts that comprise the external shell of the phylogenetically oldest part of the forebrain the paleocortex, and the underlying nuclei such as the amygdala and the neurohumoral brain stem nuclei (Panksepp, 1998) with which the cortex is interconnected. These components suffice to support identifiable patterns of intentional behavior in animals, when all of the newer parts of the forebrain have been surgically removed (Goltz, 1892) or chemically inactivated by spreading depression (Bures et al., 1974). Intentional behavior is severely altered or lost following major damage to these parts. Phylogenetic evidence comes from observing intentional behavior in salamanders, which have the simplest of the existing vertebrate forebrains (Herrick, 1948; Roth, 1987) comprising only the limbic system. Its three cortical areas are sensory (which is predominantly the olfactory bulb), motor (the pyriform cortex), and associational (Figure 1).

The latter has the primordial hippocampus interconnected with the septal, amygdaloid and striatal nuclei. It is identified in higher vertebrates as the locus of the functions of spatial orientation (the "cognitive map") and temporal orientation ("short term memory") in learning. These integrative frameworks are essential for intentional action into the world, because even the simplest actions, such as observation, searching for food, or evading predators require an animal to coordinate its position in the world with that of its prey or refuge, and to evaluate its progress during evaluation, attack or escape. These limbic structures in the medial temporal lobe appear to be the principal controllers of the neurohumoral nuclei in the hypothalamus, periacqueductal gray matter, and brain stem that are essential for the elaboration deployment and maintenance of states of readiness to act, which we identify with emotion (James, 1893; Panksepp, 1998).

The crucial question for neuroscientists is, how are the patterns of neural activity that sustain intentional behavior constructed in brains? A route to an answer is provided by studies of the electrical activity of the primary sensory cortices of animals that have been trained to identify and respond to conditioned stimuli. An answer appears in the capacity of the cortices to construct novel patterns of neural activity by virtue of their self-organizing dynamics.

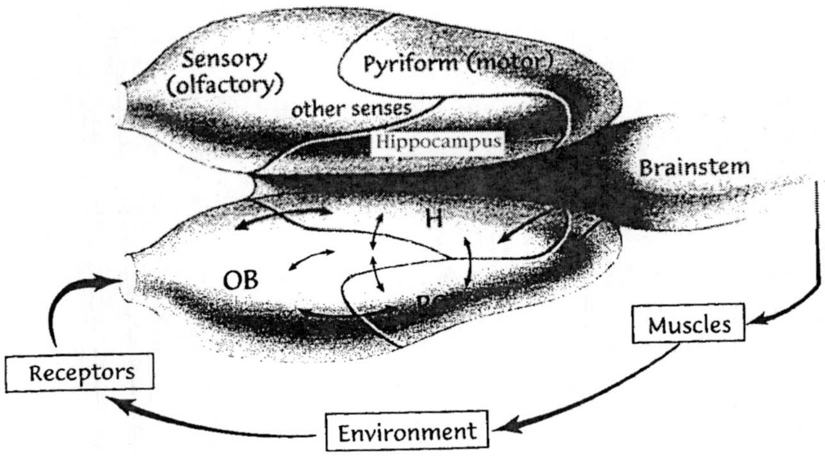

Figure 1. The schematic shows the dorsal view of the right cerebral hemisphere of the salamander (adapted from Herrick 1948). The cortical interactions are demarcated by arrows between the sensory area (olfactory bulb) with a 'transitional zone' (Tr) for all other senses and a motor area (Pir, pyriform cortex with descending connections to the corpus striatum (CS), amygdaloid (A) and septum, S), and both with the primordial hippocampus (Hip). This primitive forebrain suffices as an organ of intentionality, comprising the limbic system.

Two approaches to the study of sensory cortical dynamics are in contrast. One is based in linear causality. An experimenter identifies a neuron in sensory cortex by recording its action potential with a microelectrode, and then determines the sensory stimulus or motor action with which that neuron is most closely correlated. The pulse train of the neuron is treated as a symbol to 'represent' that stimulus as the 'feature' of an object, for example the color, contour, or motion of an eye or a nose in a face, or a ‘command’ for an action. The pathway of activation from the sensory receptor through relay nuclei to the primary sensory cortex and then beyond is described as a series of maps, in which successive representations of the stimulus are activated. The firings of the feature detector neurons must then be synchronized or 'bound' together to represent the object, such as a moving colored ball, as it is conceived by the experimenter. Neurobiologists postulate that this representation is transmitted to a higher cortex, where it is compared with representations of previous objects that are retrieved from memory storage. A solution to the 'binding problem' is still being sought (Gray, 1994; Hardcastle, 1994; Singer and Gray, 1995).

The other approach is based in circular causality. In this view the experimenter trains a subject to cooperate with him or her through use of positive or negative reinforcement, thereby inducing a state of expectancy and search for a stimulus, as the subject conceives it. When the expected stimulus arrives, the activated receptors transmit pulses to the sensory cortex, where they elicit the construction by nonlinear dynamics of a macroscopic, spatially coherent oscillatory pattern that covers the entire primary sensory cortex (Freeman 1975, 1991). Such patterns are observed by means of the electroencephalogram (EEG) from electrode arrays on any or all of the sensory cortices (Freeman 1975, 1992, 1995; Freeman and Schneider, 1978; Barrie et al., 1996; Kay and Freeman 1998). They are not seen in recordings from single neuronal action potentials, because the fraction of the variance in the typical single neuronal pulse train that is covariant with the neural mass is far too small, on the order of 0.1%.

The emergent pattern is not a representation of a stimulus, nor is it a ringing as when a bell is struck, nor a resonance as when one string of a guitar vibrates when another string does so at its natural frequency. It is a state transition that is induced by a stimulus, followed by a construction of a spatial pattern of amplitude modulation (AM) of the rapid oscillations in potential. The AM pattern is shaped by the synaptic modifications among cortical neurons from prior learning. It is also dependent on the brain stem nuclei that bathe the forebrain in neuromodulatory chemicals. It is a dynamic action pattern that creates and carries the meaning of the stimulus for the subject. It reflects the individual history, present context, and expectancy, corresponding to the unity and the wholeness of intentionality. Owing to dependence on history, the AM patterns created in each cortex are unique to each subject. The first event in neocortex upon stimulus arrival maintains information that relates to the stimulus directly, but the events thereafter reflect the category of the stimulus, its value, significance and meaning (Ohl, Scheich and Freeman 2001). This is because the mechanism of construction derives from destabilization of the cortex by input, which increases the density of excitatory interactions among neurons in the cortex, so that their activity reflects predominantly the synaptic modifications in cortex from previous learning, not the activity driven by the input (Freeman 1991; Freeman and Barrie 2000). These properties have been simulated in models both in software (Kozma and Freeman 2001) and in VLSI hardware (Principe et al. 2001). They demonstrate the difference between the passive representation of stimuli and the active engagement of the brain in the construction of the meanings of stimuli.

The visual, auditory, somesthetic and olfactory cortices serving the distance receptors all transmit their constructions through the entorhinal cortex from whence they converge into the limbic system, where they are integrated with each other over time. Clearly they must have similar dynamics, in order that the messages be combined into Gestalts. The resultant integrated meaning is transmitted back to the cortices in the processes of selective attending, expectancy, and the prediction of future inputs (Freeman 1995; Kay and Freeman 1998; Ohl, Scheich and Freeman 2001). The same waveforms of EEG activity as those found in the sensory cortices are found in various parts of the limbic system. This similarity indicates that the limbic system also has the capacity to create its own spatiotemporal patterns of neural activity. The patterns are embedded in past experience and convergent multisensory input, and they are self-organized. The limbic system provides interconnected populations of neurons that, according to the hypothesis being proposed, generate continually the patterns of neural activity that form goals and direct behavior toward them.

EEG evidence shows that the process in the various areas of cortex occurs in discontinuous steps, like frames in motion pictures on multiple screens (Freeman 1975; Barrie, Freeman and Lenhart, 1996). Being intrinsically unstable, the limbic system continually transits across states that emerge, transmit to other parts of the brain, and then dissolve to give place to new ones. Its output controls the brain stem nuclei that serve to regulate its excitability levels, implying that it regulates its own neurohumoral context, enabling it to respond with equal facility to changes both in the body and the environment that call for arousal and adaptation or rest and recreation. Again by inference, it is the neurodynamics of the limbic system, with contributions from other parts of the forebrain such as the frontal lobes and basal ganglia, that initiates the novel and creative behaviors seen in search by trial and error.

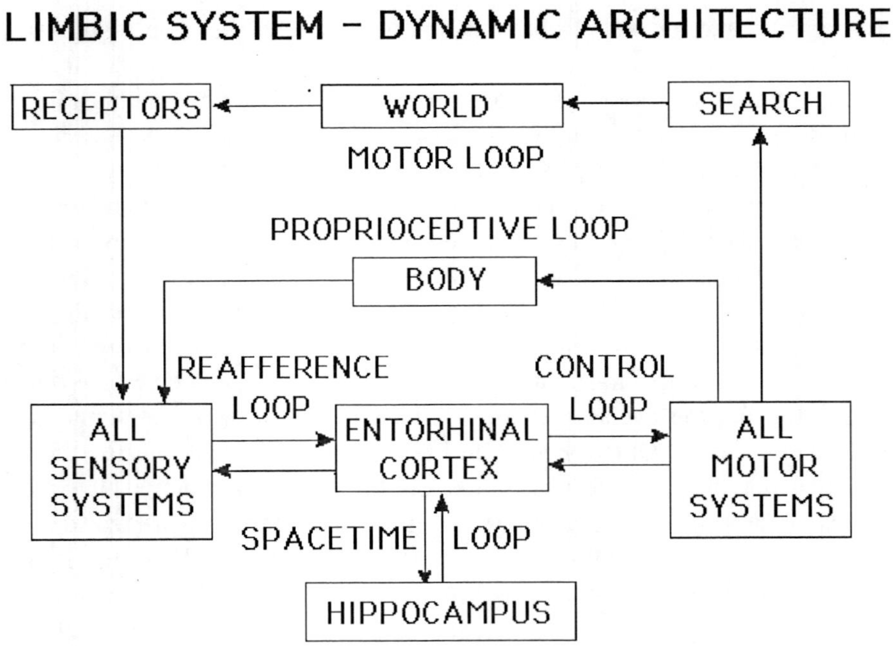

Figure 2. The limbic architecture is formed by multiple loops. The mammalian entorhinal cortex receives from and transmits to all sensory areas. It provides the main input for the hippocampus and is the main target for hippocampal output. The hypothesis is proposed that intentional action is by flow of activity around the loops that extend into the body and the world, and that awareness and consciousness are engendered by the flows within brains that, are described by circular causality.

The limbic activity patterns of directed arousal and search are sent into the motor systems of the brain stem and spinal cord (Figure 2). Simultaneously, patterns are transmitted to the primary sensory cortices, preparing them for the consequences of motor actions. This process has been called "reafference" (von Holst and Mittelstädt 1950; Freeman 1995), "corollary discharge" (Sperry 1950), and "preafference" (Kay and Freeman 1998). It compensates for the self-induced changes in sensory input that follow the actions organized by the limbic system, and it selectively sensitizes sensory systems to anticipated stimuli prior to their expected times of arrival.

The concept of preafference began with an observation by Helmholtz (1872) on patients with paralysis of lateral gaze, who, on trying and being unable to move an eye, reported that the visual field appeared to move in the opposite direction. He concluded that "an impulse of the will" that accompanied voluntary behavior was unmasked by the paralysis. He wrote: "These phenomena place it beyond doubt that we judge the direction of the visual axis only by the volitional act by means of which we seek to alter the position of the eyes.". J. Hughlings Jackson (1931) repeated the observation but postulated alternatively that the phenomenon was caused by "an in-going current", which was a signal from the non-paralyzed eye that moved too far in the attempt to fixate an object, and which was not a recursive signal from a "motor centre". Edward Titchener (1907) and, unfortunately, William James (1893) joined in this interpretation, thus delaying deployment of the concepts of neural feedback and re-entrant cognitive processes until late in the 20th century.

The sensory cortical constructions consist of staccato messages to the limbic system, which convey what is sought and the result of the search. After multisensory convergence, the spatiotemporal activity pattern in the limbic system is up-dated through temporal integration in the hippocampus. Accompanying sensory messages there are return up-dates from the limbic system to the sensory cortices, whereby each cortex receives input that has been integrated with the input from all others, reflecting the unity of intentionality. Everything that a human or an animal knows comes from the circular causality of action, preafference, perception and assimilation. Successive frames of self-organized activity patterns in the sensory and limbic cortices embody the cycle. This is the full program that was implicit in James' pragmatism, before the electrophysiological techniques of brain imaging made explicit the preconscious neural operations of intentionality.

5. Level 1 - Microscopic: Circular Causality among Neurons and Neural Masses

The "state" of the brain is a description of what it is doing in some specified time period. A state transition occurs when the brain changes and does something else. For example, locomotion is a state, within which walking is a rhythmic pattern of activity that involves large parts of the brain spinal cord, muscles and bones. The entire neuromuscular system changes almost instantly with the transition to a pattern of jogging or running: Similarly, a sleeping state can be taken as a whole, or divided into a sequence of slow wave and REM stages. Transit to a waking state can occur in a fraction of a second, whereby the entire brain and body shift gears, so to speak. The state of a neuron can be described as active and firing or as silent, with sudden changes in patterns of firing constituting state transitions. Populations of neurons also have a range of states, such as slow wave, fast activity, seizure, or silence. The science of dynamics describes states and their transitions.

The most critical question to ask about a state is its degree of stability or resistance to change or dissolution. Stability is evaluated by perturbing an object or a system (Freeman 1975). For example, an egg on a flat surface is unstable, but a coffee mug is stable. A person standing on a moving bus and holding on to a railing is stable, but someone walking in the aisle is not. If a person regains his chosen posture after each perturbation no matter in which direction the displacement occurs, that state is regarded as stable, and it is said to be governed by an attractor. This is a metaphor to say that the system goes (" is attracted") to the steady state through interim transience. The range of displacement from which recovery can occur defines the basin of attraction in analogy to a ball rolling to the bottom of a bowl. If a perturbation is so strong that it causes concussion or a broken leg, so that the person cannot stand up again, then the system has been placed outside the basin of attraction, and a new state supervenes with its own attractor and basin of attraction.

Stability is always relative to the time duration of observation and the criteria for what the experimentalist chooses to observe. In the perspective of a lifetime, brains appear to be highly stable, in their numbers of neurons, their architectures and major patterns of connection, and in the patterns of behavior they produce, including the character and identity of the individual that can be recognized and followed for many years. A brain undergoes repeated state transitions from waking to sleeping and back again coming up refreshed with a good night or irritable with insomnia, but still, giving arguably the same person as the night before. But in the perspective of the short term, brains are highly unstable. Thoughts go fleeting through awareness, and the face and body scintillate with the flux of emotions. Glimpses of the internal states of neural activity reveal patterns that are more like hurricanes than the orderly march of symbols in a computer, with the difference that hurricanes don't learn. Brain states and the states of populations of neurons that interact to give brain function are highly irregular in spatial form and time course. They emerge, persist for a small fraction of a second, then disappear and are replaced by other states.

Neuroscientists aim to describe and measure these states and tell what they signify for observations of behavior and experiences with awareness. The approach of dynamics is by defining three kinds of stable state, each with its type of attractor. The simplest is the point attractor. The system is at rest unless perturbed, and it returns to rest when allowed to do so. As it relaxes to rest, it has a brief history that is lost upon convergence to rest. Examples of point attractors are neurons or neural populations that have been isolated from the brain and also the brain that is depressed into inactivity by injury or a strong anesthetic, to the point where the EEG has gone flat. A special case of a point attractor is noise. This state is observed in populations of neurons in the brain of a subject at rest, with no evidence of overt behavior or awareness. The neurons fire continually but not in concert with each other. Their pulses occur in long trains at irregular times. Knowledge about the prior pulse trains from each neuron and those of its neighbors up to the present fails to support the prediction of when the next pulse will occur. The state of noise has continual activity with no history of how it started, and it gives only the expectation that its average amplitude and other statistical properties will persist unchanged.

A system that gives periodic behavior is said to have a limit cycle attractor. The classic example is the clock. When it is viewed in terms of its ceaseless motion it is regarded as unstable until it winds down runs out of power, and goes to a point attractor. If it resumes its regular beat after it is re-set or otherwise perturbed, it is stable as long as its power lasts. Its history is limited to one cycle, after which there is no retention of its transient approach in its basin to its attractor. Neurons and populations rarely fire periodically, and when they appear to do so, close inspection shows that the activities are usually irregular and unpredictable in detail, and when periodic activity does occur, it is likely to be either intentional, as in rhythmic drumming, or pathological, as in nystagmus and Parkinsonian tremor.

The third type of attractor gives aperiodic oscillation of the kind that is observed in recordings of EEGs and of physiological tremors. There is no one or small number of frequencies at which the system oscillates. The system behavior is therefore unpredictable, because performance can only be projected far into the future for periodic behavior. This type was first called "strange"; it is now widely known as "chaotic". The existence of this type of oscillation was known to mathematicians a century ago, but systematic study was possible only recently after the full development of digital computers. The best-known simple systems with chaotic attractors have a small number of components and a few degrees of freedom, as for example, the double-hinged pendulum. These model systems are noise-free, stationary and autonomous, meaning that they do not interact with their environments. Large and complex systems such as neurons and brains are noisy, unstable, and engaged with their environments, so the low-dimensional models do not apply. The hallmark of chaos is the capacity for rapid switching between states despite high dimensionality, plus the capacity for creating information in novel patterns. These are the properties that make chaotic dynamics interesting and applicable to behavior, even though proof is not yet possible at the present level of developments in mathematics.

The discovery of chaos has profound implications for the study of brain function (Skarda and Freeman 1987). A dynamic system has a collection of attractors, each with its basin which forms an 'attractor landscape' with all three types. The state of the system can jump from one to another in an itinerant trajectory (Tsuda 1991). Capture by a point or limit cycle attractor wipes clean the history upon asymptotic convergence, but capture in a chaotic basin engenders continual aperiodic activity, thereby creating novel, unpredictable patterns that retain its recent history. Although the trajectory is not predictable, the statistical properties such as the mean and standard deviation of the state variables of the system serve as measures of its steady state. Chaotic fluctuations carry the system endlessly around in the basin. However, if energy is fed into the system so that the fluctuations increase in amplitude, or if the landscape of the system is changed so that the basin shrinks or flattens, a microscopic fluctuation can carry the trajectory across the boundary between basins to another attractor.

In the dynamic space of each sensory cortex there are multiple chaotic attractors with basins corresponding to previously learned classes of stimuli, including those for learned background stimulus configurations. These multiple basins and attractors constitute an attractor landscape. Chaotic prestimulus states of expectancy establish the sensitivity of the cortex by warping the landscape, so that a small number of sensory action potentials driven by an expected stimulus (the "figure"), accompanied by a large number of action potentials from irrelevant stimuli (the "background"), can carry the cortical trajectory into the basin of an appropriate attractor. Circular causality enters in the following way. The state of a neural population in an area of cortex is a macroscopic event that arises through the interactions of the microscopic activity of the neurons comprising the neuropil. The global state is upwardly generated by the microscopic neurons, and simultaneously the global state downwardly organizes ("enslaves" — Haken 1983) the activities of the individual neurons.

Each cortical state transition requires this circularity. It is preceded by a conjunction of antecedents. A stimulus is sought by the limbic brain through orientation of the sensory receptors in sniffing, looking, and listening. The landscape of the basins of attraction is shaped by limbic preafference, which facilitates access to an attractor by expanding its basin for the reception of a desired class of stimuli as with Voronoi tesselations. Preafference provides the ambient context by multisensory divergence from the limbic system. The web of synaptic connections modified by prior learning maintains the basins and attractors. Pre-existing chaotic fluctuations are enhanced by input, forcing the selection of a new macroscopic state that then engulfs the stimulus-driven microscopic activity.

There are two reasons that all the sensory systems (visual, auditory, somatic and olfactory) operate this way. First, the brain faces the infinite complexity of the world with a finite capacity for understanding. The solution for the brain is to create its own information in the form of hypotheses, which it tests by acting into the environment. In olfaction for example, a significant odorant may consist of a few molecules mixed in a rich and powerful background of undefined substances, and the odorant the brain seeks may be continually changing in age, temperature, and concentration. Each sniff in a succession with the same chemical activates a different subset of equivalent olfactory receptors, so the microscopic input is unpredictable and unknowable in detail. Detection and tracking require abstraction to an invariant pattern over trials. The attractor provides the abstraction, and the basin provides the generalization over equivalent receptors. The attractor determines the response, not the particular stimulus. Unlike the view proposed by stimulus-response reflex determinism, the dynamics gives no linear chain of cause and effect from stimulus to response that can lead to the necessity of environmental determinism. The second reason that all sensory systems operate in the same way is the requirement that the neural outputs of all sensory systems have the same basic form, so that they can be combined into Gestalts, as they are converged by integration and extension over time in the entorhinal-hippocampal system (Freeman. 1995).

6. Circular Causality in Awareness

Circular causality underlies all state transitions in sensory cortices and the olfactory bulb, when fluctuations in microscopic activity exceed a certain threshold, such that a new macroscopic oscillation emerges that forces cooperation on the very neurons that have brought the pattern into being. EEG measurements show that multiple patterns self-organize independently in overlapping time frames in the several sensory and limbic cortices. The patterns emerge and coexist briefly with stimulus-driven activity in specialized areas of the neocortex that receive the projections of sensory pathways. On transmission of cortical output, the created patterns are broadcast widely, whereas the stimulus-driven activity, having done its work, is deleted. The overlapped cortical output patterns are combined into a hemisphere-wide pattern by the neocortex, which structurally is an undivided sheet of neuropil in each hemisphere .

Circular causality can serve as the framework for explaining the operation of awareness in the following way. The multimodal macroscopic patterns converge simultaneously into the limbic system, and the results of integration over time and space are simultaneously returned to all of the sensory systems. Here comes into play James' "... organ added for the sake of steering a nervous system grown too complex to regulate itself ...", though it is not another organ but, instead, another hierarchical level in brain function: a hemisphere-wide attractor, for which the local mesoscopic activity patterns are the components. The forward limb of the loop provides the bursts of oscillations converging into the limbic system that destabilize it to form new patterns. The feedback limb incorporates the limbic and sensory cortical patterns into a global activity pattern or order parameter that enslaves all of the components (Haken 1983). The enslavement enhances the coherence among all of them, which dampens the chaotic fluctuation instead of enhancing it, as the receptor input does in the sensory cortices.

A global operator of this kind must exist, for the following reason. The synthesis of sense data first into cortical wave packets and then into a multimodal packet takes time. After a Gestalt has been achieved through embedding in past experience, a decision is required as to what the organism is to do next. This also takes time (Libet 1994) for an evolutionary trajectory through a sequence of attractors constituting the attractor landscape of possible goals and actions (Tsuda 1991). The triggering of a state transition in the motor system may occur at any time, if the fluctuations in its multiple inputs are large enough, thereby terminating the search trajectory. In some emergent behavioral situations an early response is most effective: action without reflection. In complex situations with unclear ramifications into the future, precipitate action may lead to disastrous consequences. More generally, the forebrain appears to have developed in phylogenetic evolution as an organ taking advantage of the time provided by distance receptors for the interpretation of raw sense data. The quenching function of a global operator to delay decision and action can be seen as a necessary complement on the motor side, to prevent premature closure of the process of constructing and evaluating alternative courses of possible action.

Action without the deferral that is implicit in awareness can be found in so-called 'automatic' sequences of action in the performance of familiar complex routines such as driving a car, and in thoughtless and potentially self-destructive actions that Davidson (1980) described as "incontinent". Actions can "flow" without cautionary awareness. Implicit cognition is continuous, and it is simply unmasked in the conditions that lead to 'blindsight'. In this view, emotion is defined as the impetus for action, more specifically, for impending action. Its degree is proportional to the amplitudes of the chaotic fluctuations in the limbic system, which appear in the modulation depths of the carrier waves of limbic neural activity patterns (Lesse 1957). In accordance with the James-Lange theory of emotion (James 1893), it is experienced through awareness of the activation of the autonomic nervous system in preparation for and support of overt action, as described by Cannon (1939). It is observed in the patterns of behavior that social animals have acquired through evolution (Darwin 1872). Emotion is not in opposition to reason. Behaviors that are seen as irrational and incontinent result from premature escape of the chaotic fluctuations from the leavening and smoothing of the awareness operator. The most intensely emotional behavior, as it is experienced in artistic creation scientific discovery, and religious awe, occurs as the intensity of restraining awareness rises in concert with the strength of the fluctuations (Freeman 1995). As with all other difficult human endeavors, self-control is achieved through long and arduous practice. As James correctly surmised (James 1893), the contents of moments of awareness are severely limited, giving rise to his metaphor of a "penumbra" around a moving spot of light to describe the seamless texture of intertwined meanings, knit together by the awareness operator, with continuous influence of contents well outside the penumbra constituting the unconscious.

Evidence for the existence of the postulated global operator is found in the high level of covariance in the EEGs simultaneously recorded from the bulb and the visual, auditory, somatic and limbic (entorhinal) cortices of animals, and from the scalp of humans (Lehmann and Michel 1990; Miltner et al. 1999; Rodriguez et al. 1999; Müller 2000; Csibra et al. 2000; Tallon-Baudry et al. 1998; Haig et al. 2000). The magnitude of the shared activity can be measured in limited circumstances by the largest component in principle components analysis (PCA). Even though the waveforms of the several sites vary independently and unpredictably, the first component has 50-70% of the total variance (Smart et al., 1997; Gaál and Freeman 1998; Freeman, Gaál and Jörsten, 2003). These levels are lower than those found within each area of 90-98% (Barrie, Freeman and Lenhart, 1996), but they are far greater than can be accounted for by any of a variety of statistical artifacts or sources of correlation such as volume conduction pacemaker driving, or contamination by the reference lead in monopolar recording. The high level of coherence holds for all parts of the EEG spectrum and for aperiodic as well as near-periodic waves. The maximal coherence appears to have zero phase lag over distances up to several centimeters between recording sites and even between hemispheres (Singer and Gray, 1995). Attempts are being made to model the observed zero time lag among the structures by cancellation of delays in bidirectional feedback transmission (König and Schillen 1991; Traub et al. 1996; Roelfsma et al., 1997).

7. Consciousness Viewed as a System Parameter Controlling Chaos

An unequivocal choice can be made now between the three meanings of causality proposed in the Introduction. Consciousness and neural activity are not acausal processes operating in parallel, nor does either make or move the other as an agency in temporal sequences. Circular causality is a form of explanation that can be applied at several hierarchical levels without recourse to agency. This formulation provides the sense or feeling of necessity that is essential for human satisfaction, by addressing the elemental experience of cause and effect in acts of observation, even though logically it is very different from linear causality in all aspects of temporal order, spatial contiguity, and invariant reproducibility. The phrase is a cognitive metaphor. It lacks the attribute of agency, unless and until the loop is broken into the forward (microscopic) limb and the recurrent (macroscopic) limb, in which case the agency that is so compelling in linear causality can be re-introduced. This move acquiesces to the needs of the human observers, who use it in order to seek the quale of certainty in causation by studies of dynamic events and processes in the world.

I propose that the globally coherent brain activity, which can be recorded from scalp sensors noninvasively in human subjects, and which can be interpreted as an order parameter, may be an objective correlate of awareness through preafference, comprising expectation and attention, which are based in prior proprioceptive and exteroceptive feedback of the sensory consequences of previous actions, after they have undergone limbic integration to form Gestalts, and in the goals that are emergent in the limbic system. In this view, awareness is basically akin to the intervening state variable in a homeostatic mechanism, which is a physical quantity, a dynamic operator, and the carrier of influence from the past into the future that supports optimizing the relation between a desired set point and an existing state. The content of the awareness operator may be found in the spatial pattern of amplitude modulation of the shared waveform component, which is comparable to the amplitude modulation (AM patterns) of the carrier waves in the primary sensory receiving areas.

What is most remarkable about this operator is that it appears to be antithetical to initiating action. It provides a pervasive neuronal bias that does not induce state transitions, but defers them by quenching local fluctuations (Prigogine 1980). It alters the attractor landscapes of the lower order interactive masses of neurons that it enslaves. In the dynamic view, intervention by states of awareness in the process of consciousness organizes the attractor landscape of the motor systems, prior to the instant of its next state transition: the moment of choosing in the limbo of indecision when the global dynamic brain activity pattern is increasing its complexity and fine-tuning the guidance of overt action. This state of uncertainty and unreadiness to act may last a fraction of a second, a minute, a week, or a lifetime. Then when a contemplated act occurs, awareness follows the onset of the act and does not precede it.

In that hesitancy, between the last act and the next, comes the window of opportunity, when the breaking of symmetry in the next limbic state transition will make apparent what has been chosen. The observer of the self intervenes by awareness that organizes the attractor landscape, just before the instant of the next transition. The causal technology of self-control is familiar to everyone: hold off fear and anger; defer closure; avoid temptation; take time to study; read and reflect on an opportunity with its meaning and consequences; take the long view as it has been inculcated in the educational process. According to Mill (1843): "We cannot, indeed, directly will to be different from what we are; but neither did those who are supposed to have formed our characters directly will that we should be what we are. Their will had no direct power except over their own actions. ... We are exactly as capable of making our own character, *if we will*, as others are of making it for us" (p. 550).

There are numerous unsolved problems with this hypothesis. Although strong advances are being made in analyzing the dynamics of the limbic system and its centerpieces, the entorhinal cortex and hippocampus (Boeijinga and Lopes da Silva, 1988; O'Keefe and Nadel, 1978; Rolls et al., 1989; McNaughton 1993; Wilson and McNaughton 1993; Buzsaki, 1996; Eichenbaum, 1997; Traub et al., 1996), their self-organized spatial patterns, their precise intentional contents and their mechanisms of formation in relation to intentional action are still unknown. The prepyriform cortex to which the bulb transmits is strongly driven by its input, and it lacks evidence for self-organizing state transitions (Freeman and Barrie 2000) comparable to those of the sensory cortices. Whether the hippocampus has those capabilities or is likewise a driven structure is unknown. The neural mechanisms by which the entire neocortical neuropil in each hemisphere maintains spatially coherent activity over a broad spectrum with nearly zero time lag are still undefined. In order to establish the significance of this coherent activity for behavior, it will be necessary to find and classify the mental correlates of the spatial patterns of brain electrical activity. If those correlates are meanings (Freeman, 2003), then the subjects must be asked to make representations of the meanings in their minds, in order to communicate them to observers. If the subjects are animals, their representations of meanings are restricted to gestures and goal-directed actions. If the subjects are humans, they can speak, write, draw, and make music. Given the new techniques for brain imaging now available to us, knowledge of human brain function may well be within the present reach of neurodynamics, in particular the Jamesian operator that has evolved to match and manage the complexity of our brains.

8. Conclusion

Consciousness in the neurodynamic view is a global internal state variable and self-regulating operator acting in a sequence of momentary states of awareness. It is essential for incorporating each new frame of awareness in the life history of an individual, which is the wholeness of intentionality. Its regulatory role is minimally comparable to that of the operator in a thermostat, that instantiates the difference between the sensed temperature and a set point, and that initiates corrective action by turning a heater on or off. The machine state variable has little history and no capacities for learning or determining its own set point, but the principle is the same: the internal state is a form of energy, an operator, a predictor of the future, and a carrier of information that is available to the system as a whole. A thermostat is a prototype, an evolutionary precursor, not to be confused with awareness, any more than tropism in plants and bacteria is to be confused with intentionality. In the Jamesian framework consciousness is the utilitarian organizer of whatever working parts the brain can provide at its present level of evolution or devolution, to use the terms of J. Hughlings Jackson (1931). In humans, the operations and informational contents of the global state variable, which are sensations, images, feelings, thoughts and beliefs, constitute the experience of cause and effect.

To deny this comparability and assert that humans are not machines is to miss the point. Two things distinguish humans from all other beings. One is the form and function of the human body, including the brain, which has been given to us by three billion years of biological evolution. The other is the heritage given to us by two million years of cultural evolution. Our mental attributes have been characterized for millennia as the soul or spirit or consciousness that makes us not-machines. The uniqueness of the human condition is not thereby explained, but the biological foundation forged by James is enhanced by the concept of circular causality. It provides a tool for intervention when something has gone wrong, because the circle can be broken into forward and feedback limbs. Each of the limbs, physical and mental, can be explained by linear causality, which can enable us to understand where and how we might intervene. The only error would be to assign causal agency to the parts of the brain instead of to ourselves as physicians and psychiatrists, as we act in the belief of our efficacy as causal agents, yet still with the humility expressed in the epitaph of Ambroise Paré, 16th century French surgeon extraordinary: "Je le pansay, Dieu le guarit" ("I bound his wounds, God healed him").

Acknowledgments

This research was supported by grants from the National Institutes of Health MH-06686 and the Office of Naval Research N00014-93-1-0938. The essay appeared in an earlier version in the **Journal of Consciousness Studies** 6 Nov/Dec: 143-172, 1999 and in book form in "Reclaiming Cognition" edited by R. Núñez and W. J. Freeman.

References

Aquinas, St. Thomas (1272) *The Summa Theologica.* Translated by Fathers of the English Dominican Province. Revised by Daniel J Sullivan. Published by William Benton as Volume 19 in the Great Books Series. Chicago: Encyclopedia Britannica, Inc., 1952.

Barrie JM, Freeman WJ & Lenhart M (1996) Modulation by discriminative training of spatial patterns of gamma EEG amplitude and phase in neocortex of rabbits. *J. Neurophysiol.* **76**: 520-539.

Boeijinga, P.H. & Lopes da Silva, F.H. (1988) Differential distribution of beta and theta EEG activity in the entorhinal cortex of the cat. *Brain Res.* **448**: 272-286, 1988.

Brentano FC (1889) *The Origin of our Knowledge of Right and Wrong*. Chisolm RM and Schneewind EH (trans.). New York: Humanities Press (1969).

Bures J, Buresová O, Krivánek J (1974) *The Mechanism and Applications of Leão's Spreading Depression of Electroencephalographic Activity*. New York: Academic Press.

Buzsaki G. (1996) The hippocampal-neocortical dialogue. *Cerebral Cortex* **6**: 81-92.

Cannon WB (1939) *The Wisdom of the Body.* New York: WW Norton

Cartwright N (1989) *Nature's Capacities and their Measurement.* Oxford UK: Clarendon Press.

Chalmers DJ (1996) *The Conscious Mind. In Search of a Fundamental Theory*. New York: Oxford University Press.

Crick F (1994) *The Astonishing Hypothesis: The Scientific Search for the Soul*. New York: Scribner.

Csibra G, Davis G, Spratling MW, Johnson MH (2000) Gamma oscillations and object processing in the infant brain. *Science* **290**: 1582-1585.

Darwin C (1872) *The Expression of the Emotions in Man and Animals*. London UK: J. Murray.

Davidson D (1980) Actions, reasons, and causes. In: *Essays on Actions & Events.* Oxford UK: Clarendon Press.

Dennett DH (1991) *Consciousness Explained*. Boston: Little, Brown.

Dewey J (1914) Psychological doctrine in philosophical teaching. *J. Philosophy* **11**: 505-512.

Eccles JC (1994) *How the Self Controls Its Brain*. Berlin: Springer-Verlag.

Eichenbaum H. (1997) How does the brain organize memories? *Science* **277**: 330-332.

Foucault M (1976) *The History of Sexuality: Vol. 1. An Introduction* (R Hurley, trans.). New York: Random House (1980).

Freeman WJ (1975) *Mass Action in the Nervous System*. New York: Academic

Freeman WJ (1984) Premises in neurophysiological studies of learning. Ch 13 in Lynch G, McGaugh JL, Weinberger NM (eds.) *Neurobiology of Learning and Memory*. New York: Guilford Press.

Freeman WJ (1991) The physiology of perception. *Scientific American* **264**: 78-85.

Freeman WJ (1992) Tutorial in Neurobiology: From Single Neurons to Brain Chaos. *Int. J. Bifurc. Chaos* **2**: 451-482.

Freeman WJ (1995) *Societies of Brains. A Study in the Neuroscience of Love and Hate.* Mahwah NJ: Lawrence Erlbaum Associates.

Freeman WJ. (2003) A neurobiological theory of meaning in perception. Part 1. Information and meaning in nonconvergent and nonlocal brain dynamics. *Int. J. Bifurc. Chaos* **13**: in press.

Freeman WJ, Schneider W (1982) Changes in spatial patterns of rabbit olfactory EEG with conditioning to odors. *Psychophysiol.* **19**: 44-56.

Freeman WJ, Barrie JM (2000) Analysis of spatial patterns of phase in neocortical gamma EEGs in rabbit. *J. Neurophysiol.* **84**: 1266-1278.

Freeman WJ, Gaál G and Jörsten R. (2003) A neurobiological theory of meaning in perception. Part 2. Multiple cortical areas synchronize without loss of local autonomy. *Int. J. Bifurc. Chaos* **13**: in press.

Gibson JJ (1979) *The Ecological Approach to Visual Perception*. Boston: Houghton Mifflin.

Gloor P (1997) *The Temporal Lobe and the Limbic System*. New York: Oxford University Press.

Goltz FL (1892) Der Hund ohne Grosshirn. Siebente Abhandlung über die Verrichtungen des Grosshirns. *Pflügers Archiv* **51**: 570-614.

Gray CM (1994) Synchronous oscillations in neuronal systems: mechanisms and functions. *J. Comp. Neurosci.* **1**: 11-38.

Grossberg S (1982) *Studies of Mind and Brain: Neural Principles of Learning, Perception Development, Cognition and Motor Control*. Boston: D. Reidel.

Haken H (1983) *Synergetics: An Introduction*. Berlin: Springer.

Hardcastle VG (1994) Psychology's binding problem and possible neurobiological solutions. *J. Consciousness Studies* **1**: 66-90.

Hebb DO (1949) *The Organization of Behavior*. New York: Wiley.

Helmholtz HLF von (1872) *Handbuch der physiologischen Optik. Vol. III.* Leipzig: L. Voss (1909).

Herrnstein RJ & Murray C (1994) *The Bell Curve*. New York: The Free Press.

Herrick CJ (1948) *The Brain of the Tiger Salamander*. Chicago IL: University of Chicago Press.

Hull CL (1943) *Principles of Behavior, An Introduction to Behavior Theory*. New York: Appleton-Century

Hume D (1739) *Treatise on Human Nature*. London: J Noon.

Jackson JH (1931) *Selected writings of John Hughlings Jackson*. Taylor J, Walshe FMR, Holmes G (eds.). London: Hodder & Stoughton.

James W (1879) Are we automata? *Mind* **4**: 1-21.

James W (1893) *The Principles of Psychology*. New York. H. Holt.

James, W (1890) The Principles of Psychology p. 1264 in: *The Works of William James*, **2**, p. 1264, Burkhardt FH (ed. 19 vols). Cambridge MA: Harvard University Press.

Kay LM, Freeman WJ (1998) Bidirectional processing in the olfactory-limbic axis during olfactory behavior. *Behav. Neurosci.* **112**: 541-553.

König P, Schillen TB (1991) Stimulus-dependent assembly formation of oscillatory responses: I. synchronization. *Neural Comp.* **3**, 155-166.

Kozma R, Freeman WJ (2001) Chaotic resonance: Methods and applications for robust classification of noisy and variable patterns. *Int. J. Bifurc. Chaos* **10**: 2307-2322.

Lehmann D, Michel CM (1990) Intracerebral dipole source localization for FFT power maps. *Electroenceph. Clin. Neurophysiol.* **76**: 271-276.

Lesse H (1957) Amygdaloid electrical activity during a conditioned response. *Proc. 4th Intern. Congress EEG clin. Neurophysiol.*, Brussels, 99-100.

Libet B (1994) *Neurophysiology of Consciousness: Selected Papers and New Essays*. Boston MA: Birkhauser.

Lorente de Nó R (1934) Studies on the structure of the cerebral cortex. I The area entorhinalis. *J. für Psychologie und Neurologie* **45**: 381-438.

McCulloch WS (1967) *Embodiments of Mind.* Cambridge MA: MIT Press.

McNaughton BL (1993) The mechanism of expression of long-term enhancement of hippocampal synapses: Current issues and theoretical implications. *Ann. Rev. Physiol.* **55**: 375-96.

Menand L (2001) *The Metaphysical Club*. New York: Farrar, Straus & Giroux, 2001.

Merleau-Ponty M (1942) *The Structure of Behavior* (AL Fischer, Trans.). Boston: Beacon Press (1963).

Mill JS (1843) *Of Liberty and Necessity, Ch. II, Book VI. A System of Logic.* London UK: Longmans, Green 18th ed. (1965)

Mill JS (1873) *Autobiography*. New York: Columbia University Press (1924).

Miltner WHR, Barun C, Arnold M, Witte H, Taub E (1999) Coherence of gamma-band EEG activity as a basis for associative learning. *Nature* **397**: 434-436.

Müller MM (2000) Hochfrequente oszillatorische Aktivitäten im menschlichen Gehirn. *Zeitschrift für Exper. Psychol.* **47**: 231-252.

Ohl, F.W., Scheich, H., Freeman W.J. (2001) Change in pattern of ongoing cortical activity with auditory category learning. *Nature* **412**: 733-736.

O'Keefe J & Nadel L (1978) *The Hippocampus as a Cognitive Map*. Oxford UK: Clarendon.

Panksepp, J. (1998) *Affective Neuroscience.* Oxford University Press.

Penrose R (1994) *Shadows of the Mind.* Oxford UK: Oxford University Press.

Piaget J (1930) *The child's conception of physical causality*. New York: Harcourt, Brace.

Popper KR (1990) *A World of Propensities*. Bristol UK: Thoemmes.

Prigogine I (1980) *From Being to Becoming: Time and Complexity in the Physical Sciences*. San Francisco: Freeman.

Principe, J.C., Tavares, V.G., Harris, J.G. and Freeman W.J. (2001) Design and implementation of a biologically realistic olfactory cortex in analog VLSI. *Proceedings IEEE* **89**: 1030-1051.

Putnam H (1990) *Realism With a Human Face*. Cambridge MA: Harvard University Press.

Rodriguez E, George N Lachaux J-P, Martinerie J, Renault B, Varela F (1999) Perception's shadow: long-distance synchronization of human brain activity. *Nature* **397**: 430-433.

Roelfsema, P.R., Engel, A.K., König, P. & Singer, W. (1997) Visuomotor integration is associated with zero time-lag synchronization among cortical areas. *Nature* **385**, 157-161.

Rolls ET, Miyashita Y, Cahusac PBM, Kesner RP, Niki H, Feigenbaum JD & Bach L (1989) Hippocampal neurons in the monkey with activity related to the place in which the stimulus is shown. *J. Neurosci.* **9**: 1835-1845.

Roth G (1987) *Visual Behavior in Salamanders*. Berlin: Springer-Verlag

Searle JR (1992) *The Rediscovery of Mind.* Cambridge MA: MIT Press.

Sherrington CS (1940) *Man on his Nature.* Oxford UK: Oxford University Press.

Singer W, Gray CM (1995) Visual feature integration and the temporal correlation hypothesis. *Ann. Rev. Neurosci.* **18**: 555-586.

Skarda CA, Freeman WJ (1987) How brains make chaos in order to make sense of the world. *Behav. Brain Sci.* **10**: 161-195.

Skinner BF (1969) *Contingencies of Reinforcement; A Theoretical Analysis*. New York: Appleton-Century-Crofts

Smart A, German P, Oshtory S, Gaál G, Barrie JM, Freeman WJ (1997) Spatio-temporal analysis of multi-electrode cortical EEG of awake rabbit. *Soc. Neurosci. Abstracts* 189.13.

Sperry RW (1950) Neural basis of the spontaneous optokinetic response. *J. Comp. Physiology* **43**: 482-489.

Tallon-Baudry C, Bertrand O, Peronnet F, Pernier J (1998). Induced gamma-band activity during the delay of a visual short-term memory task in humans. *J. Neurosci.* **18**: 4244-4254.

Thelen E, Smith LB (1994) *A Dynamic Systems Approach to the Development of Cognition and Action.* Cambridge MA: MIT Press.

Titchener EB (1907) *An Outline of Psychology.* New York: Macmillan.

Traub RD, Whittington MA, Colling SB, Buzsaki G, Jefferys JGR (1996) A mechanism for generation of long-range synchronous fast oscillations in the cortex. *Nature* **383**: 621-624.

Tsuda I. (2001) Toward an interpretation of dynamics neural activity in terms of chaotic dynamical systems. *Behav. & Brain Sci.* **24**: 793-847.

von Holst E & Mittelstädt H (1950) Das Reafferenzprinzip (Wechselwirkung zwischen Zentralnervensystem und Peripherie). *Naturwissenschaften* **37**: 464-476.

von Neumann J (1958) *The Computer and the Brain*. New Haven CT: Yale University Press.

Whitehead AN (1938) *Modes of Thought*. New York: Macmillan.

Wilson MA & McNaughton BL (1993) Dynamics of the hippocampal ensemble code for space. *Science* **261**: 1055-1058.

In: Chaos and Complexity Research Compendium
Editors: F. Orsucci and N. Sala, pp. 47-60
ISBN: 978-1-60456-787-8

Chapter 5

THE STRUCTURAL EQUATIONS TECHNIQUE FOR TESTING HYPOTHESES IN NONLINEAR DYNAMICS: CATASTROPHES, CHAOS, AND RELATED DYNAMICS

Stephen J. Guastello*
Marquette University

Abstract

This article summarizes the central analysis points for testing hypotheses concerning catastrophe models, chaos, and related dynamics as they are encountered in the social sciences. Two model classes are considered: the catastrophe models for discontinuous change processes, and the exponential series for continuous change. Some frequently asked questions concerning the types of and amount of data that are required for viable analyses are parenthetically answered here.

One of the gripping problems in nonlinear dynamics is difficulty in testing hypotheses with real data, especially the type of data that are collected in the social sciences. In spite of controversies and tales of woe that persist in discussions on listservers, reliable techniques exist, and indeed some have been in use for more than twenty years. The techniques described in this article are predicated on the assumption that the researcher has a viable hypothesis concerning one or more specific models that require explicit testing. A basic understanding of attractors, bifurcations, chaos, catastrophes, and self-organization is necessarily assumed here.

It is a good idea, nonetheless, to review why anyone would bother studying nonlinear dynamic systems (NDS) at all. There are four basic reasons: (a) NDS theory provides concepts that explain changes that occur over time. (b) NDS theory allows for structural comparisons of models across situations that may be very different in their outward appearances. (c) NDS solutions to problems, when they have been adopted, provide better

* E-mail address: stephen.guastello@marquette.edu. Tel: 414-288-6900, Send correspondence concerning the manuscript to:Stephen J. Guastello, Ph.D., Dept. Psychology, Marquette University, P.O. Box 1881, Milwaukee, WI 53201-1881 USA

explanations of data if and when they are expressed by R^2, or percentage of variance accounted for by a nonlinear model (Guastello, 1992a, 1995, 2002). (d) NDS provides effective solutions to theoretical questions; see Guastello (2001a) for a survey of progress on topics in psychology.

Overview

The structural equations technique begins be defining a model in the form of an equation, then testing it statistically with real (as opposed to simulated) data. The analysis separates the deterministic portion of the data from noise. "Noise" here denotes that portion of the data variance that is not explained by the deterministic equation. Social scientists will recognize this model-versus-noise approach as "business as usual." This technique contrasts, however, with a prevailing habit in the physical sciences, which works in the opposite fashion: Separate the noise first, then make calculations on what remains (e.g. Kanz & Schreiber, 1997).

Two series of hierarchical models are considered here. The first is the catastrophe models for discontinuous change. The catastrophes, which were originally introduced by Thom (1975) have received renewed attention because of their relevance to self-organized systems (Kelso, 1995; Zhang, 2002). The second set involved exponential models for continuous change and includes a test for the Lyapunov exponent which distinguishes between chaos and non-chaotic dynamics. The latter set was introduced by Guastello (1995) and built on previous work by May and Oster (1976), Wiggins (1988) and numerous other contributors to the field of NDS.

Each model in a hierarchy subsumes properties of the simpler models. Each progressively complex model adds a new dynamical feature. This article covers models involving only one order parameter (dependent measure). Two-parameter models can be tested as well, but the reader is directed to Guastello (1995) to see how those extrapolations can be accomplished.

The following sections of this article address the type of data and amounts that are required; probability functions, location, and scale; the structure of behavioral measurements; the catastrophe model group, which can be tested through power polynomial regression; the exponential series of models, which can be tested through nonlinear regression; and catastrophe models that are testable as static probability functions through nonlinear regression.

Types and Amount of Data

The procedures that follow require dependent measures (order parameters) that are measured at two points in time at least. One may have many entities that are measured at two points in time, or one long time series of observations from one entity. Alternatively, one may have an ensemble of shorter time series taken from several entities.

In general it is better to have a smaller number of observations that cover the full dynamical character of a phenomenon than to have a large number of observations that cover the underlying topology poorly. Because these are statistical procedures, all the usual rules and caveats pertaining to statistical power apply. The simplest models can be tested with 50 data points, and sometimes fewer of them, if (a) a good model of the phenomenon in question, (b) reliable measurements, (c) only one or two regression parameters to estimate. In all cases, more data is better than less data so long as the data are actually covering all the nonlinear dynamics that are thought to exist in the system.

Statistical Power

The calculation of statistical power for ordinary multiple regression depends on the intended effect size, overall sample R^2, population R^2, the number of independent variables, the degree of correlation among the independent variables, and occasionally the assumption that all independent variables are equally weighted. Not surprisingly there are different rubrics for determining proper sample size.

According to the Cohen (1988), the appropriate sample size to detect a "medium" effect size of .15 for one of the independent variables with a power of .80 is 52 plus the number of estimated parameters (Maxwell, 2000). Thus the required sample size would be 58 observations for a six-parameter model. According to a similar rubric by Green (1991), a sample size of 110 should to detect an effect size of .075 for one independent variable with a power of .80. The current sample sizes would thus detect effect sizes of .07 and .04, respectively, with a power of .80. Neither rubric takes into account that the odds of finding a smaller partial correlation increase to the extent that the overall R^2 is large. According to Maxwell (2000), the odds of detecting one of the effects within a multiple regression model drops sharply as the correlations among independent variables increases.

One should bear in mind, however, that the calculation of statistical power for nonlinear regression is still generally uncharted territory. It is thus necessary to rely on the rubrics for linear models. If there is sufficient statistical power for the linear comparison models which, in the past, have been generally weaker in overall effect size than nonlinear models when the nonlinear model was held true, there should not be much concern with the statistical power of the nonlinear models. On the other hand, the power for specific effects within a nonlinear model probably depends on whether the regression parameter is associated with an additive, exponential, multiplicative, or other type of mathematical operator.

Optimal Time Lag

Put simply, the time lag between observations is optimal if it reflects the real time frame in which data points are generated. For instance, catastrophe models are usually lagged "before" and "after" a discrete event. Macroeconomic variables might be studied best at lags equal to an economic quarter of the year (e.g. Guastello, 1999a).

Probability Density Functions

It is convenient that any differential function can be transformed into a probability density function pdf using the Wright-Ito transformation. The variable y in Eq. 1. is a dependent measure that exhibits the dynamical character under study; y is then transformed into z with respect to *location* (λ) and *scale* (σ_s, Eq. 2).

$$\text{pdf}(z) = \xi \exp[-I f(z)]; \quad (1)$$

$$z = (y - \lambda) / \sigma_s . \quad (2)$$

Location

In most discussions of probability functions, "location" refers to the mean of the function. In dynamics the pdf is a member of an exponential family of distributions and is asymmetrical, unlike the so-called normal distribution. Thus the location parameter for Eq. 2 is the lower limit of the distribution, which is the lowest observed value in the series. The transformation in Eq. 2 has the added advantage of fixing a zero point and thus transforming measurements with interval scales (common in the social sciences) to ratio scales. A fixed location point defines where the nonlinear function is going to start.

Scale

The scale parameter in common discussions of pdfs is the standard deviation of the distribution. The standard deviation is used here also. The use of the scale parameter later on while testing sturctural equations serves the purpose of eliminating bias between two or more variables that are multiplied together. Although the results of linear regression are not affected by values of location and scale, nonlinear models are clearly affected by the transformation.

Occasionally one may obtain a better fit using the alternative definition of scale in Eq. 3, which measures variability around statistical modes rather than around a mean:

$$\sigma_s = \left[\sqrt{\Sigma\Sigma\left(y_{m-}^{m+} - m\right)^2} \right] / N - M \quad \text{y=1 m=1} \quad (3)$$

To use it, the distribution must be broken into sections, each section containing a mode or an antimode. The values of the variable around a mode will range from *m*- to *m*+ as depicted in Eq. 3.

Corrections for location and scale should be made on control variables as well as for dependent measures. The ordinary standard deviation is a suitable measure of scale for control variables.

Structure of Behavioral Measurements

In the classic definition, a measurement consists of true scores (T) plus error (e). The variance structure for a population of scores is thus:

$$\sigma^2(X) = \sigma^2(T) + \sigma^2(e). \quad (4)$$

The classical assumption is that all errors are independent of true scores and all other errors.

In nonlinear dynamics our true score is the result of a linear (L) and nonlinear deterministic process (NL), dependent error (DE), and independent error (IIDE):

$$\sigma^2(X) = \sigma^2(L) + \sigma^2(NL\text{-}L) + \sigma^2(DE) + \sigma^2(IIDE) \quad (5)$$

"Independent error" in conventional psychometrics is known as independently and identically distributed (IID) error in the NDS literature. Importantly the non-IID error is a result of the nonlinear deterministic process (Brock, Hseih, & LeBaron, 1991).

Catastrophe Models

The analysis that follows requires the polynomial form of multiple linear regression. The analysis can be performed with most any standard statistical software package. Several concepts for hypothesis testing carry through to subsequent analyses of other dynamics.

The set of catastrophe models was the result of the classification theorem by Thom (1975): Given certain constraints, all discontinuous changes in events can be described by one of seven elementary models. Four of the models contain one order parameter; this is the cuspoid series: fold model which has one control parameter, cusp model with two control parameters, swallowtail model with three control parameters, butterfly model with four control parameters. The remaining three models, known as the umbilic group, contain two order parameters. The instructions that follow pertain to the cuspoid group.

The process of hypothesis testing begins by choosing a model that appears to be closest to the phenomenon under investigation. Because the cusp is the most often used model, the following remarks are framed in terms of the cusp model. The cusp (Fig. 1) depicts two stable states of behavior and requires two control variables. The two stable states are separated by a bifurcation manifold or separatrix. The asymmetry parameter governs how close the system is to a sudden discontinuous change of events. The bifurcation parameter governs how large the change will be. Two or more experimental variables may be hypothesized for each control parameter without changing the basic model or analytic procedure.

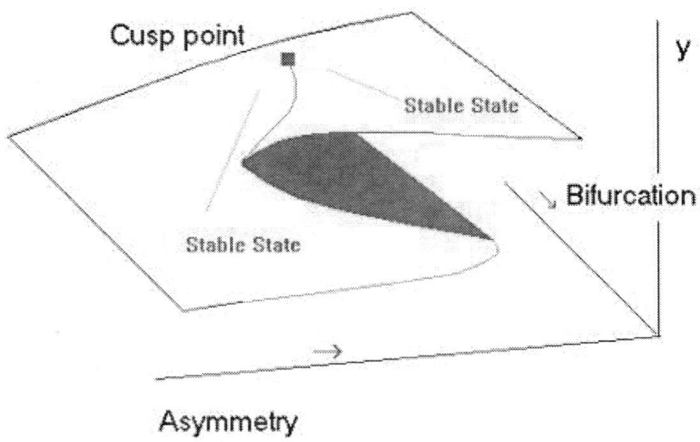

Figure 1. Response surface for the cusp catastrophe model.

Nonlinear Statistical Model

The deterministic equation for the cusp is shown in Eq. 6, and followed by its probability density function using the Wright-Ito transformation in Eq. 7:

$$\delta f(y)/\delta y = y^3 - by - a, \tag{6}$$

$$\text{pdf}(z) = \xi\, e^{[-z_4/4 + bz_2/2 + az]}. \tag{7}$$

Figure 2 shows an example of what a cusp pdf could look like using real data (Guastello, 2002, p. 136).

Next, we take the deterministic equation for the cusp response surface, insert regression weights, and a quadratic term:

$$\Delta z = \beta_0 + \beta_1 z_1^3 + \beta_2 z_1^2 + \beta_3 b z_1 + \beta_4 a, \tag{8}$$

The quadratic term is an additional correction for location. The dependent measure Δz denotes a change in behavior over two subsequent points in time.

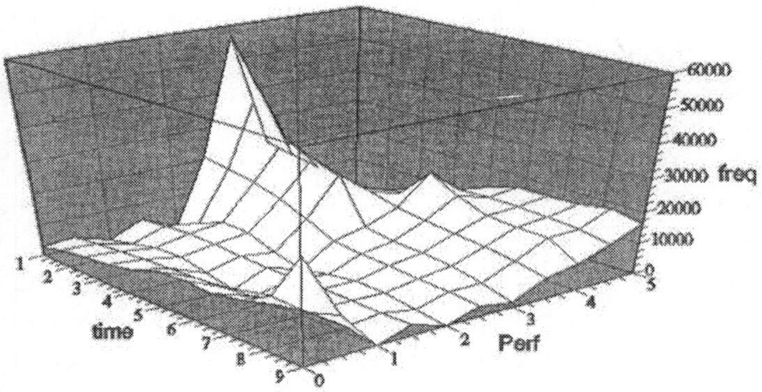

Figure 2. Example of a cusp pdf for a personnel selection and turnover problem (Guastello, 2002). Reprinted with permission from Lawrence Erlbaum Associates.

Several hypotheses are being tested in the power polynomial equation (Eq. 8). There is the *F* test for the model overall; retain the R^2 coefficient and save it for later use. There are *t* tests on the beta weights; they denote which parts of the model account for unique portions of variance.

Some model elements are more important than other elements. The cubic term expresses whether the model is consistent with cusp structure; the correct level of complexity for a catastrophe model is captured by the leading power term. If there is a cusp structure, then one must identify a bifurcation variable as represented by the $\beta b z_1$ term. A cusp hypothesis is not complete without a bifurcation term; shabby results may be expected otherwise. The asymmetry term βa is important in the model, but failing to find one does not negate the cusp structure if the cubic and bifurcation elements are present. The lack of an asymmetry term only means that the model is not complete.

The quadratic term is the most expendable. It is not part of the formal deterministic cusp structure. Rather it is an additional correction for location (Cobb, 1981a). In the event that unique weights are not obtained for all model components, delete the quadratic term and test the remaining elements again.

Note the procedural contrast with linear regression analysis: In common linear regression, when a variable does not attain a significant weight, we simply delete that variable. In NDS, we delete variables based on their relative importance to the structural model. In linear analyses, there is only a linear structure under consideration, so particular variables are then kept or discarded. In nonlinear analyses, different variables may be playing different structural roles.

Linear Comparison Models

Next construct Eqs. 9 and 10 and compare their R^2 coefficients against the R^2 that was obtained for the cusp:

$$y_2 = B_0 + B_1 y_1 + B_2 a + B_3 b, \tag{9}$$

$$\Delta y = B_0 + B_1 a + B_2 b. \tag{10}$$

Next evaluate the elements of the cusp model. If all the necessary parts of the cusp are significant, and the R^2 coefficients compare favorably, then a clear case of the cusp has been obtained.

Exponential Model Series

This section describes a series of models that exhibit continuous but nevertheless interesting change. The model structures are functions of the Naperian constant e. They produce, among other things, the Lyapunov exponent, which is a test for chaos and a value comparable to the fractal dimension.

Nonlinear regression is required the test this series of models. Nonlinear regression may be familiar to biologists, but is probably much less familiar to social scientists at the present time. The hierarchical series of models ranges from simplest to complex as follows: (a) simple Lyapunov exponent, (b) Lyapunov with additional fitting constants, (c) May-Oster model with the bifurcation parameter unknown, (d) model with an explicitly hypothesized bifurcation model, and (e) models with two or more order parameters.

Lyapunov Models

The simplest model predicts behavior z_2 from a function of z_1. Note that the corrections for location and scale apply here as well:

$$z_2 = e^{(\theta_1 z_1)} \tag{11}$$

The nonlinear regression weight θ_1 is located in the exponent. θ_1 is also the Lyapunov exponent. It is a measure of turbulence in the time series. If θ_1 is positive, then chaos is occurring. If θ_1 is negative, then a fixed point or periodic dynamic is occurring. D_L is an approximation of the fractal dimension (Ott, Sauer, & Yorke, 1994):

$$D_L = e^{\theta_1}. \tag{12}$$

The second model in the series is the same as the first except that two constants have been introduced to absorb unaccounted variance. The Lyapunov exponent is now designated as θ_2:

$$z_2 = \theta_1 e^{(\theta_2 z_1)} + \theta_3 . \tag{13}$$

In nonlinear regression it is necessary to specify the placement of constants in a model. Unlike the general linear model, constants in nonlinear models can appear anywhere at all. Hence θ_1 and θ_3 are introduced in Eq. 13.

The suggested strategy here is to start with the second model (Eq. 13). If statistical significance is not obtained for all three weights, delete θ_1 and try again. If that result is not good enough, drop the additive constant θ_3 and return to the simplest model of the series (Eq. 11).

Bifurcation Models

The third level of model is shown in Eq. 14. Note the introduction of z_1 between θ_1 and e:

$$z_2 = \theta_1 z_1 e^{\theta_2 z_1} + \theta_3. \tag{14}$$

Eq. 14 tests for the presence of a variable that is possibly changing the dynamics of a model. For instance, some learning curves could be sharper than others. A positive test for the model indicates that a variable is present, but its identity is not yet known.

The computation of dimension is similar to that of previous exponential models, except that a value of 1 must be added to account for the presence of a bifurcation variable:

$$D_L = \mathrm{e}^{\theta_2} + 1. \tag{15}$$

At the fourth level of complexity, the researcher has a specific hypothesis for the bifurcation variable, which is designated as c in Eq. 16:

$$z_2 = \theta_1 c\, z_1 e^{\theta_2 z_1} + \theta_3. \tag{16}$$

Linear Contrasts

As in the case of the catastrophes, we test the R2 for the nonlinear regression model against that of the linear alternatives, such as

$$y_2 = B_0 + B_1 y_1 \quad (17)$$

or

$$y_2 = B_0 + B_1 t, \quad (18)$$

where *t* is time.

Tips for Using Nonlinear Regression

On the model parameter command line (or comparable command in statistical packages), the names of the regression weights are specified with initial values. Use either the initial values of 0.5 or pick your own. When in doubt, give the initial weights equal value. Often it does not matter whether the iterative computational procedure states off with equal weights or not. If the model results are not affected by the initial values, then the resulting model is more robust than what would be the case otherwise.

If there is an option to choose constrained versus unconstrained nonlinear regression, use the unconstrained option, which is typically the default. Constraints indicate that the researcher expects the values of parameters to remain within numerical boundaries that have been pre-determined. Occasionally there may be a good rationale for containing parameters, but they would be specific to the problem when they exist.

If there is an option to choose least squares or maximum likelihood error term specification be forewarned: Maximum likelihood is more likely to capitalize on chance aspects of the pdf, and is thus more likely to return a significant result. I use least squares.

If the results of a nonlinear regression analysis are so poor that they produce a negative R^2, do not be alarmed. Just treat the negative R^2 as if it were .00.

When testing for significance, the tests on the weights are very important. Some researchers value them more than the overall R^2. Tests for weights are made using the principle of confidence intervals. An alpha level of $p < .05$ is regarded as unilaterally sufficient. A nonsignificant weight with a high overall R^2 could be the result of a high correlation among the parameter estimates; this condition is akin to multicollinearity in ordinary linear models.

Testing Catastrophes through Nonlinear Regression

The two strategies previously delineated can now be combined for some special circumstances. Sometimes one might obtain a pdf that bears a strong resemblance to that of an elementary catastrophe, and it is logical to frame a hypothesis as to whether that association is true or false. In another situation, there may be a catastrophe process occurring, but all the time-1 measurements are the same value of 0.00.

In both types of situations it would be good to test a hypothesis concerning the catastrophe distribution. This type of test is described below for a situation that involves that swallowtail catastrophe model. The swallowtail response surface is shown in Fig 3. Because the response surface is four-dimensional, it must be presented in two 3-dimensional sections.

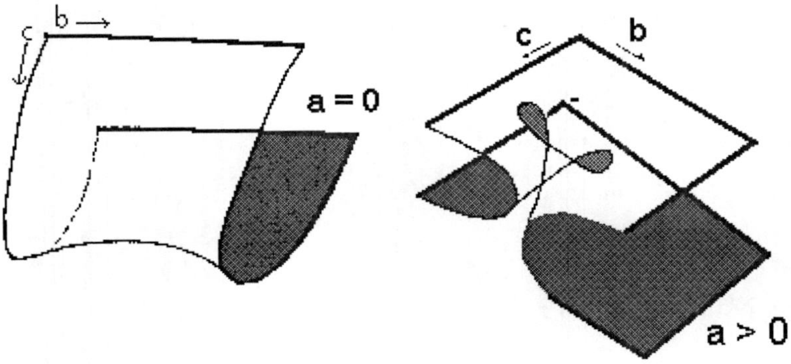

Figure 3. Swallowtail catastrophe response surface (Guastello, 2002). Reprinted with permission from Lawrence Erlbaum Associates.

The equation for the response surface is shown in Eq. 19:

$$\delta f(y)/\delta y = y^4 - cy^2 - by - a, \tag{19}$$

The swallowtail pdf is shown in Eq. 20 and Fig. 4:

$$\text{pdf}(z) = \xi \exp[-z^5/5 + z^4/4 + cz^3/3 + bz^2/2 + az]. \tag{20}$$

The pdf is tested as a nonlinear regression model in Eq. 21:

$$\text{pdf}(z) = \xi \exp[-\theta_1 z^5 + \theta_2 z^4 - \theta_3 cz^3 - \theta_4 bz^2 - \theta_5 az]; \tag{21}$$

Note where the regression weights are inserted in Eq. 21. θ_i is also treated as a regression weight. Pdf(z) is the cumulative probability of *z* within the distribution.

If the control variables are not known yet (it is necessary to have hypotheses about them in the polynomial regression models), variables *a*, *b*, and *c*, in Eq. 21 can be ignored. One would thus be treating them essentially as constants (or part of the regression weight).

An actual (slice of a) swallowtail PDF appears in Fig. 5. The jagged contour is real data. The smooth contour is based on estimated values that were obtained from an analysis using Eq. 22 (Guastello, 1998a):

$$\text{freq}(y) = \beta_0 + \beta_1 y + \beta_2 y^2 + \beta_3 y^3 + \beta_4 y^4 \tag{22}$$

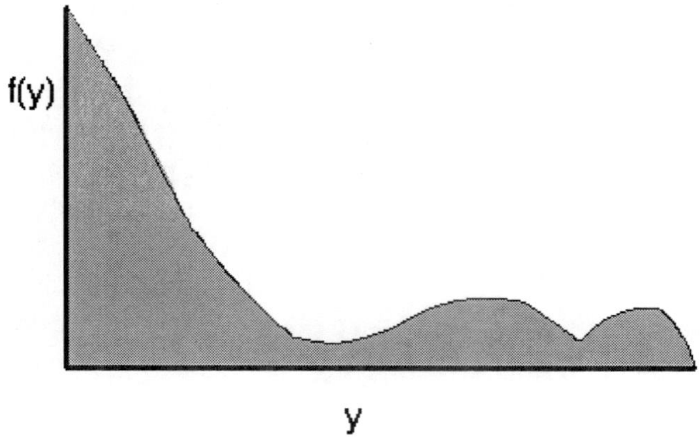

Figure 4. Slice of the swallowtail pdf.

Eq. 22 is a polynomial regression model where the *frequency of y* is a function of the value of *y*. This is the easiest way to make the smooth contour and identify critical points that correspond to statistical modes and antimodes. The jagged contour in Fig. 5 is the actual pdf of leadership ratings from an emergent leadership experiment. The modes are located at values of 0, 9, and 12, with antimodes at 7 and 11. The estimated values in the smooth contour denote modes at values of 0, 9, and 16, with antimodes at 5, and 13.

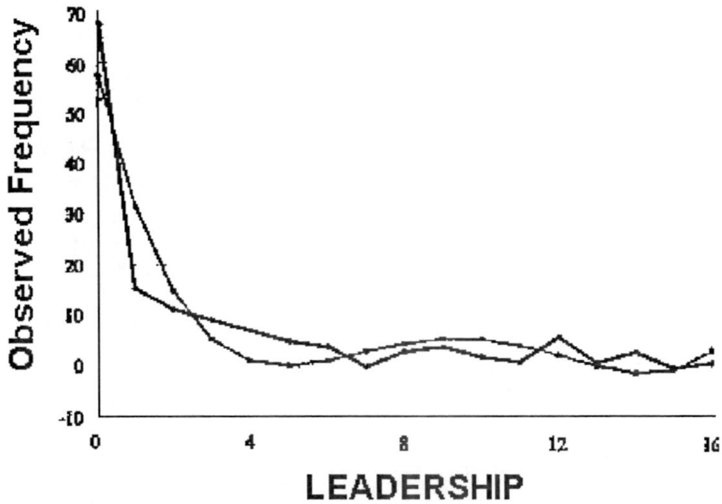

Figure 5. Comparison of actual and predicted pdf data from a swallowtail catastrophe problem (Guastello, 1998a). Reprinted with permission from Kluwer Academic Publishers.

Examples on Record

Examples of analyses using the polynomial regression method for catastrophes date back to Guastello (1982). Recent examples, however, can be found in Guastello (1995, 2002),

Guastello, Gershon, & Murphy (1999), Clair, (1998); Lange (1999), and Byrne, Mazanov, and Gregson (2001).

Examples of analyses using the nonlinear regression method for chaos and related exponential models date back to Guastello (1992b). More recent examples can be found in Guastello (1995, 1998b, 1999a, 1999b, 2001b, 2002), Guastello and Philippe (1997), Guastello and Guastello (1998). Guastello and Johnson (1999), Guastello, Johnson, and Rieke (1999), Guastello and Bock (2001), Rosser, Rosser, Guastello, and Bond (2001), and Guastello and Bond (in press). For examples that compared dimensionality estimates made through nonlinear regression with values obtained by other means, see Johnson and Dooley (1996) and Guastello and Philippe (1997).

Examples of analyses using the nonlinear regression method for testing catastrophe pdfs are sparse, although the method was proposed by Cobb (1981a, 1981b). More recent examples can be found in Hanges, Braverman, & Rentch (1991), Guastello (1998a, 2002), and Zaror and Guastello (2000).

References

Brock, W. A., Hseih, D. A., LeBaron, B. (1991). *Nonlinear dynamics, chaos, and instability: Statistical theory and economic evidence*. Cambridge, MA: MIT Press.

Byrne, D. G., Mazanov, J., & Gregson, R. A. M. (2001). A cusp catastrophe analysis of changes to adolescent smoking behavior in response to smoking prevention programs. *Nonlinear Dynamics, Psychology, and Life Sciences,* **5**, 115-138.

Clair, S. (1998). A cusp catastrophe model for adolescent alcohol use: An empirical test. *Nonlinear Dynamics, Psychology, and Life Sciences,* **2**, 217-241.

Cobb, L.(1981a). Multimodal exponential families of statistical catastrophe theory. In C. Taillie, G. P. Patel, & B. Baldessari (Eds.), *Statistical distributions in scientific work* (vol. 6, pp. 67-90). Hingam, MA: Reidel.

Cobb, L. (1981b). Parameter estimation for the cusp catastrophe model. *Behavioral Science,* **26**, 75-78.

Cohen, J. (1988). *Statistical power analysis for the behavioral sciences*. Hillsdale, NJ: Lawrence Erlbaum Associates.

Green, B. F. (1991). How many subjects does it take to do a regression analysis*? Multivariate Behavioral Research,* **12**, 263-288.

Guastello, S. J. (1982a). Color matching and shift work: An industrial application of the cusp-difference equation. *Behavioral Science,* **27**, 131-137.

Guastello, S. J. (1992a). Clash of the paradigms: A critique of an examination of the polynomial regression technique for evaluating catastrophe theory hypotheses. *Psychological Bulletin,* **111**, 375-379.

Guastello, S. J. (1992b). Population dynamics and workforce productivity. In M. Michaels (Ed.), *Proceedings of the annual conference of the Chaos Network: The second iteration* (pp. 120-127). Urbana, IL: People Technologies.

Guastello, S. J. (1995). *Chaos, catastrophe, and human affairs: Applications of nonlinear dynamics to work, organizations, and social evolution*. Mahwah, NJ: Lawrence Erlbaum Associates.

Guastello, S. J. (1998a). Self-organization and leadership emergence. *Nonlinear Dynamics,*

Psychology, and Life Sciences, **2**, 303-316.

Guastello, S. J. (1998b). Creative problem solving groups at the edge of chaos. *Journal of Creative Behavior*, **32**, 38-57.

Guastello, S.J. (1999a). Hysteresis, bifurcation, and the natural rate of unemployment. In E. Elliott & L. D. Kiel, (Eds.). *Nonlinear dynamics, complexity and public policy* (pp. 31-46). Commack, NY: Nova Science.

Guastello, S. J. (1999b). Hierarchical dynamics affecting work performance in organizations. In W. Tschacher & J-P. Dauwaulder (Eds.), *Dynamics, synergetics and autonomous agents* (pp. 277-302). Singapore: World Scientific.

Guastello, S. J. (2001a). Nonlinear dynamics in psychology. *Discrete Dynamics in Nature and Society,* **6**, 11-29.

Guastello, S. J. (2001b). Attractor stability in unemployment and inflation rates. In Y. Aruka (Ed.) *Evolutionary controversies in economics: A new transdiscipinary approach* (pp. 89-99). Tokyo: Springer-Verlag.

Guastello, S. J. (2002). Managing emergent phenomena: Nonlinear dynamics in work organizations. Mahwah, NJ: Lawrence Erlbaum Associates.

Guastello, S. J., & Bock, B. R. (2001). Attractor reconstruction with principal components analysis: Application to work flows in hierarchical organizations. *Nonlinear Dynamics, Psychology, and Life Sciences,* **5**, 175-192.

Guastello, S. J., Gershon, R. M., & Murphy, L. R. (1999). Catastrophe model for the exposure to blood-borne pathogens and other accidents in health care settings. *Accident Analysis and Prevention,* **31**, 739-750.

Guastello, S. J., & Guastello, D. D. (1998). Origins of coordination and team effectiveness: A perspective from game theory and nonlinear dynamics. *Journal of Applied Psychology,* **83**, 423-437.

Guastello S. J., & Johnson, E. A. (1999). The effect of downsizing on hierarchical work flow dynamics in organizations. *Nonlinear Dynamics, Psychology, and Life Sciences,* **3**, 347-378.

Guastello, S. J., Johnson, E. A., & Rieke, M. L. (1999). Nonlinear dynamics of motivational flow. *Nonlinear Dynamics, Psychology, and Life Sciences,* **3**, 259-273.

Guastello, S. J., & Philippe, P. (1997). Dynamics in the development of large information exchange groups and virtual communities. *Nonlinear Dynamics, Psychology, and Life Sciences,* **1**, 123-149.

Hanges, P. J., Braverman, E. P., Rentch, J. R. (1991). Changes in raters' perception of subordinates: A catastrophe model. *Journal of Applied Psychology,* **76**, 878-888.

Johnson, T. L., & Dooley, K. J. (1996). Looking for chaos in time series data. In W. Sulis and A. Combs (Eds.), *Nonlinear dynamics in human behavior* (pp. 44-76). Singapore: World Scientific.

Kanz, H., & Schreiber, T. (1997). *Nonlinear time series analysis*. New York: Cambridge University Press.

Kelso, J. A. S. (1995). *Dynamic patterns: Self-organization of brain and behavior*. Cambridge, MA: MIT Press.

Lange, R. (1999). A cusp catastrophe approach to the prediction of temporal patterns in the kill dates of individual serial murderers. *Nonlinear Dynamics, Psychology, and Life Sciences,* **3**, 143-159.

Maxwell, S. E. (2000). Sample size and multiple regression analysis. *Psychological Methods,* **5**, 343-458.

May, R. M., & Oster, G. F. (1976). Bifurcations and dynamic complexity in simple ecological models. *American Naturalist,* **110**, 573-599.

Ott, E., Sauer, T., & Yorke, J. A. (Eds.). (1994). *Coping with chaos*. New York: Wiley.

Rosser, J. B., Rosser, M. V., Guastello, S. J., & Bond, R. W. Jr. (2001). Chaotic hysteresis and systemic economic transformation: Soviet investment patterns. *Nonlinear Dynamics, Psychology, and Life Sciences,* **5**, 345-368.

Thom, R. (1975). *Structural stability and morphegenesis*. New York: Benjamin-Addison-Wesley.

Wiggins, S. (1988). *Global bifurcations and chaos.* New York: Springer-Verlag.

Zaror, G., & Guastello, S. J. (2000). Self-organization and leadership emergence: A cross-cultural replication. *Nonlinear Dynamics, Psychology, and Life Sciences,* **4**, 113-119.

Zhang, W-B. (2002). Theory of complex systems and economic development. *Nonlinear Dynamics, Psychology, and Life Sciences,* **6**, 83-102.

In: Chaos and Complexity Research Compendium
Editors: F. Orsucci and N. Sala, pp. 61-84
ISBN: 978-1-60456-787-8

Chapter 6

SYNCHRONIZATION OF OSCILLATORS IN COMPLEX NETWORKS

***Louis M. Pecora*[1] *and Mauricio Barahona*[2]**
[1]Code 6343, Naval Research Laboratory, Washington, DC 20375, USA
[2]Department of Bioengineering, Mech. Eng. Bldg., Imperial College of STM, Exhibition Road, London SW7 2BX, UK

Abstract

We introduce the theory of identical or complete synchronization of identical oscillators in arbitrary networks. In addition, we introduce several graph theory concepts and results that augment the synchronization theory and tie is closely to random, semirandom, and regular networks. We then use the combined theories to explore and compare three types of semirandom networks for their efficacy in synchronizing oscillators. We show that the simplest *k*-cycle augmented by a few random edges or links appears to be the most efficient network that will guarantee good synchronization.

I. Introduction

In the past several years interest in networks and their statistics has grown greatly in the applied mathematics, physics, biology, and sociology areas. Although networks have been structures of interest in these areas for some time recent developments in the construction of what might be called structured or semirandom networks has provoked increased interest in both studying networks and their various statistics and using them as more realistic models for physical or biological systems. At the same time developments have progressed to the point that the networks can be treated not just as abstract entities with the vertices or nodes as formless place-holders, but as oscillators or dynamical systems coupled in the geometry of the network. Recent results for such situations have been developed and the study of dynamics on complex networks has begun.

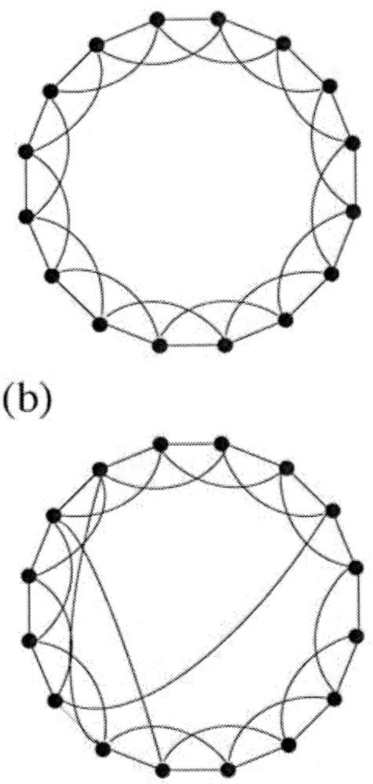

Figure 1. Example of a cycle and a semiregular cycle (smallworld a la Watts&Strogatz).

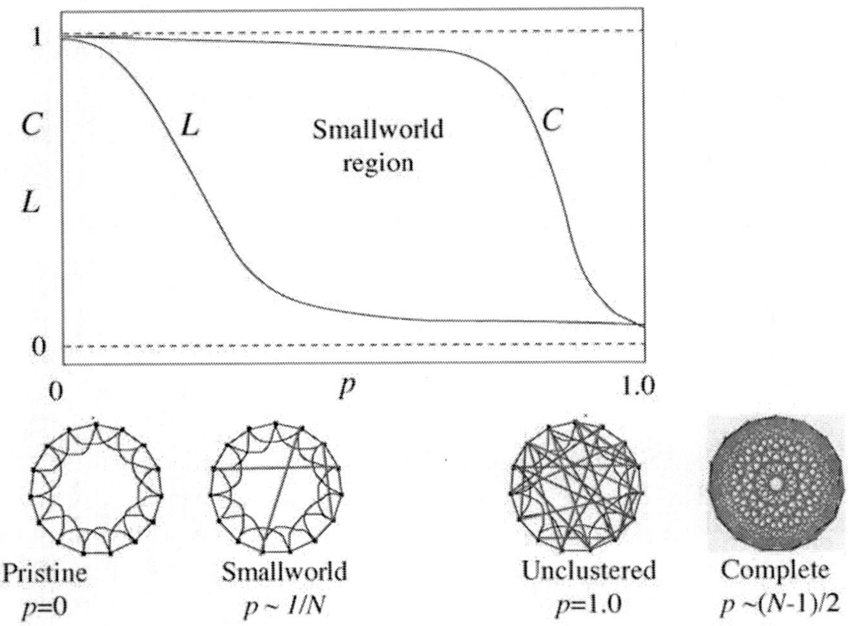

Figure 2. Plot of $L=L(\mathrm{p})/L(0)$ (normalized average distance between nodes) and $C=C(\mathrm{p})/C(0)$ (normalized clustering) vs. p. Shown at the bottom are typical graphs that would obtain at the various p values including the complete graph. Note in the smallworld region we are very far from a complete graph.

In 1968 Watts and Strogatz [1] showed that simple cyclical networks called ***k*-cycles** (nodes connected to each other in circles, see Fig. 1 for an example) make the transition from networks where average distances between nodes is large to short average distance networks with the addition of surprisingly few edges randomly rearranged and reattached at random to other nodes in the network. At the same time the network remained highly **clustered** in the sense that nodes were connected in clumps. If we think of connected nodes as friends in a social network, highly clustered would mean that friends of a particular node would, with high probability be friends of each other. Thus with only a few percent or less of rearranged edges the network shrank in size, determined by average distance, but stayed localized in the clustering sense. These networks are referred to as **smallworld** networks. See Fig. 2 for a plot of fractional change in average distance and clustering vs. probability of edge rearrangement.

Such smallworld networks are mostly regular with some randomness and can be referred to as semirandom. The number of edges connecting to each node, the **degree** of the node, is fairly uniform in the smallworld system. That is, the distribution of degrees in narrow, clustered around a well-defined mean. The Watts and Strogatz paper stimulated a large number of studies [2] and is seminal in opening interest in networks to new areas of science and with a new perspective on modeling actual networks realistically.

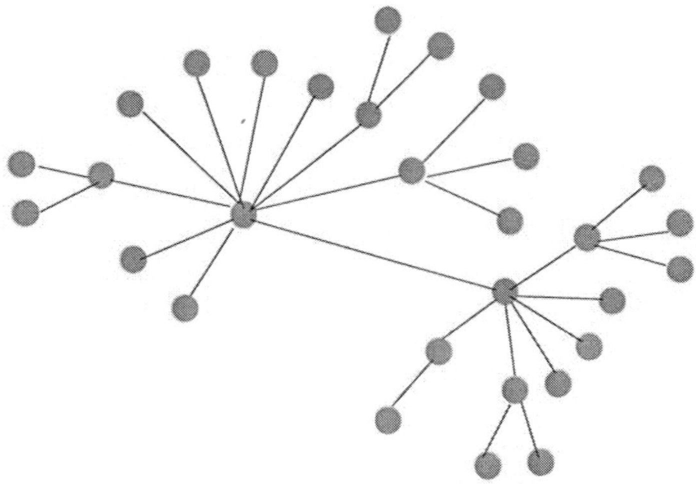

Figure 3. Example of SFN with m=1. Note the hub structure.

A little later in a series of papers Barabasi, Albert, and Jeong showed how to develop scale-free networks which closely matched real-world networks like co-authorship, protein-protein interactions and the world-wide web in structure. Such networks were grown a node at a time by adding a new node with a few edges connected to existing nodes. The important part of the construction was that the new node was connected to existing nodes with preferences for connections to those nodes already well-connected, i.e. with high degree. This type of construction led to a network with a few highly connected hubs and more lower connected hubs (see Fig. 3). In this network the **degree** of the node, has no well-defined average. The distribution of node degrees is a power law and there is no typical degree size in the sense that the degrees are not clustered around some mean value as in the smallworld case. The network is referred to as **scale-free**. The natural law of growth of the scale-free network, the rich get richer in a sense, seems to fit well into many situations in many fields. As a result interest is this network has grown quickly along with the cycle smallworld.

In the past decade in the field of nonlinear dynamics emphasis on coupled systems, especially coupled oscillators has grown greatly. One of the natural situations to study in arbitrarily connected identical oscillators is that of **complete synchronization** which in a discrete system is the analog of the uniform state in continuous systems like fluids where the uniform state would be laminar flow or chemical reactions like the BZ reaction where there is no spatial variation although temporal evolution can be very complex and/or chaotic. That is, in a completely synchronized system all the oscillators would be doing the same thing at the same time. The stability of the uniform state is of great interest for it portends the emergence of patterns when its stability is lost and it amounts to a coherent state when the stability can be maintained. The uniform or completely synchronized state is then a natural first choice to study in coupled systems.

In the last several years a general theory has been developed for the study of the stability of the synchronized state of identical oscillators in arbitrary coupling topologies [3,4]. A natural first step in the study of dynamics on complex or semirandom networks is the study of synchronization in cycle smallworlds and scale-free networks of oscillators. In the next section we develop the formalism of synchronization stability in arbitrary topologies and presents some ideas from networks and graph theory that will allow us to make some broad and generic conclusions.

II. Formal Development

A. Synchronization in Arbitrary, Coupled Systems

Here we present a formal development of the theory of stability of the synchronous state in any arbitrary network of oscillators. It is this theory which is very general that allows us to make broad statements about synchronous behavior in classes of semirandom networks.

We start with a theory based on linear coupling between the oscillators and show how this relates to an important quantity in the structure of a network. We then show how we can easily generalize the theory to a broader class of nonlinearly coupled networks of oscillators and iterated maps.

Let's start by assuming all oscillators are identical (that is, after all, how we can get identical or complete synchronization). This means the uncoupled oscillators have the following equation of motion,

$$\frac{d\mathbf{x}^i}{dt} = \mathbf{F}(\mathbf{x}^i) \tag{1}$$

where the superscript refers to the oscillator number (i=1,...,N) and subscripts on dynamical variables will refer to components of each oscillators, viz., x_j^i, j=1,...,m. We can linearly couple the N oscillators in some network by specifing a connection matrix, **G**, that consists of 1's and 0's to specify what oscillators is coupled to which other ones. We restrict our study to symmetric connections since our networks will have non-directional edges, hence, **G** is symmetric. Generalizations to non-symmetric couplings can be made (see Refs [4,5]).

We also assume all oscillators have an output function, **H**, that is a vector function of dimension m of the dynamical variables of each oscillator. Each oscillator has the same output function and its output is fed to other oscillators to which it is coupled. For example, **H**, might be an $m \times m$ matrix that only picks out one component to couple to the other oscillators.

The coupled equations of motion become [6],

$$\frac{d\mathbf{x}^i}{dt} = \mathbf{F}(\mathbf{x}^i) - \sigma \sum_{j=1}^{N} G_{ij}\mathbf{H}(\mathbf{x}^j), \qquad (2)$$

where σ is the overall coupling strength and note that **G** acts on each oscillator as a whole and only determines which are connected and which are not. **H** determines which components are used in the connections. Since we want to examine the case of identical synchronization, we must have the equations of motion for all oscillators be the same when the system is synchronized. We can assure this by requiring that the sum $\sum_{j=1}^{N} G_{ij}\mathbf{H}(\mathbf{x}^j)$ is a constant when all oscillators are synchronous. The simplest constant is zero which can be assured by restricting the connection matrix **G** to have zero row sums. This works since all $\mathbf{H}(\mathbf{x}^j)$ are them same at all times in the synchronous state. It means that when the oscillators are synchronized they execute the same motion as they do when uncoupled (Eq. (1)), except all variables are equal at all times. Generalization to non-zero constants can be done, but it unnecessarily complicates the analysis. A typical connection matrix is shown in the next equation,

$$\mathbf{G} = \begin{pmatrix} 2 & -1 & 0 & \ldots & 0 & -1 \\ -1 & 2 & -1 & 0 & \ldots & 0 \\ 0 & -1 & 2 & -1 & \ldots & 0 \\ & & \vdots & & & \\ 0 & \ldots & 0 & -1 & 2 & -1 \\ -1 & 0 & \ldots & 0 & -1 & 2 \end{pmatrix}, \qquad (3)$$

for nearest neighbor, diffusive coupling on a ring or cycle.

Our central question is, for what types of oscillators (**F**), output functions (**H**), connection topologies (**G**), and coupling strengths (σ) is the synchronous state stable? Or more generally, for what classes of oscillators and networks can we get the oscillators to synchronize? The stability theory that emerges will allow us to answer these questions.

In the synchronous state all oscillators' variables are equal to the same dynamical variable: $\mathbf{x}^1(t) = \mathbf{x}^2(t) = \ldots = \mathbf{x}^N(t) = \mathbf{s}(t)$, where $\mathbf{s}(t)$ is a solution of Eq. (1). The subspace defined by the constraint of setting all oscillator vectors to the same, synchronous, vector is called the **synchronization manifold**. We test whether this state is stable by considering small, arbitrary perturbations ξ^j to each $\mathbf{x}^j$ and see whether all the perturbations ξ^j die out or grow. This is accompllished by generating an equation of motion for each ξ^j and determining a set of **Lyapunov exponents** which tell us the stability of the state. The use of Lyapunov exponents is the weakest condition for the stability of the synchronous state. Although other stability criteria can be used [5] we will use the Lyapunov exponents here.

To generate an equation of motion for the set of ξ^j we start with the full equations of motion for the network (Eq. (2)) and insert the perturbed value of the dynamical variables $\mathbf{x}^j(t) = \mathbf{s}(t) + \xi^j$ expanding all functions (**F** and **H**) in Taylor series to 1st order (we are only interested in small ξ^j values). This gives,

$$\frac{d\xi^i}{dt} = \sum_{j=1}^{N} \left[D\mathbf{F}(\mathbf{s})\delta_{ij} - \sigma G_{ij} D\mathbf{H}(\mathbf{s}) \right] \cdot \xi^j, \tag{4}$$

where $D\mathbf{F}$ and $D\mathbf{H}$ are the Jacobians of the vector field and the output function.

Eq. (4) is referred to as a **variational equation** and is often the starting point for stability determinations. This equation is rather complicated since given arbitrary coupling **G** it can be quite high dimensional. However, we can simplify the problem by noting that the equations are organized in block form. The blocks correspond to the (*ij*) indices of **G** and we can operating on them separately from the components within each block. We use this structure to diagonalize **G**. The first term with the Kronecker delta remains the same. This results in variational equations in eigenmode form:

$$\frac{d\zeta^l}{dt} = \left[D\mathbf{F}(\mathbf{s}) - \sigma \gamma_l D\mathbf{H}(\mathbf{s}) \right] \cdot \zeta^l, \tag{5}$$

where γ_l is the *l*th eigenvalue of **G**. We can now find the Lyapunov exponents of each eigenmode which corresponds to a "spatial" pattern of desynchronization amplitudes and phases of the oscillators. It would seem that if all the eigenmodes are stable (all Lyapunov exponents are negative), the synchronous state is stable, but as we will see this is not quite right and we can also simplify our analysis and not have to calculate the exponents of each eigenblock separately. We note that because of the zero-sum row constraint $\gamma=0$ is always an eigenvalue with eigenmode the major diagonal vector (1,1,...,1). We denote this as the first eigenvalue γ_1 since by design it is the smallest. The first eigenvalue is associated with the synchronous state and the Lyapunov exponents associated with it are those of the isolated oscillator. Its eigenmode represents perturbations that are the same for all oscillators and hence do not desynchronize the oscillators. The first eigenvalue is therefore not considered in the stability analysis of the whole system.

Next notice that Eq. (5) is the same form for all eigenmodes. Hence, if we solve a more generic variational equation for a range of couplings, then we can simply examine the exponents for each eigenvalue for stability. This is clearer if we show the equations. Consider the generic variational equation,

$$\frac{d\zeta}{dt} = \left[D\mathbf{F}(\mathbf{s}) - \alpha\, D\mathbf{H}(\mathbf{s})\right]\cdot\zeta, \qquad (6)$$

where α is a real number (**G** is symmetric and so has real eigenvalues). If we know the maximum Lyapunov exponent $\lambda_{max}(\alpha)$ for α over a range that includes the Lyapunov spectrum, then we automatically know the stability of all the modes by looking at the exponent value at each $\alpha=\sigma\gamma_l$ value. We refer to the function $\lambda_{max}(\alpha)$ as the **master stability function**.

For example, Fig. 4 shows an example of a typical stability curve plotting the maxium Lyapunov exponent vs. α. This particular curve would obtain for a particular choice of vector field (**F**) and output function (**H**). If the spectrum $\{\gamma_l\}$ all falls under the negative part of the stability curve (the deep well part), then all the modes are stable. In fact we need only look to see whether the largest γ_{max} and smallest γ_2, non-zero eigenvalues fall in this range. If there exists a continuous, negative λ_{max} regime in the stabilty diagram, say between α_1 and α_2, then it is sufficient to have the following inequality to know that we can always tune σ to place the entire spectrum of **G** in the negative area:

$$\frac{\gamma_{max}}{\gamma_2} < \frac{\alpha_2}{\alpha_1}, \qquad (7)$$

We note two important facts: (1) We have reduced the stability problem for an oscillator with a particular stability curve (say, Fig. 4) to a simple calculation of the ratio of the extreme, non-zero eigenvalues of **G**; (2) Once we have the stability diagram for an oscillator and output function we do not have to re-calculate another stability curve if we reconfigure the network, i.e. construct a new **G**. We need only recalculate the largest and smallest, non-zero eigenvalues and consider their ratio again to check stability.

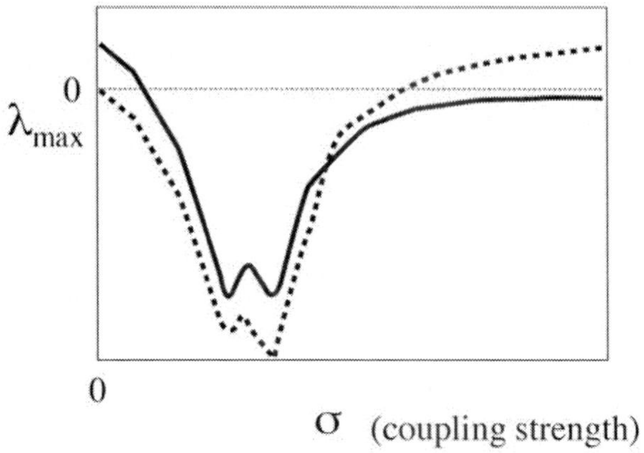

Figure 4. Stability curve for a generic oscillator. The curve may start at (1) $\lambda_{max}=0$ (regular behavior), (2) $\lambda_{max}>0$ (chaotic behavior) or (3) $\lambda_{max}<0$ (stable fixed point) and asymptotically ($\sigma\to\infty$) go to (a) $\lambda_{max}=0$, (b) $\lambda_{max}>0$, (c) $\lambda_{max}<0$. Of course the behavior of λ_{max} at intermediate σ values is somewhat arbitrary, but typical stability curves for simple oscillators have a single minium. Shown are the combinations (1)-(b) and (2)-(c) for some generic simple oscillators.

Finally, we remark that stability curves like Fig. 4 are quite common for many oscillators in the literature, especially those from the class arising from an unstable focus. Appendix A gives a heuristic reason for this common shape. For the rest of this article, we assume that we are dealing with the class of oscillators which have stability curves like Fig. 4. They may or may not be chaotic. If they are then at $\alpha=0$ $\lambda_{max} > 0$, otherwise $\lambda_{max} = 0$ at $\alpha=0$. And their values for large α may go positive or not for either chaotic or periodic cases. We will assume the most restrictive case that there is a finite interval where $\lambda_{max} < 0$ as in Fig. 4. This being the most conservative assumption will cover the largest class of oscillators including those which have multiple, disjoint α regions of stability as can happen in Turing pattern-generating instabilities [7]. Several other studies of the stability of the synchronous state have chosen weaker assumptions, including the assumption that the stability curve $\lambda_{max}(\alpha)$ becomes negative at some threshold (say, α_1) and remains negative for all $\alpha > \alpha_1$. Conclusions of stability in these cases only require the study of the first non-zero eigenvalue γ_2, but cover a smaller class of oscillators and are not as general as the broader assumption of Eq. (7).

B. Beyond Linear Coupling

We can easily generalize the above situation to one that includes the case of nonlinear coupling. If we write the dynamics for each oscillator as depending, somewhat aribtrarily on it's input from some other oscillators, then we will have the equation of motion,

$$\frac{d\mathbf{x}^i}{dt} = \mathbf{F}^i(\mathbf{x}^i, \mathbf{H}\{\mathbf{x}^j\})\}, \tag{8}$$

where here $\mathbf{F}^i$ is different for each i because it now contains arbitrary couplings. $\mathbf{F}^i$ takes N+1 arguments with $\mathbf{x}^i$ in the first slot and $\mathbf{H}\{\mathbf{x}^j\}$ in the remaining N slots. $\mathbf{H}\{\mathbf{x}^j\}$ is short for putting in N arguments which are the result of the output function $\mathbf{H}$ applied in sequence to all N oscillator vectors $\mathbf{x}^j$, viz., $\mathbf{H}\{\mathbf{x}^j\}= (\mathbf{H}(\mathbf{x}^1), \mathbf{H}(\mathbf{x}^2), \ldots \mathbf{H}(\mathbf{x}^N))$. We require the constraint,

$$\mathbf{F}^i(\mathbf{s},\mathbf{H}\{\mathbf{s}\})\} = \mathbf{F}^j(\mathbf{s}, \mathbf{H}\{\mathbf{s}\})\}, \tag{9}$$

for all i and j so that identical synchronization is possible.

The variational equation of Eq. (8) will be given by,

$$\frac{d\xi^i}{dt} = \sum_{j=1}^{N}\left[D_0\mathbf{F}^i(\mathbf{s}, \mathbf{H}\{\mathbf{s}\})\delta_{ij} + D_j\mathbf{F}^i(\mathbf{s},\mathbf{H}\{\mathbf{s}\})\cdot D\mathbf{H}(\mathbf{s})\right]\cdot\xi^j, \tag{9}$$

where D_0 is the partial derivative with respect to the first argument and D_j, j=1,2,...N, is the derivative with respect to the remaining N argument slots. Eq. (9) is almost in the same form as Eq. (4). We can regain Eq. (4) form by restricting our analysis to systems for which the partial derivatives of the second term act simply like weighting factors on the outputs from each oscillator. That is,

$$D_j \mathbf{F}^i(\mathbf{s},\mathbf{H}\{\mathbf{s}\}) = -\sigma G_{ij}\mathbf{1}_m, \tag{10}$$

where G_{ij} is a constant and $\mathbf{1}_m$ is the $m \times m$ unit matrix. Now we have recovered Eq. (4) exactly and all the analysis that led up to the synchronization criterion of Eq. (7) applies. Note that Eq. (10) need only hold on the synchronization manifold. Hence, we can use many forms of nonlinear coupling through the $\mathbf{F}^i$ and/or $\mathbf{H}$ functions and still use stability diagrams and Eq. (7).

C. Networks and Graph Theory

A **network** is synonomous with the definition of a **graph** and when the nodes and/or edges take on more meaning like oscillators and couplings, respectively, then we should properly call the structure a network of oscillators, etc. However, we will just say network here without confusion. Now, what is a graph? A graph U is a collection of **nodes** or **vertices** (generally, structureless entities, but oscillators herein) and a set of **connections** or **edges** or **links** between some of them. See Fig. 1. The collection of vertices (nodes) are usually denoted as $V(U)$ and the collection of edges (links) as $E(U)$. We let N denote the number of vertices, the **cardinality** of U written as $|U|$. The number of edges can vary between 0 (no vertices are connected) and $N(N-1)/2$ where every vertex is connected to every other one.

The association of the synchronization problem with graph theory comes through a matrix that appears in the variational equations and in the analysis of graphs. This is the connection matrix or $\mathbf{G}$. In graph theory it is called the **Laplacian** since in many cases like Eq. (3) it is the discrete version of the second derivative $\Delta = \nabla^2$. The Laplacian can be shown to be related to some other matrices from graph theory.

We start with the matrix most studied in graph theory, the **adjacency** matrix $\mathbf{A}$. The is given by the symmetrix form where $A_{ij}=1$ if vertices i and j are connected by an edge and $A_{ij}=0$ otherwise. For example, for the graph in Fig. 5 has the following adjacency matrix,

$$\mathbf{A} = \begin{pmatrix} 0 & 1 & 1 & 0 & 0 \\ 1 & 0 & 1 & 0 & 1 \\ 1 & 1 & 0 & 1 & 0 \\ 0 & 0 & 1 & 0 & 0 \\ 0 & 1 & 0 & 0 & 0 \end{pmatrix}, \tag{11}$$

Much effort in the mathematics of graph theory has been expended on studying the adjacency matrix. We will not cover much here, but point out a few things and then use $\mathbf{A}$ as a building block for the Laplacian, $\mathbf{G}$.

The components of the powers of $\mathbf{A}$ describe the number of steps or links between any two nodes. Thus, the non-zero components of $\mathbf{A}^2$ show which nodes are connected by following exactly two links (including traversing back over the same link). In general, the mth power of $\mathbf{A}$ is the matrix whose non-zero components show which nodes are connected by m steps. Note, if after $N-1$th step $\mathbf{A}$ still has a zero, off diagonal component, then the graph must be **disconnected**. That is, it can be split into two subgraphs each of whose nodes have no edges connecting them to the other subgraph. Given these minor observations and the fact that a matrix must satisify its own characteristic equation, one can appreciate that much work in graph theory has gone into the study of the eigenspectrum of the adjacency matrix. To pursue this further, we recommend the book Ref. [8].

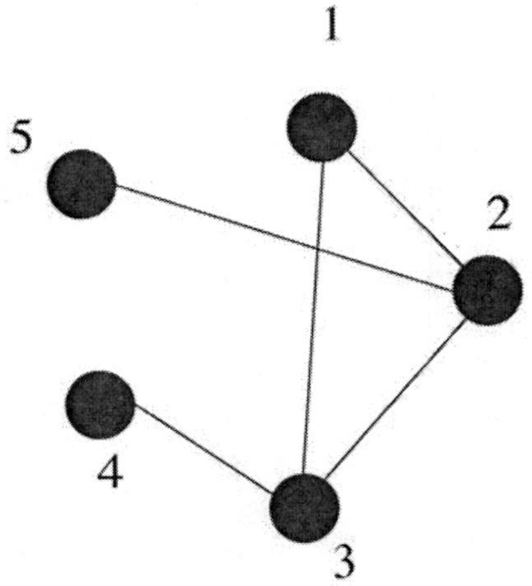

Figure 5. Simple graph generating the adjacency matrix in Eq. (11).

The **degree** of a node is the sum of the number of edges connecting it to other nodes. Thus, in Fig. 5, node 2 has a degree = 3. We see that the degree of a node is just the row sum of the row of **A** associated with that node. We form the degree matrix or **valency** matrix **D** which is a diagonal matrix whose diagonal entries are the row sums of the corresponding row of **A**, viz., $D_{ij}=\Sigma_k A_{ik}$. We now form the new matrix, the Laplacian **G**=**D**–**A**. For example, for the diffusive coupling of Eq. (3), **A** would be a matrix like **G**, but with 0 replacing 2 on the diagonal and **D** would be the diagonal matrix with 2's on the diagonals. The eigenvalues and vectors of **G** are also studied in graph theory [8,9], although not as much as the adjacency matrix. We have seen how **G**'s eigenvalues affect the stability of the synchronous state so some results of graph theory on the eigenspectrum of **G** may be of interest and we present several below.

The Laplacian is a postive, semi-definite matrix. We assume its N eigenvalues are ordered as $\gamma_1<\gamma_2<...<\gamma_N$. The Laplacian always has 0 for the smallest eigenvalue associated with the eigenvector $\mathbf{v}_1=(1,1,...,1)$, the major diagonal. The latter is easy to see since **G** must have zero row sum by construction. The next largest eigenvalue, γ_2 is called the **algebraic connectivity**. To see why, consider a graph which can be cleanly divided into two disconnected subgraphs. This means by rearranging the numbering of the nodes **G** would be divided into two blocks with no non-zero "off-diagonal" elements since they are not connected,

$$\mathbf{G}=\begin{pmatrix}\mathbf{G}_1 & 0\\ 0 & \mathbf{G}_2\end{pmatrix}, \tag{12}$$

where we assume $\mathbf{G}_1$ is $n_1 \times n_1$ dimensional and $\mathbf{G}_2$ is $n_2 \times n_2$ dimensional. In this latter case we now have two zero eigenvalues each associated with unique eigenvectors mutually orthogonal, namely, $\mathbf{v}_1=(1,1,...,1,0,0,...,0)$ with n_1 1's and $\mathbf{v}_2=(0,0,...,0,1,1,...,1)$ with n_2 1's. The degeneracy of the zero eigenvector is one greater than the connectivity. If $\gamma_2>0$, the graph is connected. We immediately see that this has consequences in synchronization stabilty, since a disconnected set of oscillators cannot synchronize physically and mathematically this shows up as a zero eigenvector.

Much work has gone into obtaining bounds on eigenvalues for graphs [8,9]. We will present some of these bounds (actually deriving a few ourselves) and show how they can lead to some insight into synchronization conditions in general. In the following the major diagonal vector (1,1,...,1) which is the eigenvector of γ_1 is denoted by $\boldsymbol{\theta}$.

We start with s few well-known min-max expressions for eigenvalues of a real symmetric matrix which is $\mathbf{G}$ in our case, namely,

$$\gamma_1 = \min\left\{\frac{\langle \mathbf{Gv}, \mathbf{v}\rangle}{\langle \mathbf{v}, \mathbf{v}\rangle} \middle| \mathbf{v} \neq 0, \mathbf{v} \in \mathbf{R}^N\right\}, \tag{13}$$

$$\gamma_2 = \min\left\{\frac{\langle \mathbf{Gv}, \mathbf{v}\rangle}{\langle \mathbf{v}, \mathbf{v}\rangle} \middle| \mathbf{v} \neq 0, \mathbf{v} \in \mathbf{R}^N, \mathbf{v}\perp\boldsymbol{\theta}\right\}, \tag{14}$$

and,

$$\gamma_{\max} = \max\left\{\frac{\langle \mathbf{Gv}, \mathbf{v}\rangle}{\langle \mathbf{v}, \mathbf{v}\rangle} \middle| \mathbf{v} \neq 0, \mathbf{v} \in \mathbf{R}^N\right\}, \tag{15}$$

Now with the proof of two simple formulas we can start deriving some inequalities.

First we show that $\sum_{i,j=1}^{N} A_{ij}\left(v_i - v_j\right)^2 = 2\langle \mathbf{Gv}, \mathbf{v}\rangle$, where A_{ij} are the components of the adjacency matrix. By symmetry of $\mathbf{A}$, we have

$$\sum_{i,j=1}^{N} A_{ij}\left(v_i - v_j\right)^2 = 2\sum_{i,j=1}^{N} (A_{ij}v_j^2 - A_{ij}v_j v_i),$$

The first term in the sum is the row sum of $\mathbf{A}$ which is the degree D_{ii} making the term in parenthesis an matrix product with the Laplacian so that we have $\sum_{i,j=1}^{N} v_i\left(D_{ii}\delta_{ij} - A_{ij}\right)v_j = 2\langle \mathbf{Gv}, \mathbf{v}\rangle$, thus proving the first formula.

Second, we show that $\sum_{i,j=1}^{N}\left(v_i - v_j\right)^2 = 2N\langle \mathbf{v}, \mathbf{v}\rangle$. Because we are using $\mathbf{v}\perp\boldsymbol{\theta}$ we can easily show that $\sum_{i,j=1}^{N}\left(v_i - v_j\right)^2 = 2\sum_{i,j=1}^{N}\left(v_i^2 - v_i v_j\right) = 2\sum_{i,j=1}^{N}\left(v_i^2 - \langle \mathbf{v}.\boldsymbol{\theta}\rangle^2\right) = 2N\langle \mathbf{v}, \mathbf{v}\rangle$ since $\mathbf{v}\perp\boldsymbol{\theta}$.

Noting that the $(v_i - v_j)$ terms are not affected if we add a constant to each component of **v**, that is, $(v_i - v_j) = ((v_i + c) - (v_j + c))$. We can now write Eq.(14) and Eq. (15) as

$$\gamma_2 = N \min \left\{ \frac{\sum_{i,j=1}^{N} A_{ij}(v_i - v_j)^2}{\sum_{i,j=1}^{N} (v_i - v_j)^2} \middle| \mathbf{v} \in \mathbf{R}^N, \mathbf{v} \neq c\boldsymbol{\theta}, c \in \mathbf{R} \right\}, \tag{16}$$

and,

$$\gamma_{\max} = N \max \left\{ \frac{\sum_{i,j=1}^{N} A_{ij}(v_i - v_j)^2}{\sum_{i,j=1}^{N} (v_i - v_j)^2} \middle| \mathbf{v} \in \mathbf{R}^N, \mathbf{v} \neq c\boldsymbol{\theta}, c \in \mathbf{R} \right\}, \tag{17}$$

If δ is the minimum degree and Δ the maximum degree of the graph then the following inqualities can be derived:

$$\gamma_2 \leq \frac{N}{N-1}\delta \leq \frac{N}{N-1}\Delta \leq \gamma_{\max} \leq \max\{D_{ii} + D_{jj}\} \leq 2\Delta, \tag{18}$$

where $D_{ii}+D_{jj}$ is the sum of degrees of two nodes that are connected by an edge. We can prove the first inequality by choosing a particular **v**, namely the standard basis $\mathbf{e}^{(i)}$= all zeros except for a 1 in the *i*th position. Putting $\mathbf{e}^{(i)}$ into Eq. (16) we get $\gamma_2 \leq \frac{ND_{ii}}{N-1}$. This last relationship is true for all D_{ii} and so is true for the minimum, δ. A similar argument gives the first $\gamma_{\max}$ -Δ inequality. The remaining inequalities bounding $\gamma_{\max}$ from above are obtained through the Perron-Frobenius theorem [9].

There remains another inequality that we will use, but not derive here. It is the following:

$$\gamma_2 \geq \frac{4}{N\,\mathrm{diam}(U)}, \tag{19}$$

where diam(U) is the diameter of the graph. The distance between any two nodes is the minimum number of edges transversed to get from one node to the other. The diameter is the maximum of the distances [8]. There are other inequalities, but we will only use what is presented up to here.

The synchronization criterion is that the ratio $\gamma_{\max}/\gamma_2$ has to be less than some value (α_2/α_1) for a particular oscillator node. We can see how this ratio will scale and how much we can bound it using the inequalities. First, we note that the best we can do is $\gamma_{\max}/\gamma_2$=1 which occurs only when each node is connected to all other nodes. We can estimate how much we are bounded away from 1 by using the first inequalities of Eq. (18) . This gives,

$$\frac{\Delta}{\delta} \leq \frac{\gamma_{max}}{\gamma_2}, \tag{20}$$

Hence, if the degree distributions are not narrow, that is, there is a large range of distributions, then the essential eigenratio γ_{max}/γ_2 cannot be close to 1 possibly making the oscillators difficult to synchronize (depending on α_2/α_1). Note that this does not mean necessarily that if δ/Δ is near 1 we have good synchronization. We will see below a case where $\delta/\Delta=1$, but γ_{max}/γ_2 is large and not near 1.

To have any hope in forcing the eigenratio down we need an upper bound. We combine Eq. (18) with Eq. (19) . This gives,

$$\frac{\gamma_{max}}{\gamma_2} \leq \frac{N \operatorname{diam}(U) \max\{D_{ii} + D_{jj} \mid i \text{ and } j \text{ are connected}\}}{4}, \tag{21}$$

This inequality is not very strong since even if the network has small diameter and small degrees the eigenratio still scales as N. We want to consider large networks and obviously this inequality will not limit the eigenratio as the network grows. However, it does suggest that if we can keep the degrees low and lower the diameter we can go from a high-diameter, possibly hard to synchronize network to an easier to synchronize one. We will see this is what happens in smallworld systems.

III. Synchronization in Semirandom Smallworlds (Cycles)

A. Generation of Smallworld Cycles

The generation of smallworld cycles starts with a k-cycle (defined shortly) and then either rewires some of the connections randomly or adds new connections randomly. We have chosen the case of adding more, random connections since we can explore the cases from pristine k-cycle networks all the way to compete networks (all nodes connected to all other nodes).

To make a k-cycle we arrange N nodes in a ring and add k connections to each of the k nearest neighbors to the right. This gives a total of kN edges or connections. Fig. 1 shows a 2-cycle. We can continue the construction past this initial point by choosing a probability p and going around the cycle and at each node choosing a random number r in (0,1) and if $r \leq p$ we add an edge from the current node to a randomly chosen node currently not connected to our starting node. Note that we can guarantee adding an edge to each node by choosing p=1. We can extend this notion so we can reach a complete neetwork by allowing probabilities greater than 1 and letting the integer part specify how many times we go around the cycle adding edges with probability p=1 with the final loop around the cycle using the remainig fraction of probability. For example, if we choose p=2.4 we go twice around the cycle adding an edge at each node (randomly to other nodes) and then a third time around adding edges with probability 0.4. With this contruction we would use a probability $p=N(N-2k-1)/2$ to make a complete network.

We refer to the k-cycle without any extra, random edges as the pristine network. It is interesting to examine the eigenvalues of the pristine Laplacian especially as they scale in ratio.

B. Eigenvalues of Pristine Cycles

The Laplacian for the pristine k-cycle is **shift invariant** or **circulant** which means that the eigenvalues can be calculated from a discrete Fourier transform of a row of the Laplacian matrix. This action gives for the lth eigenvalue,

$$\gamma_l = 2\left[k - \sum_{j=1}^{k} \cos\left(\frac{2\pi(l-1)j}{N}\right)\right]. \tag{22}$$

For N even, l=1 is gives γ_1 (nondegenerate), l=N/2 is gives γ_{max} (nondegenerate), with the remaining eigenvalues being degenerate where $\gamma_l = \gamma_{N-l+1}$. A plot of the eigenvalues for k=1, 2, 3, and 4 is shown in Fig. 6. The maximum eigenvalue (ME) γ_{max} occurs at different l values for different k values. The second or first-non-zero eigenvalue (FNZE) always occurs for l=2.

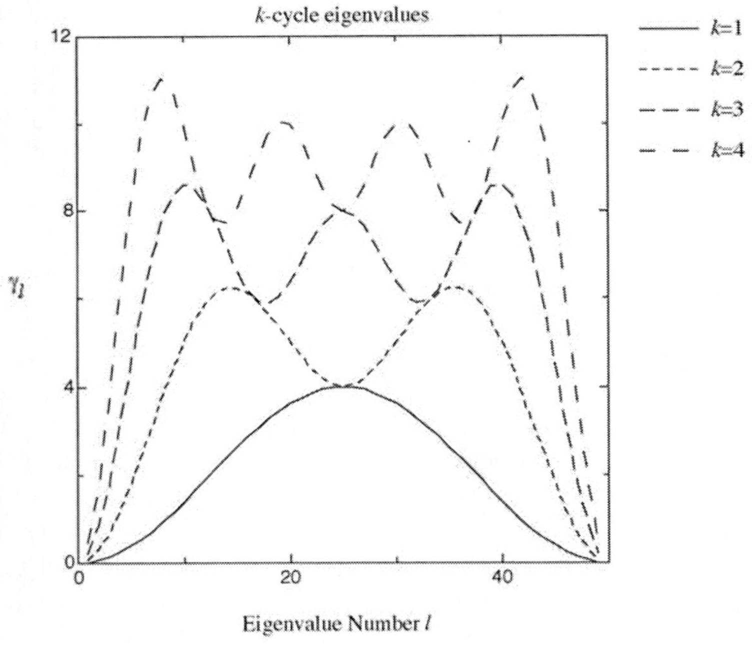

Figure 6. Eigenvalues of the pristine k-cycle for k=1, 2, 3, and 4.

What we are mostly interested in here is the eigenratio γ_{max}/γ_2. We might even question whether this ratio is sufficiently small for the pristine lattices that we might not even need extra, random connections. Fig. 7 shows the eigenratios as a function of N for k=1, 2, 3 and 4. The log-log axes of Fig. 7 shows that γ_2 scales roughly as $(2k^3+3k^2+k)/N^2$, γ_{max} values are constant for each k, and, therefore, the eigenratio γ_{max}/γ_2 scales as $N^2/(2k^3+3k^2+k)$. The scaling of the eigenratio is particularly bad in that the ratio gets large quickly with N and given that for many oscillators the stability region (α_2/α_1) is between 10 and a few hundred we see that pristine k-cycles will generally not produce stable synchronous behavior. Hence, we must add random connections in the hope of reducing γ_{max}/γ_2.

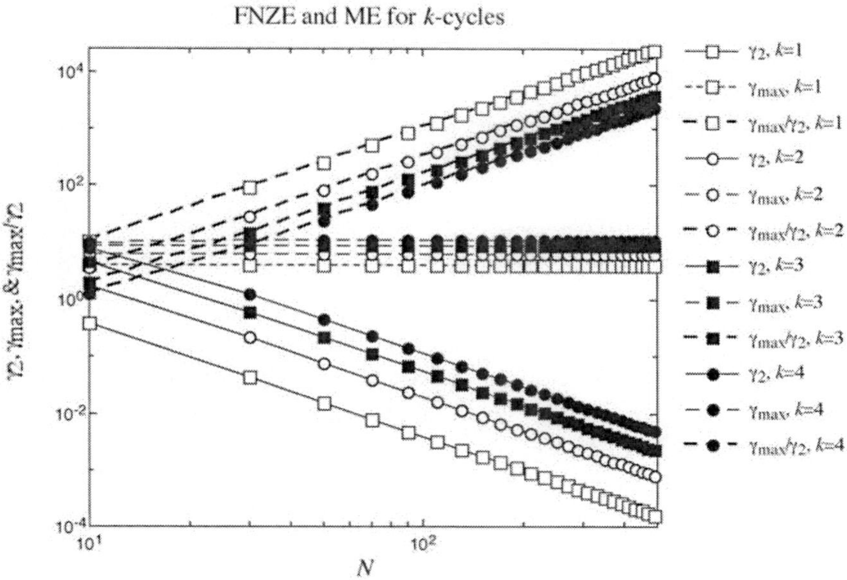

Figure 7. Eigenratios of the pristine k-cycle as a function of N for k=1, 2, 3 and 4.

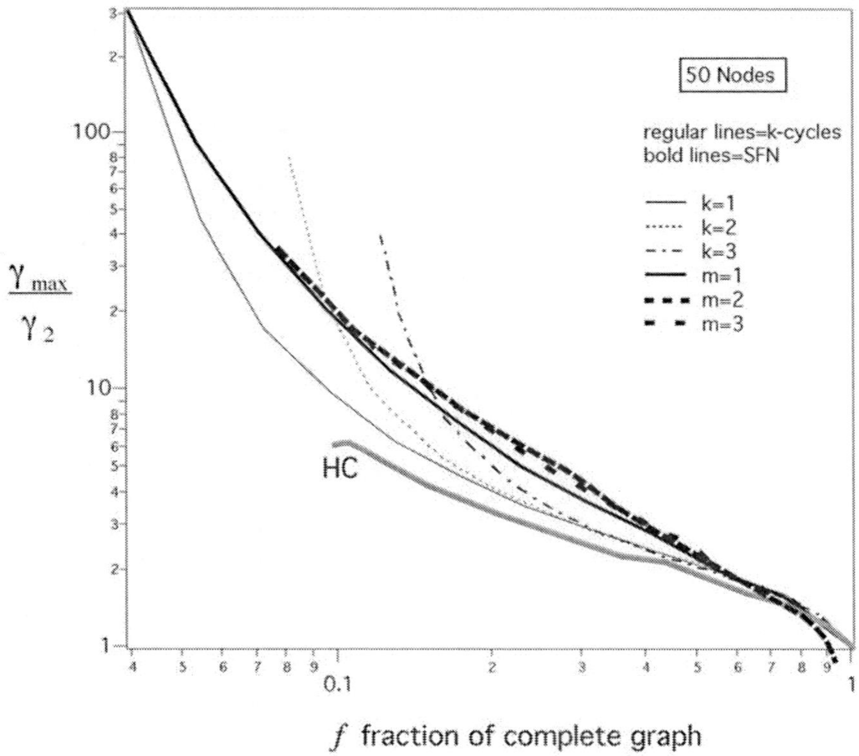

Figure 8. Eigenratio for 50 node k-cycle semirandom network as a function of fraction of complete network.

C. Eigenvalues of Smallworld Cycles

In this section and the next on SFNs we look at the trends in eigenratios as new edges are randomly added increasing the coupling between oscillators. If these new edges are added easily in a system, then surely the best choice is to form the complete network and have the best situation possible, $\gamma_{max}/\gamma_2=1$. However, we take the view that in many systems there is some expense or cost in adding extra edges to pristine networks. For biological systems the cost will be in extra intake of food and extra energy which will be fruitful only if the new networks are a great improvement and provide the organisms some evolutionary advantages. In human-engineered projects, the cost is in materials, time, and perhaps maintenance. Therefore, the trend in γ_{max}/γ_2 as edges are added randomly will be considered important.

Fig. 8 shows the eigenratio for k-cycle networks as a function of f the fraction of the complete lattice that obtains when edges are added randomly for probabilities from p=0 to values of p yielding complete graphs for N=50 nodes. Fig. 9 shows the same for 100 nodes. For now we ignore the plots for SFNs and concentrate only on the trends in the k-cycle, semirandom networks.

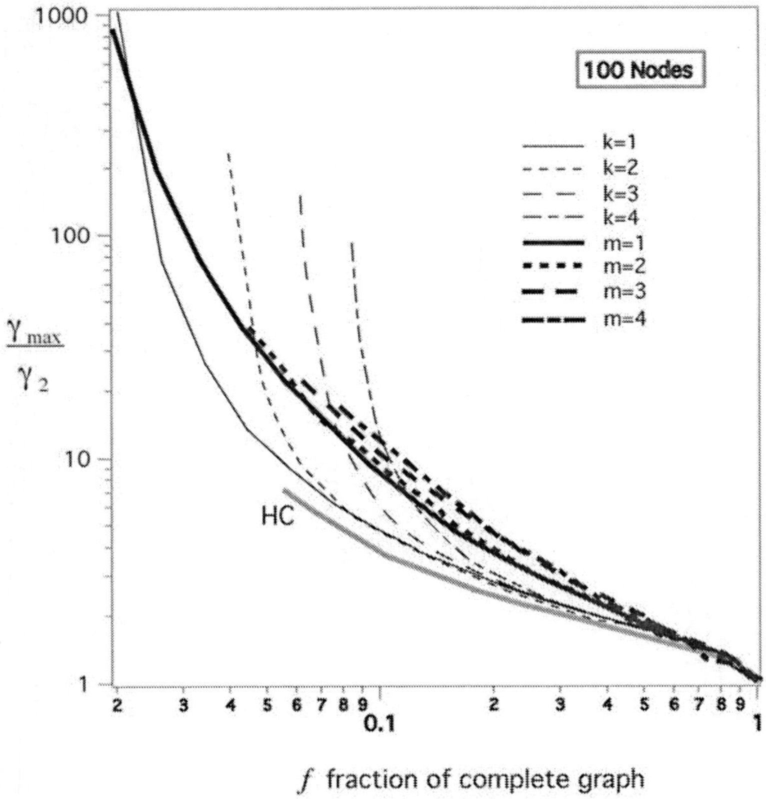

Figure 9. Eigenratio for 100 node k-cycle semirandom network as a function of fraction of complete network.

We saw in the previous section that the worse case for pristine k-cycles is when k=1. However, Fig. 8 and Fig. 9 show that this turns into the best case of all the k-cycles when random edges are added if we compare the networks at the same fraction of complete graph values where the total number of edges or cost to the system is the same. Although each case for k>1 starts at lower γ_{max}/γ_2 values for the pristine lattices, at the same fraction of complete graph values the k=1 network has lower values. This immediately suggests that a good way to generate a synchronizable network is to start with a k=1 network and randomly add edges. We will compare this to starting with other networks below.

In terms of graph diameter we can heuristically understand the lowering of γ_{max}/γ_2. Recall inequality Eq. (21) . This strongly suggests that as diam(U) decreases the eigenratio must decrease. From the original Watts and Strogatz paper we know the average distance between nodes decreases early upon addition a few random edges. Although average distance is not the same as diameter, we suspect that since the added edges are randomly connected the diameter also decreases rapidly. Note that the k-cycles are well-behaved in regard to the other inequality Eq. (20) . In fact δ/Δ starts out with a value of 1 in the pristine lattices. Of course, the eigenratio is not forced to this value from above since the diameter is large for the pristine networks – on the order of $N/(4k)$ which means the upper bound grows as N^2 just the same trend as the actual eigenratio for the pristine networks. However, adding edges does not change δ/Δ by much since the additions are random, but it does force the upper bound down.

IV. Synchronization in Scale-Free Networks

A. Generation of Scale-Free Networks

Scale-free networks (SFNs) get their name because they have no "typical" or average degree for a node. The distribution of degrees follows an inverse power law first discovered by Albert and Barabási [10] who later showed that one could derive the power law in the "continuum" limit of large N. Several methods have been given to generate SFNs, but here we use the original method [11].

The SFN is generated by an iterative process. We start with an initial number of nodes connected by single edges. The actual initial state does not affect the final network topology provided the initial state is a very small part of the final network. We now add nodes one at a time and each new node is connected to m existing nodes with no more than one connection to each (obviously the initial number of nodes must be equal to or greater than m). We do this until we have the desired number of nodes, N. The crucial part is in how we choose the m nodes to connect each new node to. This is done by giving each existing node a probability P_i which is proportional to the degree of that node. Normalizing the probabilities gives,

$$P_i = \frac{d_i}{\sum_{j \in \substack{\text{existing} \\ \text{nodes}}} d_j}, \tag{23}$$

where d_j is the degree of the jth node. Using this probability we can form the cumulative set of intervals, $\{ P_1, P_1+P_2, P_1+P_2+P_3,...\}$. To choose an existing node for connection we randomly pick a number from the interval (0,1), say c, and see which interval it falls into. Then we pick the node of that interval with the highest index. A little thought will show that the ordering of the nodes in the cumulative intervals will not matter since c is uniformly distributed over (0,1).

Such a network building process is often referred to as "the rich get richer." It is easy to see that nodes with larger degrees will be chosen preferentially. The above process forms a network in which a few nodes form highly connected hubs, more nodes form smaller hubs, etc. down to many single nodes connected only through their original m edges to the network. This last sentence can be quantified by plotting the distribution or count of the degrees vs. the degree values. This is shown in Fig. 10.

We note that since we are judging synchronization efficiency of networks we want to compare different networks (e.g. cycles and SFNs) which have roughly the same number of edges, where we view adding edges as adding to the "cost" of building a network. We can easily compare pristine cycles and SFNs since for $k=m$ we have almost the exact same number of edges (excepting the initial nodes of the SFN) for a given N.

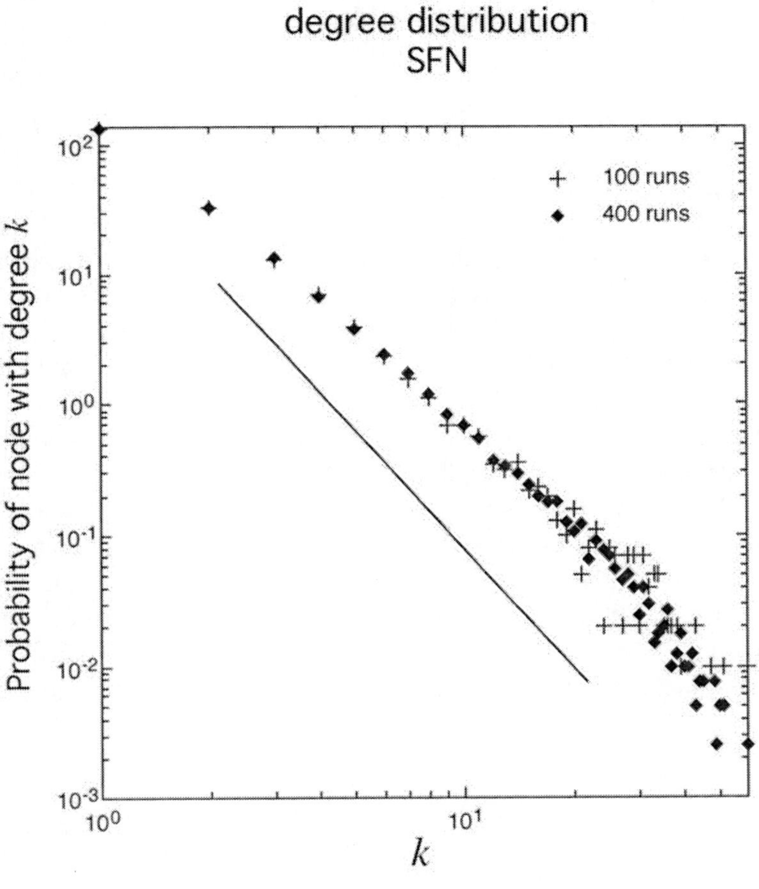

Figure 10. Distribution of node degrees for SFNs.

B. Eigenvalues of Pristine Scale-Free Networks

Although we do not start with a fully deterministic network in the SFN case, we can still characterize the eigenvalues and, to a lesser extenct, the eigenvectors of the original SFN. This is useful when comparing the effect of adding more, randomly chosen edges as we do in the next section.

Fig. 11 shows the FNZE, ME, and their ratio vs. N for various m values of 1, 2, 3, and 4. For m=1 we have (empirically) $\gamma_{max} \sim \sqrt{N}$ and $\gamma_2 \sim 1/N$. Therefore the ratio scales as $\gamma_{max}/\gamma_2 \sim N^{\frac{3}{2}}$. Recall that the pristine cycle's ratio scaled as $\gamma_{max}/\gamma_2 \sim N^2$, hence for the same number of nodes and k=m the pristine eigenratio of cycles grows at a rate $\sim\sqrt{N}$ faster then the SFN eigenratio. Thus, for large networks it would seem that SFNs would be the network of choice compared to cycles for obtaining the generic synchronization condition $\gamma_{max}/\gamma_2 \leq \alpha_2/\alpha_1$ for the same number of edges. However, as we add random edges we see that this situation does not maintain. For values of $m \geq 2$, the situation is not as clear.

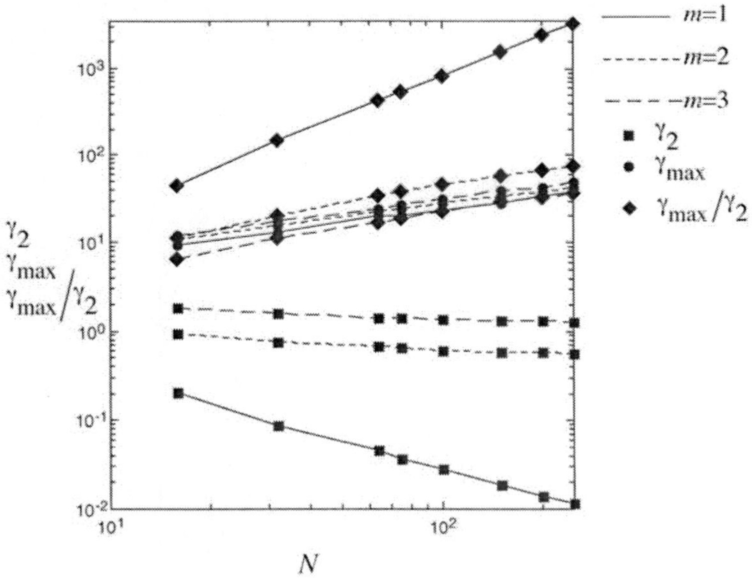

Figure 11. FNZE, ME, and their ratio vs. N for SFNs.

C. Eigenvalues of Scale-Free Networks

Fig. 8 and Fig. 9 also show the eigenratio for SFNs as a function of f the fraction of complete graph. Note that we can directly compare SW and SFNs when m=k for which they have the same f value. We are now in a position to make several conclusions regarding the comparison of SW networks and SFNs. Such comparisons seem to hold across several values of N. We tested cases for N=50, 100, 150, 200, and 300.

First SW semirandom k cycles start out in their pristine state with a larger eigenratio than SFNs with $m=k$. However, with the addition of a small number of edges – barely changing f the SW k cycle eigenratio soon falls below its SFN counterpart with $m=k$. The eigenratios for all SFNs fall rapidly with increasing f, but they do not reach the same low level as the SW networks until around f=0.5. This implies that the SWs are more efficient than SFNs in reducing eigenratio or, equivalently, synchronizing oscillators.

An interesting phenomenon for SFNs is that changing m only appears to move the eigenratio up or down the m=1.curve. That is, curves for SFNs with different m values fall on top of each other with their starting point being higher f values for higher m values. This suggests an underlying universal curve which the m=1 curve is close to. The connection to the previous section where level spacing for the SFN eigenvalues begins to appear random-like as we go to higher m values is mirrored in the present section by showing that higher m values just move the network further along the curve into the randomly added edge terroritory.

At f=1 all networks are the same. They are complete networks with $N(N-1)/2$ edges. At this point the eigenratio is 1. As we move back away from the f=1 point by removing edges, but preserving the underlying pristine network, the networks follow a "universal" curve down to about f=0.5. Larger differences between the networks don't show up until over 50% of the edges have been removed which implies that the underlying pristine skeleton does not control the eigenratio beyond f=0.5. This all suggests that an analytic treatment of networks beyond f=0.5 might be possible. We are investigating this.

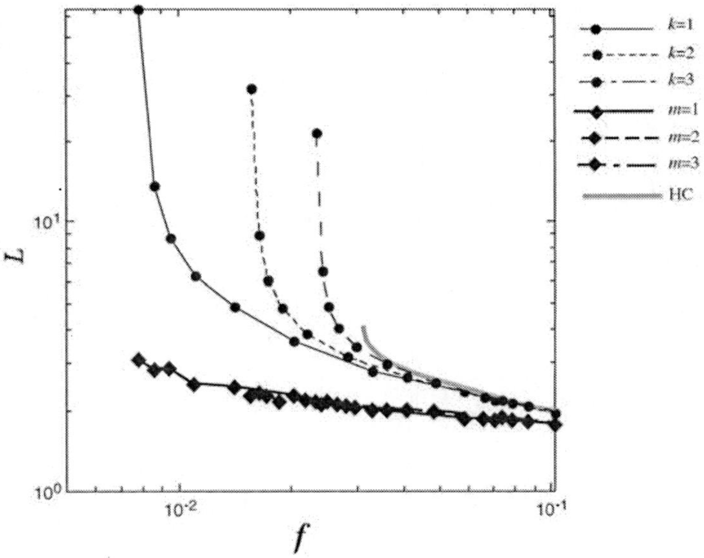

Figure 12. Average distances between nodes on SW, SFN, and HC networks.

In their original paper Watts and Strogatz [12] suggested that the diminishing average distances would allow a smallworld network of oscillators to couple more efficiently to each other and therefore synchronize more readily. In Fig. 12 we plot the average distance in each network as a function of fraction of completed graph f. In Fig. 13 the same type of plot is given for the clustering coefficient. What we immediately notice is that there seem to be no obvious relationship to eigenratios as a function of f. SFNs start out with very small average distance and have very low clustering coeffieicnts, at least until many random edges are added to beyond f=0.1 which is beyond the smallworld regime. Thus, it appears that neither average distance nor clustering affect the eigenratio. What network statistics, if any, do affect γ_{max}/γ_2?

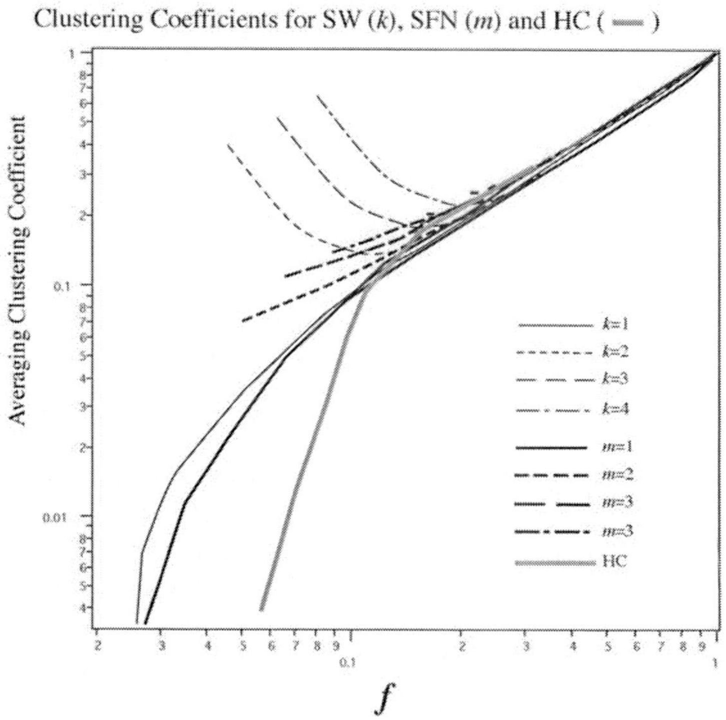

Figure 13. Clustering coefficient for SW, SFN, and HC networks.

From the graph theory bounds in section II.C. Eqs. (20) and (21) we can explain some of the phenomena we find numerically for eigenratios. We can immediately explain the higher eigenratio of SFNs using Eq. (20) . The eigenratio is bounded away from 1 (the best case) by the ratio of largest to smallest degrees (Δ/δ). For smallworld k-cycles this ratio starts at 1 (since all nodes have degree $2k$) and does not change much as edges are added randomly. Hence, it is at least possible for the eigenratio to approach 1, although this argument does not require it. Conversely, in the SFN the degree distribution is wide with many nodes with degree=m and other nodes with degree on the order of some power of N. Hence, for SFNs Δ/δ is large, precluding the possibility that γ_{max}/γ_2 can get close to 1 until many random edges are added.

We can bound the eigenratio from above using Eq. (21) . Thus, γ_{max}/γ_2 is sandwiched between Δ/δ and $\frac{1}{4}N\,\mathrm{diam}(U)\max\{D_{ii}+D_{jj}\}$. We show this schematically in Fig. 14. The ratio Δ/δ provides a good lower bound, but the upperbound of Eq. (21) is not tight in that it scales with N. Hence, even in small networks the upper bound is an order of magnitude or more larger than the good lower bound. At this stage it does not appear that graph theory can give the full answer as to why smallworld k-cycles are so efficient for synchronizing oscillators.

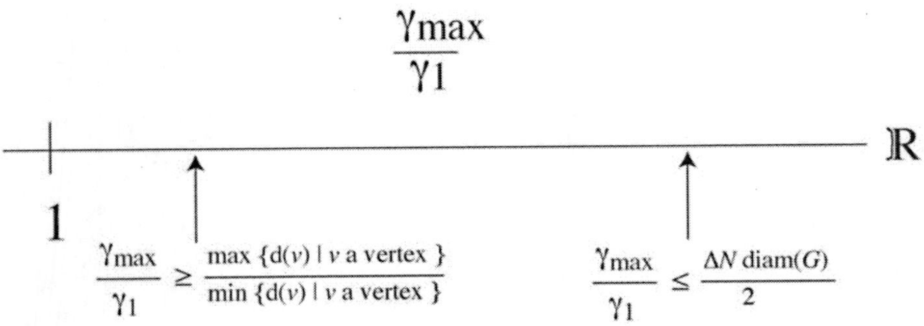

Figure 14. The bounded interval containing the eigenratio γ_{max}/γ_2.

V. Hypercube Networks

As a simple third comparison we examine the eigenratio of the **hypercube** (HC) network which we showed in Ref. [13] was a very efficient network for obtaining low eigenratios. HC networks are built by connecting $N=2^D$ nodes in a topology so that all nodes form the corners of a D-dimensional hypercube with degree D. Hence, the 1st graph theory inequality (Eq. (20)) allows the eigenratio to be the minimum allowed value of 1 although this is not mandatory. Furthermore, it is not hard to show that the maximum eigenvalue scales as $2D=2\log_2(N)$ and the smallest eigenvalue is always 2 so that $\gamma_{max}/\gamma_2=D$ for the pristine HC network and this increases very slowly with N. We therefore expect the eigenratio for any HC network to start off small and decrease to 1.

Fig. 8 and Fig. 9 contain plots of the HC network eigenratio vs. f from pristine state to complete network. Indeed, the initial eigenratios are low, but this comes with a price. The price contains two contributions. One is that we need to start at a much higher f value, that is, HC networks in their pristine state require more edges than the SW $k=1$ network. And the other contribution is that the HC must be constructed carefully since the pristine state is apparently more complex than any other network.

In the end one gets the same performance by just connecting all the nodes in a loop and then adding a small number of random edges. The construction is simpler and the synchronization as good as one of the best networks, the HC.

VI. Conclusions

Our earlier work [13] showed that many regular networks were not as efficient in obtaining synchronization as the smallworld network or fully random networks. Here we added two other semirandom networks, the SFN+random edges and the HC+random edges to our analysis. The results appear to be that elaborately constructed networks offer no advantage and perhaps even disadvantages over the simple *k*-cycle+random edges, especially in the smallworld regime.

The fully random network [14] can also have ratios γ_{max}/γ_2 which are similar to the smallworld case for large enough probabilities for adding edges. However, the random networks are never guaranteed to be fully connected. When unconnected, as we noted earlier, synchronization is impossible. The observation we can make then is that the 1-cycle smallworld network may be the best compromise between fully random and fully regular networks. The smallworld network maintains the fully connectedness of the regular networks, but gains all the advantages of the random networks for efficient synchronization.

References

[1] D.J. Watts, *Small Worlds*. (Princeton University Press, Princeton, NJ, 1999).

[2] M.E.J. Newman and D.J. Watts, *Physics Letters* **A 263** (4-6), 341 (1999); M.E.J. Newman, C. Moore, and D.J. Watts, *Physical Review Letters* **84** (14), 3201 (2000); C. Moore and M.E.J. Newman, *Physical Review* **E 61** (5), 5678 (2000); L.F. Lago-Fernandez, R. Huerta, F. Corbacho et al., *Physical Review Letters* **84** (12), 2758 (2000); Rémi Monasson, *European-Physical-Journal-B* **12** (4), 555 (1999); R.V. Kulkami, E. Almaas, and D. Stroud, *Physical Review* **E 61** (4), 4268 (2000); M. Kuperman and G. Abramson, *Physical Review Letters* **86**, 2909 (2001); S. Jespersen, I. Sokolov, and A. Blumen, *J. Chem. Phys.* **113**, 7652 (2000); H. Jeong, B. Tombor, R. Albert et al., *Nature* **407**, 651 (2000); M. Barthelemy and L. Amaral, *Physical-Review-Letters* **82** (15), 3180 (1999).

[3] Y. Chen, G. Rangarajan, and M. Ding, *Physical Review* **E 67**, 026209 (2003); A.S. Dmitriev, M. Shirokov, and S.O. Starkov, *IEEE Transactions on Circuits and systems I. Fundamental Theory and Applications* **44** (10), 918 (1997); P. M. Gade, *Physical Review* **E 54**, 64 (1996); J.F. Heagy, T.L. Carroll, and L.M. Pecora, *Physical Review* **E 50** (3), 1874 (1994); Gang Hu, Junzhong Yang, and Winji Liu, *Physical Review* **E 58** (4), 4440 (1998); L. Kocarev and U. Parlitz, *Physical Review Letters* **77**, 2206 (1996); Louis M. Pecora and Thomas L. Carroll, *International Journal of Bifurcations and Chaos* **10** (2), 273 (2000); C. W. Wu, presented at the 1998 *IEEE International Symposium of Circuits and Systems*, Monterey, CA, 1998 (unpublished); Chai Wah Wu and Leon O. Chua, *International Journal of Bifurcations and Chaos* **4** (4), 979 (1994); C.W. Wu and L.O. Chua, I*EEE Transactions on Circuits and Systems* **42** (8), 430 (1995).

[4] L.M. Pecora and T.L. Carroll, *Physical Review Letters* **80** (10), 2109 (1998).

[5] K. Fink, G. Johnson, D. Mar et al., *Physical Review* **E 61** (5), 5080 (2000).

[6] Unlike our earlier work we use a minus sign for the coupling terms so the definition of matrices from graph theory agree with the graph theory literature.

[7] A. Turing, *Philosophical Transactions* **B 237**, 37 (1952).

[8] D.M. Cvetkovi´c, M. Doob, and H. Sachs, *Spectra of Graphs: Theory and Applications*. (Johann Ambrosius Barth Verlag, Heidelberg, 1995).

[9] B. Mohar, in *Graph Symmetry: Algebraic Methods and Applications, NATO ASI Ser. C*, edited by G.Hahn and G.Sabidussi (Kluwer, 1997), Vol. 497, pp. 225.

[10] Réka Albert and Albert-László Barabási, *Physical Review Letters* **84**, 5660 (2000).

[11] Réka Albert and Albert-László Barabási, *Reviews of Modern Physics* **74**, 47 (2002).
[12] Duncan J. Watts and Steven H. Strogatz, *Nature* **393** (4 June), 440 (1998).
[13] M. Barahona and L. Pecora, *Physical Review Letters* **89**, 054101 (2002).
[14] P. Erdös and A. Renyi, *Publicationes Mathematicae (Debrecen)* **6**, 290 (1959); P. Erdös and A. Renyi, *Bull. Inst. Internat. Statist.* **38**, 343 (1961).

In: Chaos and Complexity Research Compendium
Editors: F. Orsucci and N. Sala, pp. 85-104
ISBN: 978-1-60456-787-8

Chapter 7

CTML: A Mark Up Language for Holographic Representation of Document Based Knowledge

Graziella Tonfoni[*]
DSLO University of Bologna
Bologna 40126, Italy

Abstract

This chapter is specifically meant to illustrate an articulated framework for information visualization based upon a set of conceptual tools meant to promote qualitative reasoning in documentation management.

The complexity of our information world today poses the immediate and urgent need for powerful conceptual tools, enabling individuals to cope with daily communicative activities which have become more and more complex.

On the other side decision makers in our interconnected world need to reach the top level of accuracy and be able minimize the risk of fuzziness and misinterpretation which may be caused by lack of originating context, where a certain information was first produced or by continuously shifting contexts throughout a diversity of media and cultures.

CTML, the Context Transport Mark up Language, (Tonfoni,1998i, 1999) is a high power derivative language of CPP-TRS, the Communicative Positioning Program-Text Representation Systems (Tonfoni,1996i,1996ii), which is a widely tested and comprehensive methodology based upon a consistent set of dynamic visuals, to be used in combinations meant to teach individuals how to represent those cognitively different textual operations they perform, while they are writing and reading in multimedia environments.

CTML, which may definitely be viewed as a really advanced and a highly specialized sublanguage on its own; is also a mark up system meant empower already existing advanced information systems.

[*] Author of the CPP-TRS methodology and of the CTML

As a controlled language on its own it is meant to enable verification of asysmmetries which are likely to occur in interpretation of those very documents, which are most subject to continuous change, revision and further updates.

Keywords: Context identification; context transport, context reproduction, context verification

1.1. Introduction

Document management systems are crucial today. To ensure most effective performance, they need to incorporate highly specialized conceptual tools for identifying misunderstandings and for fixing interpretive mistakes. Such problems may be caused by significant changes occurring in our highly interconnected information environment, leading to unwished and unintended communication casualties and interpretation aberration phenomena.

Interpretation aberration phenomena are likely to occur for different reasons, such as information overload and loss of context and also as a consequence of uncontrolled perceptual changes, which are unintentionally triggered by too rapid data expansion and amplification produced by new information technologies.

A document may be viewed as white light, which is made out of a variety of colors.

An advanced document management system needs to provide users with special lenses that may allow various colors to all come up as the result of different kinds of refractions.

Documents and single paragraphs are most likely to absorb external energies, along with external and undesired information matter, which is part of the communicative environment, where such documents are first analyzed and then classified.

Our perceptual system is undergoing efforts of major proportion and major stress is generated as a consequence.

Information flowing and floating in huge and uncontrolled quantities is dramatically affecting our usual modes of perception. Meaning vibrations, caused by fuzzy communication waves and their order of magnitude and speed, may therefore alter dramatically an originating meaning even when just literally transported from one given context into another one.

Besides, random turbulences are generated within an information environment due to culture-bound interpretation attribution progressive waves, which may deeply affect the originating and intended meaning of each given document.

Any advanced document management system needs to be complemented by a whole set of tools and resources, which may serve the purpose of allowing information flows to come in; monitor, control, transfer information; filter, distinguish, discriminate, label information; verify, store, classify information; retrieve, declassify, package information; repackage, update, upgrade information; and finally select, reconfigure, deliver information.

Information detection may be envisioned as visualization enhancement made possible throughout a whole visual apparatus, enabling it to trace back communication patterns consistently and correctly.

Advanced document management systems will highly benefit from specifically targeted and fully enhanced information compilation capacities, making discrimination between verified and unverified documents actually possible.

A whole set of techniques for analyzing information meant to allow information capturing and reproduction of the closest approximation possible to the originating context have been derived from Tonfoni's theoretical framework and are incorporated within the CTML system, which is here synthetically illustrated.

Within such framework, information compilations of various kinds are made available and transportable as a consequence of a consensually shared set of criteria. Document management systems designed according to the highly articulated conceptual model here provided also provide a highly consistent visual screen and display for qualitative reasoning upon information throughout a highly specialised vocabulary to be shared interculturally.

More precisely: information compilations, visually enhanced, will provide: interpretive clues for understanding, parallely and synchronously supporting evidence for each statement made; interpretive cues showing that originating sources were in fact accessed and accurately referenced or referred to; interpretive labels based upon a set of preselected categories, which have been established as a consequence of an in-depth search and as the result of highly consistent knowledge accumulation, filtering and selection; interpretive labels based upon a set of paragraphs, which derive from an already summarized description of each item, and where paragraphs are complemented by comments and carry evaluation criteria along; interpretive cues indicating full continuity and consistency for further insertion of more summarized information, which is of relevance in order to better understand facts and events illustrated; interpretive cues in the form of a premise, meant to declare that more information is needed in order to know enough about a certain fact, event or phenomenon; and finally interpretive labels indicating a set of examples complemented by explanation and definition, showing to be a subset derived from a larger body of knowledge.

Representation of time actually needed for accurate perception, which will indicate clearly that time of reading will radically differ from time of writing, will also be displayed along with the document representation in progressively evolving stages.

1.1.1. On Document Based Knowledge

Continuous improvement within an organizational structure needs to be seen as strictly linked and dependent upon individuals' cognitive ability and teams' cooperative and communicative competence in monitoring continuously incoming information flows, some of which are synchronous and consistent, where others are asynchronous, contradictory up to fully fuzzy. Different information flows need to be carefully analyzed as to be made compatible.

Creating a conceptual platform for documentation sharing means being able to value each individual contribution to the production of knowledge within any organization, we know well as for most extended research in this area that a continuously shifting context of communication may result in a threatening and discouraging factor if not properly presented and then handled.

On the other hand, continuous sharing of information, coming in documental formats today, may be viewed positively and result into a unique opportunity for individuals to become active protagonists of a most significant process, such as the creation of a broader communication context.

Information managers today need to be aware all the time that they are working for the present and the future, and that what only what has been consistently designed may be built in steady ways and is therefore likely to be stored as to remain available for further use. Documentation, which is conceived as to be turned into a collective memory based repository of organizational knowledge, is most likely to be productively accessed and reused at different times and for different purposes.

Enhanced data warehousing may therefore be consistently viewed as some kind of high performance knowledge management, where knowledge is the derivative product and the end result of a set of highly articulated processes of information selection and organization, made visible throughout a whole series of precisely defined transitional states, incorporating the history and evolution of each document. Within the framework, here illustrated, individuals' contributions will be stored and will therefore remain visible in their originating context, as such they will be accessible and available all the time.

Naturally occurring instability and sudden shifts of context and priorities, due to continuous change, are compensated for, by a flexible interpretive model for information visualization, which allows changes and context shifts to be reported, traced back and triggered together with the originating context through the use of a consensually agreed upon and shared code.

To summarize: each document contributor's visibility made possible throughout an enhanced text encoding system, will significantly gain out of consistent visualization of those very conditions in which the document was first generated and then revised, upgraded, updated up to even radically transformed.

1.1.2. Change: From Threat to Challenge

Even if interpreted as challenging and rewarding, there is no doubt that change and transitioning conditions within any organizational structure are perceived as threatening and need to be appropriately monitored.

Motivation and commitment by individuals or teams to just one stage of a process, which is presented and therefore perceived to be most likely either discontinued, or passed to different individuals or teams, is hard to keep.

Evidently individuals involved with a certain process perceive and resent instability either consciously or unconsciously, even if encouraged to feel committed to what they have been asked to do temporarily, to complete those tasks they have been assigned.

Results achieved by individuals may actually not even become part of the organizational memory and may not be used at all.

CTML allows a whole set of contributions, which are likely to remain invisible if waves of new information, often contradictory and fuzzy are allowed to flow in indiscriminately, to still remain fully available and visible as to facilitate tracing back of the originating sources and clear-cut decision making as for their reliability.

Individuals are today continuously encouraged to think of change as some real opportunity to learn for themselves first and as a true challenge to cooperatively interact and work in teams.

But it is also a fact insecurity and instability are perceived as diminishing and frightening factors playing both against personal initiative and collaborative attitudes.

As a consequence of instability, information management subject to continuous change is characterized by the following features:

- increasingly wide amount of information coming in fuzzy, both synchronously and asynchronously;
- increasingly wide amount of contradictory and inconsistent data to be checked and verified continuously;
- increasingly frequent change of conditions of satisfaction and continuously occurring communicative context shifts.

Decision making based upon instable conditions is perceived as a fuzzy process linked to concepts such as arbitrariness and chance.

If decision makers may neither identify nor visualize relevant information as to be able to consequently envision a consistent model for taking action (Tonfoni,1998ii), will they not be able to make reasonable projections and predictions and will they be inclined, as a consequence, to resist full involvement and full commitment.

Decision making on most crucial and most delicate matters, when subject to change due to continuing modification of roles, played by single individuals and teams, is likely to become disturbed.

Information managers through the use of a CTML enhanced document management system arre practically enabled to foresee, decide and disclose how long and to what extent they will actually be in charge of and effectively responsible for a certain process they have initiated.

They are this way more likely to resist providing their own experience to the new comers and they are definitely more keen to provide those knowledge based interpretive clues they have themselves developed in the course of the years to their collaborators or successors as to facilitate problem solving.

New comers are infact likely to miss most relevant knowledge and to be missing the context also as they are likely to be invited to take action unexpectedly and under fully different conditions and within a significantly modified context.

Such resistance is not to be referred exclusively to personality issues which are obviously there too, as it also derives from awareness that each change of context and protagonists does create a new situation, where a certain problem solving strategy, which may have been previously suggested and considered to be valid, may simply just not work any longer.

Documentation management is strictly bound to accuracy, motivation and responsibility shown by individuals involved in the process of creation and further development throughout a whole set of transitional states.

Lack of those elements radically affects the way information flows are perceived, monitored and channeled within an organizational structure. Collective memories show different structures and require complex representation systems. Information systems designers do have to tailor complexity of the representation according to organizational scale (Bowker and Star, 2000).

There needs to be a good match between the types of information collected in the form of documents and their repository and its basic mission and scope.

The framework here proposed is aimed toward creating and keeping optimized conditions for positively coping with continuous change, and for supporting effective decision making under unstable conditions by reinforcing the decision makers' roles and responsibilities and by providing both conceptual and practical tools meant to establish most favorable conditions for taking action.

2.1. Documentation Visualization as a Teaching and Learning Opportunity

Documentation (from Latin *documentum* : to be taught) is the result of whole set of textual operations performed by an individual or more individuals, working collaboratively in groups within an organizational structure.

Daily events, currently occurring interactions, occasional conversations as well as planned meetings establish the conditions for information to flow through different media in many and various ways and also create the context for an accurate understanding of the very culture of each organization and institution.

Information may therefore be viewed as coming in various flows and waves and also be filtered, categorized and organized, as to then become accessible and reusable for different purposes at different times.

As soon as information is found to be of relevance and becomes stabilized it is then turned into a document: at this very stage the need to store information in ways which may be made fully transparent becomes a major issue.

Information visualization based on both topic continuity and context consistency is a most fundamental process upon which accurate and timely decision making resides.

Availability and accessibility of contextual information coming in documental format is well supported by an enhanced labeling system, such as CTML is, which may help speed up effective search and retrieval in significant ways.

Each document or piece of a document- where by piece of document a consistent topic and context completed section of a document is indicated, which is recognized as a whole entity by itself, ready to be linked up to other documents, according to topic continuity and context consistency - may be labeled according to qualitative reasoning criteria upon the nature of information and effectively stored and retrieved.

Not only is it important to suggest statistical methods in order to be able to identify topical words in the form of keywords, but it is equally relevant to add interpretive clues as to be able to add qualitative reasoning clues on top.

By showing explicitly which kind of information each document contains, along with the originating context in which the information was first conceived, will the document become CTML screened and the full history of each paragraph may become visible.

By showing which kind of progressive revisions have in fact produced various documental transitional states and various outputs, and by indicating explicitly those upgrading and updating operations, which have been performed all along, it will also become possible to trace back individuals' contributions in reformulating different context shifts, which have occurred at various times and annotated as such.

Reshuffling pieces of information coming in written forms and turning them into documents which are stabilized information packages meant to incorporate CTML clues for interpretation, according to continuously changing scenarios, entails a specific competence and cannot be subject to arbitrary choice.

Not only is a very specific "context sensitivity" required, but also a consensually shared framework for interpretation needs to be made available and referred to in order to label documents and pieces of documents consistently, based upon a common understanding and naming of the different textual operations performed.

This is why a derivative controlled language such as CTML comes handy.

Each document in the process of being visualized and labeled and made ready to be reused represents a real learning opportunity for those individuals discovering, defining and finally showing those very same operations, which have been performed at each given stage.

As a consequence of the fully explicit illustration of a whole variety of actions taken upon information, some of which may also look contradictory, may the same document, become a teaching opportunity for those individuals and groups which need to access the document in order to gain knowledge about its own history as to process it further, by updating or upgrading it either locally or globally, according to the new circumstancies.

According to a CTML perspective, will each single document carry its own attached history, while undergoing changes of various kinds.

Its visual annotation apparatus, such as displayed, may therefore be viewed as an enhanced attachment, keeping trace of the various transitioning stages, that may if not considered properly, affect its originating communicative value more or less radically.

Transitional states in a document are therefore to be considered those temporarily defined and stabilized states of information which provide evidence and support for a certain set of decision making processes which have occurred or are going to occur next.

Change applies to the conditions of satisfaction any communicative occurrence entails as well as to priorities and roles played by individuals.

Temporarily stabilized textual states will therefore provide context, evidence and visibility for each individual who is actively involved with the process.

Documentation organized and built up according to a CTML based perspective, may constitute a tremendously rich repository of collective memories, for each organization or institution which may want to invest into creating the optimized conditions for keeping the consistent interpretation of actions and activities available for understanding in the future as well.

Obviously, a consensually shared encoding system such as CTML is infact, needs to be made compatible with other information technologies, which are currently used by the organizational structure as to be really made transportable all throughout the different communicative situations occurring in the specific culture.

This is precisely the reason why a common visual system for encoding documentation has been found to be most flexible and handy to serve the purpose of transparency, clarity and crossucultural transportability.

2.1.1. Multiple Viewing and Complex Textual Visualization through Holographic Representation

Before becoming a reliable stabilized document, each text may be described, viewed and considered as a hologram.

Holograms are in fact a very good model to capture the various levels and layers of complexity each communicative interaction, may entail, which then also reflects upon each paragraph.

Holograms are infact three dimensional representation systems which may go far beyond geographical mapping. In some ways, they are communication representation devices, which are likely to enhance human interpretive processes, just like the wheel is meant to enhance human motion capabilities not by imitation.

A documental hologram is a three dimensional representation meant to be perceived and processed according to a highly context sensitive set of techniques. Only a high level of accuracy and precision in the processing of each paragraph, which has been identified as particularly sensitive, will ensure the possibility of visualizing multiple and complex interpretation processes according to the document producer's point of view and perspective.

Two kinds of documental holograms are presented and described. They are precisely: transmission documental holograms and reflection documental holograms, which consistently apply to the descriptions of wider corpora, based upon joint interpretive efforts made by partners as to produce a stabilized body of document based evidence at the end.

Three dimensional visualization of different communicative positioning stages within the same document would in fact explode the number of interpretation processes, if larger interpretative patterns were not first identified and then indicated as such, in order to guarantee consistent document perception to occur at each given stage and time of observation on the document perceiver's side.

Differentiated stages of meaning may create a progressively shifting and dynamically changing document visualization and three dimensional representation of complex interpretation just on one page and may well serve both for exploiting meaning potential and for eliminating undesired ambiguity, before that very page becomes a stable document on which decision making may be based.

2.1.2. Visualization of Document Based Reasoning

It is by now very much accepted in the literature that a mental model for representing a variety of phenomena that involve knowledge representation and information processing (Johnson-Laird, 1983, Chi and Glaser and Farr, 1988) may become a useful device for information seeking (Marchionini, 1995).

More specifically diagrams are abstract graphic portrayals of the subject matter they represent (Lowe, 1993).

Correspondence between the diagram observer's and perceptor's reactions and the diagram author's intentions represents an interesting topic for exploration and further research as in the analysis of different types of individuals dealing with different kinds of problem solving tasks (Hegarty and Just and Morrison,1998).

A mental model for enlightening interpretation complexity will have to entail dynamic viewing of each text, where by text a basic linguistic entity is considered together with its own consistent communicative context, which may then, according to a multiple viewing and complex information visualization, be converted into a textual holographic representation.

A same model may guarantee for both complex and multiple contexts explosion of interpretation, which is generated by multiple context attribution and for accurate definition of one specific context, whenever multiple contexts possible attribution may result in a very undesirable condition, because it may generate confusion.

According to previous research work extensively carried on (Tonfoni,1996iii) , communicative context may be viewed as the combination of communicative function, communicative intention and communicative turn taking controlling each single paragraph or page.

Complexity evidently appears as an intrinsic aspect of any kind of documentation interpretation and management, at different levels and scales of course, which need to be taken into account.

Complexity may show at different ranks, like lexical complexity, which has a wide literature on its own.

What is really on focus here, is rather that kind of complexity which is directly associated with the different types of communicative contexts, which may control each page at different stages; they are in fact precisely related to communicative function, intention and turn taking.

Since communicative context controlling each paragraph, may vary significantly as for space and time, therefore creating different documentation perception possibilities and effects, it may become extremely useful to visualize potential interpretive values and sometimes even conflicting interpretation stages which may be analyzed in their combinations as well, also proceeding toward most different directions, even if generated and derived out of the same originating body of documentation.

Previously carried on analysis in terms of communicative pattern recognition, shows clearly how some links may be more predictable than others and how guessing and qualitative reasoning about documental further evolutions may be productively pursued.

According to such perspective, visualization comes in as an extremely powerful device, both as for interpretation and monitoring and for reducing distortion and consequently eliminating undesired ambiguity.

Visualizing complexity of interpretation first requires very sharp conceptual tools and it certainly demands that a previous selection has been made, meant to first identify different layers and levels of possible meaning, which should be first isolated and then categorized, as to be consistently handled.

A documental hologram becomes therefore not only an accurate metaphor for envisioning information exchanges, and for representing interpretation complexity, progressively shifting according to space and time perception, but also a very practical device for representing and processing a document according to a complex and multiple viewing based procedure, still in highly reliable and fully precise ways.

The interactive nature of each collaborative effort poses the need to identify two different kinds of documental holograms, "transmission documental holograms" and "reflection documental holograms" respectively.

Transmission documental holograms are monitored and controlled by the originating document producer, whereas reflection documental holograms are processed by the document receiver and are the result of the observer's interpretation, monitoring and perception.

Communicative context, meaning communicative function, intention and turn taking, may in fact be missing or be very ambiguous, even in the course of a collaborative documentation passing and exchange, given the gap both in time and space and the most likely variety of media and channels involved in our highly complex and highly interconnected world of today.

Documents are naturally subject to loss of context or context shift, which may result in possible ambiguity or multiple interpretation.

Documental holograms processing is therefore meant make visualization of the many and complex ways in which communicative context would jeopardize interpretation and understanding if missing or left invisible or distorted.

3.1. On "Transmission Documental Holograms" and "Reflection Documental Holograms"

Transmission documental holograms should be organized by the originating document producers, who may want to make sure that the document is received, meaning perceived, charged by that very precise kind of communicative function, intention and turn taking, which they want themselves to assign to the document itself. If the documental hologram producer's main concern is to make sure that a stabilized and clearly defined context is conveyed together with text, the text will be turned into a three dimensional representation of each of its paragraph, charged with its specifically determined context.

Reading of the document may therefore be conceived in terms of progressive shifting by adopting a different perception, perspective and point of view at each given stage.

Transmission documental holograms may also serve the opposite purpose, which means that the document producer may decide to explode the context potential and multiple interpretation possibilities stage by stage by attributing to each paragraph a whole variety of plausible contexts.

If such process is to be undertaken, then a whole interpretation explosion will come out of it, allowing document perceivers to react in multiple ways as opposed to directing them toward a desired and preplanned interpretive direction, as in the previous case.

Reflection documental holograms derive from the need document perceivers have to define a certain context for themselves as for interpretation of each document received to be then sent back to the original producer with added comments.

By adding a context, are perceivers indicating what they think is relevant to the point.

This is meant to allow the original sender to add a new context or to modify the indicated one, therefore trying to optimize communication as much as possible by reducing the bandwidth and gap existing between producers and perceivers.

If ambiguity or high complexity of the original document is perceived, the reflection documental hologram may well represent such complexity, in order to have the same document be sent back to the original producer for selection and identification of most appropriate context for accurate interpretation.

A reflection documental hologram, in the form of a reply, may in this last case just represent the originating message together with one or many possible contexts for interpretation to be checked and confirmed or deleted by the original producer to be properly replied to by the perceiver according to the most suitable context attribution.

The whole concept of hologram is based upon effective recognition of intrinsic complexity of each written text, especially as space and time variables may play a significant role in document production and perception and temporary vacuum of context may occur up to context redundancy.

CTML based holographic techniques play an extremely relevant role in defining or selecting the most accurate context for consistent interpretation.

Keeping multiple contexts open may in some cases result in added value to the text, as we were taught to think about literary and poetic texts or under some other circumstances: it may actually constitute a quite significant problem for accurate interpretation of a very extended set of other kinds of texts.

Multiple possible contexts attribution may in fact result in noise and redundancy, therefore slowing down or even paralyzing a whole chain of consistent reasoning patterns based upon documental knowledge.

A snowball like effect on documentation interpretation and processing may actually increase the combinatorial possibilities of misinterpretation and misjudgment, and therefore dramatically alter the intended meaning of paragraph, therefore jeopardizing consistent interpretation of the overall document.

In summary: a documental hologram is a high visibility device, and its illustrative power is high too. High visibility should not in any way be misunderstood and thought of as restrictive and constraining, rather as clarifying and disambiguating.

If multiple contexts are in fact to be kept, they may be represented and equally shown in a multiple holographic documental representation as intended by either producer or perceiver or both of them.

3.1.1. Enhanced Encoding Procedures for Documentation Visualization

The process of visualizing a certain document is here achieved throughout the pervasive and massive application of a consistent interpretive system, meant to describe and define different kinds of communicative actions taken by individuals and groups, to be finally stored electronically. Documental operations are here categorized and analytically defined in the process of progressive organization, each document is likely to undergo in the course of conversations and meetings to be finally turned into a consistent attachment to the document stored.

Once a consistently interpreted and appropriately packaged document or piece of document is labeled and recognized in its own originating context, it may then be reconfigured many times and more or less radically transformed. Such process may be activated after relevant clues have been extracted and visually represented by specific icons, supported here of course by the Context Transport Mark up Language, CTML, which are of four kinds, and are precisely the following ones:

- documental signs: meant to indicate the communicative function or type of a document or piece of a document;

- documental symbols: meant to indicate the communicative style of a document or piece of a document;
- documental turn taking symbols: meant to indicate the role and interplay between the document producer and the document reader;
- documental amplifier symbols: meant to coordinate wide sets of documents, which are characterized by topical continuity and context consistency.

By topical continuity, we mean to indicate documents focusing on the same topic, which may be either literally extracted as linear sequences out of the document or abstracted as a result of accurate interpretation and further adaptation at a more conceptual level.

By contextual consistency, we mean to indicate documents showing the same communicative context, which may be explicitly declared as to be easily retrieved.

Documental signs, which represent the various communicative functions, a document may convey paragraph by paragraph, are the following ones:

Square: for an informative document or piece of a document, which carries information about a specific event or fact, to be linked up with another document or set of documents made available, in order to extend topical continuity and context consistency.

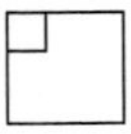

Square within the Square: for a summary of a certain document, which has been produced to reinforce contextual consistency between an original document and its own abstract.

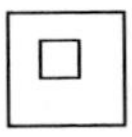

Frame: for a document or piece of document which is found to be analogous in content to other documents and previously stored cases is meant to reinforce contextual consistency between and among different documents.

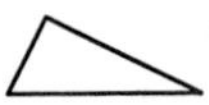

Triangle: for a memory and history generated out of a certain document meant to establish topical continuity with background information which has not been previously introduced because not available in a documental format.

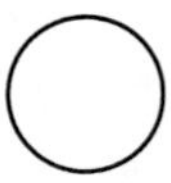

Circle: for a main concept conveyed by a certain document which has been abstracted and linked to other documents showing topical continuity. It is meant to reinforce topical keywords identification and to effectively link together documents which show the same keyword .

Grouped Semicircles: for main concepts which are abstracted out of an originating document and meant to establish both topical continuity and context consistency between the originating document and a set of topical keywords .

Semicircle: for a locally identified concept abstracted out of a piece of document and meant to reinforce context consistency by establishing further links to other documents which show the same keyword.

Inscribed Arcs: for indicating the need for an upgrade and update of a certain document; it indicates that a revision process is likely to occur, though it does not declare if such revision will be a major or a minor one.

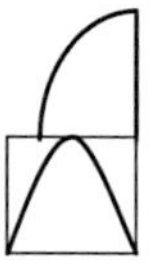

Opened Text Space: for indicating that an upgrade and update has indeed occurred within a certain document; it indicates that the document has now reached a new revision state as a consequence, though it does not declare if the revision has been a major or a minor one.

Right Triangle: for a comment made to a certain document or piece of a document coming in a non documental format where more contextual information is needed, which has to be derived from other external sources not previously available, based on topical continuity.

Documental symbols are meant to indicate communicative intentions and styles more locally within a certain document, sentence by sentence.

They are particularly useful in showing contributions made by individuals in the process of creation of a certain document and may be easily incorporated within the final document providing further interpretive clues which may significantly add to clarity and visibility.

Documental symbols, which represent different modes of information packaging activated at different times or at the same time, may be combined and used dynamically for repackaging purposes, because they effectively indicate documental transitional states by declaring explicitly the nature of those changes which have occurred or are likely to occur next.

Documental symbols are the following ones :

Describe: from Latin *describo*: write around. It means complementing the original document or piece of a document with as much information as maybe found interesting to add without any specific constraints.

It is represented by a spiral which starts from a central point –the middle point of the spiral indicating the original document- and proceeds toward expanding the document at various degrees, linking it with other documents or pieces of documents or information coming in from different sources and found to be relevant to facilitate the originating document interpretation.

In this specific information packaging mode, there is actually no need at all to follow any chronological order ; the spiral may be smaller or larger, depending on how much information the document producer - an individual or a team – may find relevant to add.

Define: from Latin *definio*: put limits. It means complementing the document with limited information about a very defined topic which has been previously selected and identified as the most relevant one, which is represented by the middle point of the square.

It indicates that there is a specific need to incorporate specific information about a relevant document or piece of a document which is made available.

Define means actually describe under specific constraints and implies accurate and most selective focusing on a very limited package of highly specific information.

Narrate: from Latin *narro*: tell the story. It means complementing the document with various facts and events, which have been referred to in the originating context by following a logical and chronological order.

It indicates a set of major points or facts representing different diachronic stages which are strictly linked up together in a sequence.

The longer the story is the more narrative points are actually added according to the document producer's decision making.

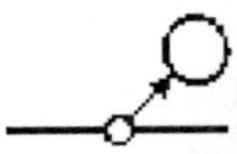

Point out: take a point out of a story chain. It means isolating a specific event or fact among those reported within a single document, focusing on just that one and adding more detailed information by expanding it significantly and linking with other documents which have been found to be of relevance to that point .

It represents the specific point chosen and does obviously entail the need to look for more extended information provided and made available.

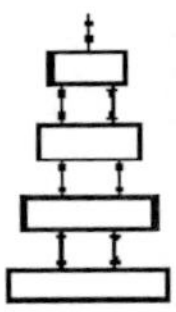

Explain: from Latin *explano*: unwrap, open up. It means that facts and reasons are given to support interpretation of a certain event within a certain document or piece of a document.

The document producer may start by indicating the originating cause and proceed toward showing the effects or start with effects and go back to the cause, according to what is found to be more significant.

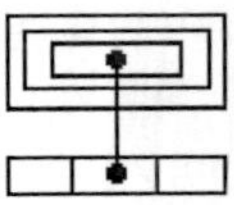

Regress: from Latin *regredior*: go back. It means that more information about a certain topic presented within the document is absolutely needed as to gain a deeper understanding.

It represents a specific topic focusing process and an in depth information expansion, which is activated only for that precise topic.

The document reader may want to consider if further information is needed and ask for availability of additional resources.

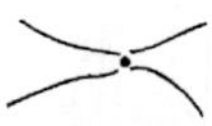

Inform: from Latin *informo*: put into shape, shape up. It means that any document is the result of some information organization and that the very specific document indicated is organized in the most unconstrained way, therefore subject to many and various kinds of reformulations.

It usually leads toward two different kinds of further specification, which are respectively conveyed by the "inform synthetically" and the "inform analytically" indication:

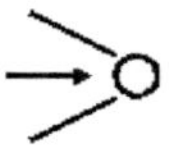

"inform synthetically" means departing from a larger document or set of documents and proceed toward a summary related to a specific topic, identified as being the most relevant one emerging from the originating document.

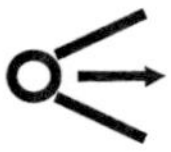

"inform analytically" means departing from a given document or limited set of documents to expand toward further documents or add more information, which needs to be previously converted into a documental format not yet available.

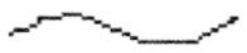

Reformulate: from Latin *reformo/reformulo*: change shape and shape again. It means changing the kind of information packaging which was adopted before and substituting a certain information request with a different one still related to the same document.

It may turn into a more or less radical transformation of the originating document according to a precisely defined request or set of requests.

Express: from Latin *exprimo*: push out and press out.It means adding personal opinions and individual feelings related to facts and events within a certain document; it indicates the most subjective mode of information organization, which is openly recognized to be bound to very personal evaluations, judgments and emotional states.

Documental turn taking symbols are meant to define the mode of accessing and reading the document, requested at each given time; they are suggested by the document producer to be followed by the document user; they are the following ones:

Major Scale: it shows that literal interpretation is needed and that those pieces of documents indicated and marked off, should be extracted and quoted literally the way they were first organized.

Minor Scale: it shows that accurate interpretation may need a further process of abstraction and that pieces of documents indicated and marked off may undergo significant reconfiguration processes up to high level conceptualization.

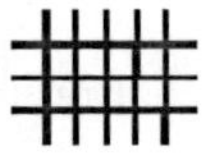

Open or Unsaturated Rhythm: it shows that accessing the document may lead the user toward incomplete interpretation of those facts and events, which are presented. It is meant to suggest the user to access more documents and various kinds of sources which are made available.

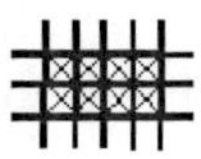

Tight or Saturated Rhythm: it shows that accessing the document will lead the user toward complete interpretation of those facts and events which are presented. It is meant to suggest the user to stick to the interpretation provided, though access to other sources is still available to support evidence.

Documentation amplifier symbols are complementary and may be added after the previously illustrated ones have been used; they apply to sets of documents and larger documentation territories and indicate specific operations, which are to be performed to connect sets of documents, which have been previously encoded and accurately stored.

They are the following ones:

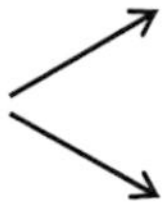

Choose: it is meant to represent the dynamic process of first identifying and then deciding between optional contexts for interpretation, which are mutually exclusive, given a certain set of documents.

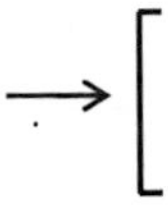

Identify:it is meant to represent definition of a more specific context within a broader context for interpretation of a set of documents; it naturally occurs before "search" and "select".

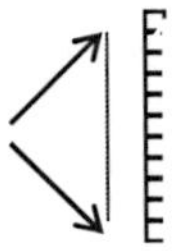

Search: it is meant to represent the dynamic process of choosing among different contexts for interpretation of a set of documents which are many and compatible as to find the most appropriate one.

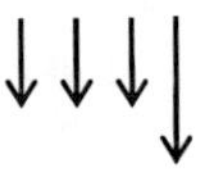

Select: it is meant to represent multiple contexts which may evolve either synchronously or asynchronously and may be modified once a certain decision making process has been performed and be stored and kept as an example.

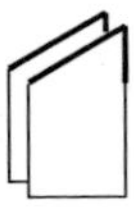

Copy/Replicate: it is meant to represent the dynamic process of duplication and repetition of a certain context, which, if lost, would radically jeopardize understanding and accurate interpretation of a set of events and facts described and explained by a set of documents.

Ahead: it is meant to represent the progression of a certain set of documents which are linked together by context consistency or harmoniously shifting contexts.

Back: it is meant to represent the need to go back to delete and replace the originating context which has radically shifted in the course of various transition states, such that, if not eliminated, would indeed affect consistent interpretation of a whole set of documents.

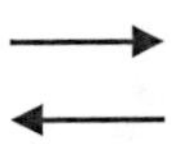

Conflict: it is meant to represent an emerging inconsistency and incompatibility between various context attributions to a set of documents which needs to be cleared as to proceed toward any further interpretation.

The documental notational system here illustrated in its various components may be applied at different layers and at various levels of complexity and is meant to underline the fundamental role and responsibility of the encoding individual and encoding team.

It is obvious that interpretive clues assigned to wide information territories will result into the production of a large amount of diversified documentation which will constitute some kind of enhanced and threedimensional geographic mapping system, carrying along its own legend, which has a direct impact upon the culture of the organizational structure it is applied to.

3.1.2. On Continuity and Change

Interpretation consistency, in spite of continuously occurring shifts of context and turn taking by individuals may still be possible. By retrieving the different layers of interpretation provided to a certain fact or set of facts, also reflecting directly upon further stages of information packaging, resulting then into a final documental format, evidence for statements made may be supported in the very form of an attachment for readers, also enhanced by a whole set of indications about the specific team members who have been put in charge and are therefore responsible for interpretation and mapping.

Just like any geographic map will only show those features which are relevant according to the nature and purpose assigned to the map itself, the same way of thinking may be extended to the domain of documentation mapping which need to be designed according to different kinds of priorities which may come out progressively as a consequence of the continuously shifting contexts.

Energy and time dedicated to quite an expensive process, such as enhanced encoding of each state of information is when packaged into a paragraph, is anyway fully cost effective because it is meant to provide an enormous amount of examples and in-house knowledge, that will remain extensively available for future generations to come.

Any further process of verification may tremendously gain and profit out of a dynamic repository which is meant to expand all the time still keeping track of the history of each given document.

as well as pride are directly connected and linked to motivation and sense of identification within

If it is a fact that not being able to foresee a future reward for a project may be a source of deep frustration for each individual involved, it is equally true that valuing each process in all of its stages, by keeping records about each individual's involvement and contribution, is a fundamental step for resourceful rethinking within a continuously changing information affluent society.

A document, which could not be completed in the past, as a certain project may have been disconnected, may turn out to be profitably triggered later on, providing therefore an example, which may serve the purpose to inform newcomers of previous work done by individuals involved.

According to such perspective, individuals initiative and authorship recognition as for documentation production is harmoniously incorporated as part of the enhanced encoding process and may reflect directly upon history of each document attached, which is meant to explicitly declare and illustrate roles played by single individuals or by highly recognizable teams.

Document management systems when CTML enhanced also represent a powerful carrier, transporting through time and space accurately filed and verified documentation viewed as the output result of individuals and teams, whose contributions are made visible according to what they decide should infact be made visible and explicit, as carrying along the context and cultural flavour in which the information packaging and repackaging did actually occur.

CTML enhanced repositories are meant to encourage active protagonism, to support consistent turn taking and to facilitate accurate decision making at each given time, ensuring individuals that each contribution provided and found to be of relevance will not be lost.

The concept of CTML enhancement in the documentation production lifecycle, resides upon the concept of harmonious integration of multiple information resources.

Rethinking of the document production lifecycle along those lines may result in a natural process very much custom friendly as memories of single contributors may be made available for further and multiple access and on line verification.

Reinventing documentation formats according to the CTML document design conceptual infrastructure is also meant to ensure continuity with the past contributions as scattered pieces of information may be made compatible and transportable.

Providing a consensually shared framework for tagging and labelling documents also represents a very constructive solution, as to cope with useless dispersion up to dissipation of relevant knowledge, which may have consolidated in the course of the years, as a consequence of the individual and collective effort of thinking and rethinking of day by day events.

4. Conclusions

Reinventing documentation formats does not mean reinventing the wheel again and again, pretending it was not there before. A CTML enhanced encoding system is based upon accurate analysis, rethinking and retracing "how the wheel was first invented and then progressively modified and refunctionalized according to continuously changing priorities and diversified needs as to best suit different kinds of evolving vehicles."

To extend the metaphor a bit further, document design innovation within any organization or institution needs to be bound to a deep understanding of the already existing tradition and of the very reasons, why such tradition was first created and evolved in various ways. New vehicles may only be conceived and designed as a consequence of observation, analysis and most accurate interpretation of the prototypes and further models which were reconfigured afterwards.

CTML, also a document annotation system, has been designed specifically as to enhance content and context visibility and intended to provide a means for information and knowledge management.

Mission of this long lasting research effort, extensive testing and verification, has been and is to be able to greatly enhance the accuracy of the process of information retrieval and of knowledge conveyance and cognition. CTML, also as a controlled language, has a whole variety of applications to information and knowledge retention, information and knowledge classification, contextual refinement of stored information, and contiguous context linking of web-based information.

Qualitative reasoning about the nature of information displayed within a document is made possible and may be further incorporated within the automation process of reviewing and redacting documents. Energy and time to be dedicated to the process of document redesigning according to the new conceptual formatting procedures have shown evidently to be fully compensated by the increased level of accuracy reached in the interpretation process.

Acknowledgments

Anytime I present my visual system, in one of its many possible applications, I cannot forget to mention those people, who have supported me, by creating the conditions for my work to become actually visible and to circulate as to be practically used.

First of all, I want to thank Marvin Minsky, Toshiba Professor at the Massachusetts Institute of Technology., who has always encouraged me to go ahead with my research and has, in the course of many years, created for me the opportunities for presenting the output results of my work and for discussing it with him and colleagues at the Massachusetts Institute of Technology, especially when I was a Research Scholar at the Artificial Intelligence Laboratory.

I am very grateful to Masoud Yazdani, Professor at the University of West England Bristol and President of Intellect for having been most supportive in having my books timely published and who has shown significant interest and a clear understanding of my work in the course of the years.

I want to thank colleagues at the School of Engineering and Applied Science at The George Washington University for having provided opportunities for my work to be fully illustrated, and particularly Chairman and Dean Lile Murphree.

I also want to thank Dean Tom Mazzuchi, and Professors Michael Stankoski and Dianne Martin also directors of the Cyberspace Policy Institute, and Professors Rachel Heller, Robert Lindeman, Richard Scotti and. Julie Ryan. Christopher Heikiman and Richard Stoler have provided me with valuable professional support as for my activities, and workshops in the Washington area.

I then want to acknowledge very valuable encouragement in pursuing this work, given to me by Dr. Pete Daniel, Curator in the Division of the History of Technology at the National Museum of American History at the Smithsonian Institution in Washington D.C., who has also carefully read this paper and keeps records of my work done in the Washington area.

References

Bowker, G.C.; Star, S.L. (2000), *Sorting things out: classification and its consequences*, Cambridge, Mass., M.I.T. Press

Chi,M.T.H.;Glaser,R., Farr,H.J. (1988), *The nature of expertise*, Hillsdale, N.J., Lawrence Erlbaum

Johnson-Laird, P.N. (1983), *Mental models: toward a cognitive science of language, inference and consciousness*, Cambridge, Cambridge University Press.

Hegarty, M.;Just, M.A., Morrison, I.R. (1988), Mental models of mechanical systems: individual differences in qualitative and quantitative reasoning, in "*Cognitive Psychology*", **20**,191-236

Lowe, R.K., (1993), Diagrammatic information: techniques for exploring its mental representation and processing, in "*Information Design Journal*", **7**.,1,3-17

Marchionini, G. (1995), *Information seeking in electronic environments*, Cambridge, Cambridge University Press

Tonfoni, G. (1996i), *Communication Patterns and Textual Forms*, Exeter, U.K., Intellect

Tonfoni, G. (1996ii), A visual news processing environment, in Artificial Intelligence Review, *Special Issue on Integration of Natural Language and Vision Processing: Recent Advances*, edited by P.McKevitt, Kluwer Academic Publishers, Amsterdam, The Netherlands, **10**,3-4,33-54

Tonfoni, G. (1996iii), On visual text planning, processing and programming, in "*Machine Learning and Perception*", edited by G.Tascini, F.Esposito, V.Roberto and P.Zingaretti, Series in Machine Perception and Artificial Intelligence, vol.23, World Scientific, Singapore,23. 99-106

Tonfoni, G. (1998i), Intelligent control and monitoring of strategic documentation: a complex system for knowledge miniaturization and text iconization, in "*Proceedings of the ISIC/CIRA/ISAS 98 Conference*", National Institute of Standards and Technology, U.S. Department of Commerce, Gaithersburg, Maryland, U.S, 869-874.

Tonfoni, G. (1998ii), *Information Design: The Knowledge Architect's Toolkit*, Scarecrow Press, Lanham, Maryland, U.S.

Tonfoni G.(1999), On augmenting documentation reliability through communicative context transport, in *The Proceedings of the 1999 Symposium on Document Image Understanding Technology*, SDIUT99, Annapolis MD, 283-286

In: Chaos and Complexity Research Compendium
Editors: F.F. Orsucci and N. Sala, pp. 105-114
ISBN: 978-1-60456-787-8

Chapter 8

SUSTAINABILITY AND BIFURCATIONS OF POSITIVE ATTRACTORS

***Renato Casagrandi*[1,*] *and Sergio Rinaldi*[1,2]**
[1]Dipartimento di Elettronica e Informazione,
Politecnico di Milano, Milano, Italy
[2]Adaptive Dynamics Network, International Institute
for Applied Systems Analysis, 2361 Laxenburg, Austria

Abstract

In this paper we show how sustainability can be rigorously defined by making reference to the positivity of the attractors of a dynamical system. Consistently, the sustainability analysis with respect to various system and policy parameters can be performed by using specialized software for the study of the bifurcations of nonlinear dynamical systems. By means of an example concerning the tourism industry, we show how the analysis can be systematically organized and how easy it is to interpret the results of the numerical bifurcation analysis.

1. Introduction

The notion of sustainability is, nowadays, one of the most pervasive (if not invasive) in all political debates. The idea of sustainability emerged in the late sixties and can be traced in some pioneering scientific work, like that of Hardin (1968) or those that inspired the Club of Rome (Forrester, 1971), and of various socio-political movements (http://greenpeace.org, http://wwf.org and http://zerogrowth.org are few among the hundreds).

A great number of studies followed the pioneering stage and gave rise to *conferences*, as the 1992 UN Commission on Environment and Development conference in Rio de Janeiro, *journals* (like International Journal of Sustainable Development, Sustainable Development and World Ecology, Environmental Modeling and Assessment, Environment and Development Economics, Ecological Economics and others), *books* (Clark and Munn, 1987; Costanza, 1991; Wackernagel and Rees, 1995; Dodds, 2000; Starke, 2002), *laws*, as the

*E-mail address: casagran@elet.polimi.it

European Directives 337/85 and 11/97 for the Environmental Impact Assessment or the 42/2001 for the Strategic Environmental Assessment, and *international agreements*, as the still unattended Kyoto protocol (see http://unfccc.int/cop3/ for details). The main result of this huge effort is that people and governments are now much more sensitive than thirty years ago to the problem of long term survival of the world. However, despite this success, the issue of sustainability is still missing a simple and clear theoretical framework. This is very unfortunate, because in the absence of unified theories and methods of analysis any issue, no matter how important it is, becomes vague and anoising and, in the long run, discourages young scientists from investing their skillness.

Since sustainability refers to the possibility of keeping alive forever all meaningfull social and natural compartments of an evolving system (from towns to continents) it is clear that any *formal* definition of sustainability must refer to the long term behavior of some appropriate dynamical system. Thus, one should *a priori* expect that the analysis of sustainability with respect to the parameters characterizing the system (e.g. latitude, resource availability, population, ...) or its government (e.g. standards on emissions, environmental taxation schemes, subsidies, ...) can be performed through the study of the bifurcations of the attractors of a mathematical model mimicking the evolution of the real system. This is, indeed, the thesis of this article, which has the ambitious target of establishing a bridge between an important issue (sustainability) and a basic chapter of modern mathematics (bifurcation analysis of dynamic systems).

The paper is organized as follows. In the next section two components of sustainability, called *profitability* and *compatibility*, are defined with reference to an abstract model of the system. The first component takes into account only the social compartment of the system, while the second is only concerned with the environmental aspects. From these definitions it follows that the study of profitability and compatibility can be carried out through the bifurcation analysis of the attractors of the model. However, not all the attractors of the system are involved, but only those which are "positive" with respect to the social or to the environmental variables. Then, in the third section, sustainability is defined by putting social and environmental aspects at the same level of importance. This definition is in line with the theory of conflict resolution in multiobjective analysis and it is not as partisan as others proposed by many economists and environmentalists. Again, from our definition it follows that a bridge can be established between sustainability and bifurcation theory. Finally, an entire section is devoted to highlight through an example the meaning of the various definitions given in the paper and to show how the bridge established with bifurcation theory can allow one to systematically and effectively discuss sustainability once a model of the system is available.

2. Profitability and Compatibility

We now assume that the time evolution of the variables involved in the problem under consideration is described by a set of ordinary differential equations (ODE). We also assume that the state vector can be partitioned in three subvectors x, y and z of dimensions n_x, n_y and n_z, respectively, and that the variables x_i, $i = 1, \ldots, n_x$ and y_j, $j = 1, \ldots, n_y$ are indicators of social and environmental value. For example, the variables x_i could be measures of employment, welfare, health or education in a given nation, while the variables

y_j could be abundances of some plant or animal species in a forest, air quality in various towns, water quality in some rivers and lakes and so on.

All these variables x_i and y_j are typically non-negative, because they represent, directly or not, densities or biomasses. The equations describing the evolution of x_i and y_j over time are simply *conservation equations* involving the balance between inputs and outputs. Moreover, the input and output rates in the balance equations are, with almost no exception, expressed in terms of net per capita rates. In other words, the rate of variation of x_i ($dx_i/dt = \dot{x}_i$) is the product of the abundance x_i and the net growth rate per capita f_i, which is a function of all variables. All this brings to the conclusion that the model can be assumed to have the following general form

$$\dot{x}_i = x_i f_i (x, y, z, p, q) \qquad i = 1, \ldots, n_x \tag{1}$$

$$\dot{y}_j = y_j g_j (x, y, z, p, q) \qquad j = 1, \ldots, n_y \tag{2}$$

$$\dot{z}_k = H_k (x, y, z, p, q) \qquad k = 1, \ldots, n_z \tag{3}$$

where f_i and g_j are net growth rates per capita, z_k are variables of no direct social and environmental interest, and p and q are constant parameters which identify the characteristics of the system (altitude, structure of trasportation networks, fleet dimension, ...) and of its management (emission standards, fishing quotas, subsidies for tourism development, ...).

The particular form of eqs. (1,2) (sometimes called Kolmogorov's form) is such that the non-negativity of the variables x_i and y_j is preserved forever if it is guaranteed at the initial time $t = 0$, i.e. the space $x_i \geqslant 0$, $y_j \geqslant 0$ for all i, j is an invariant set. For physical reasons, in the following we will always refer to this invariant set even if we do not say it explicitly. Given the pair (p, q), i.e. given system (1-3), all its attractors in the above invariant set are uniquely identified (even if not always easily computable). These attractors can be many. Some of them can be not positive – i.e. $x_i = 0$ and $y_j = 0$ for some (i, j) –, while others are positive with respect to x (i.e. the attractor is characterized by $x_i > 0 \ \forall i$) and/or with respect to y (i.e., $y_j > 0 \ \forall j$). From now on, an attractor which is positive with respect to x will be called x-positive, and similarly for y, while an attractor which is positive with respect to x and y will be called (x, y)-positive. The existence of an x-positive attractor guarantees the possibility that all compartments of social interest remain alive forever. From an economic viewpoint, this means that the system has the possibility of producing profits forever. This justifies the following definition.

Definition 1 *A pair (p, q) is profitable if system (1-3) has at least one x-positive attractor.*

From this definition it follows that the points (p^*, q^*) of the boundary of the profitability region in the space (p, q) are bifurcation points of system (1-3). In fact, given a boundary point (p^*, q^*) there exist pairs (p, q) infinitely close to (p^*, q^*) for which system (1-3) has an attractor characterized by $x_i > 0 \quad \forall i$. By varying the parameters p and q, the attractor must cease to exist or to be x-positive at (p^*, q^*). In the first case the attractor has a catastrophic bifurcation at (p^*, q^*); for example, if the attractor is an equilibrium it disappears through a saddle-node bifurcation or through a subcritical Hopf bifurcation, if it is a limit cycle it disappears through a tangent bifurcation of cycles or through a homoclinic bifurcation, and so on for more complex attractors (i.e., tori and strange attractors). In contrast, if the attractor does not disappear at (p^*, q^*) but loses its positivity with respect to x at that point, then system (1-3) has a transcritical bifurcation at (p^*, q^*). This is a consequence of the Kolmogorov's form of eq. (1) which has the constant solution $x_i = 0$ for all values of p and q Notice that transcritical bifurcations are generic in Kolmogorov's systems (Kuznetsov, 1995).

It is worth noticing that not all bifurcations of system (1-3) are involved in performing the profitability analysis of a system. In fact, bifurcations of attractors which are not positive with respect to x have nothing to do with profitability. The same is true for x-positive attractors which remain such while undergoing a non-catastrophic bifurcation (e.g. a supercritical Hopf bifurcation). Finally, if the system has multiple x-positive attractors, a catastrophic or transcritical bifurcation of one of them does not imply the loss of profitability, which is guaranteed by the remaining x-positive attractors.

It is important to notice that the persistence of all social characteristics is not always guaranteed in a profitable system. In fact, in such a system, besides an x-positive attractor $A'(p, q)$, there can also be another attractor $A''(p, q)$ which is not positive with respect to x, namely an attractor characterized by $x_i = 0$ for some i. In such a case, a sufficiently strong perturbation, like a political scandal, an epidemics or a war, can move in a very short time the state of the system from the attractor A' into the basin of attraction of the alternative attractor A''. Thus, after the perturbation has ceased, the system will tend toward A'' and lose some of its social characteristics. It is therefore lecit to distinguish between safe and risky profitability as follows.

Definition 2 *A profitable pair (p, q) is safe if all the attractors of system (1-3) are x-positive, and risky otherwise.*

Analogous considerations hold for the environmental characteristics of the system: when it is possible to preserve them forever, we say that the system is compatible.

Definition 3 *A pair (p, q) is compatible if system (1-3) has at least one y-positive attractor.*

We can also distinguish between safe and risky compatibility as follows.

Definition 4 *A compatible pair (p, q) is safe if all the attractors of system (1-3) are y-positive, and risky otherwise.*

Of course, the boundary of the compatibility region in the space (p, q) enjoyes the same properties pointed out for the boundary of the profitability region. Thus, points (p^*, q^*) belonging to that boundary are catastrophic or transcritical bifurcation points of system (1-3).

3. Sustainability

In order to give a definition of sustainability which is not as partisan as the two previous ones, we pretend that both the socio-economic and the environmental compartments of the system persist forever. In other words, our formal definition of sustainability is the following.

Definition 5 *A pair* (p, q) *is sustainable if system (1-3) has at least one* (x, y)*-positive attractor.*

Notice that a sustainable system is both profitable and compatible, while the converse is not always true. In fact, a system could have many attractors, some x-positive and some y-positive, without having one (x, y)-positive attractor.

As far as risk is concerned, in the case of sustainability one can distinguish between economic and environmental risk as follows.

Definition 6 *A sustainable pair* (p, q) *is safe if all attractors of system (1-3) are* (x, y)*-positive, and risky otherwise. Moreover, a risky sustainable pair* (p, q) *is at economic [environmental] risk if one of the attractors of system (1-3) has* $x_i = 0$ *for some* i *[*$y_j = 0$ *for some* j*].*

Of course, the boundary of the sustainability region in the space (p, q) can be determined through bifurcation analysis by looking, in particular, at the catastrophic and transcritical bifurcations of the (x, y)-positive attractors. Moreover, also the boundaries separating safe and risky systems are composed of bifurcation points.

4. An Example

The example we present in this section concerns tourism sustainability. The model is an extension of a simpler model described in detail in Casagrandi and Rinaldi (2002). The problem is rather abstract and refers to a hypothetical site characterized by four variables: number of tourists (x), environmental quality (y), and capital, subdivided into two quotas (z_1 and z_2) measuring the value (amount) of structures producing touristic services of different nature (e.g., cultural and recreational). The model has the following form

$$\dot{x} = x\left[\mu_y \frac{y}{y+\varphi_y} + \mu_z \frac{\lambda z_1 + (1-\lambda) z_2}{\lambda z_1 + (1-\lambda) z_2 + \varphi_z x + \varphi_z} - \alpha x - a\right] \quad (4)$$

$$\dot{y} = y\left[ry\left(1 - \frac{y}{K}\right) - \beta_1 z_1 - \beta_2 z_2 - \gamma x\right] \quad (5)$$

$$\dot{z}_1 = -\delta z_1, +\nu \varepsilon x \quad (6)$$

$$\dot{z}_2 = -\delta z_2 + (1-\nu)\,\varepsilon x \quad (7)$$

At the righ-hand-side of eq. (4), the first term is the flow of tourists attracted by the environmental quality of the site, while the second is the flow of tourists attracted by services (with λ and $(1-\lambda)$ representing the relative preferences for the two classes of services); the third term is negative and specifies how quickly the tourists abandon the site when it is crowded, while the last term is the basic rate at which tourists abandon the site. If the site is absolutely not attractive $(y = z_1 = z_2 = 0)$ and there is no crowd, the tourists decay as $exp(-at)$. Of course, the rate a of this exponential decay is higher if there are many other interesting touristic sites. It is therefore reasonable to assume, as done in the following, that the parameter a is a measure of the competition of the alternative touristic sites.

In eq. (5) the first term says that in the absence of touristic activities $(x = z_1 = z_2 = 0)$ the environmental quality recovers to a carrying capacity K in accordance with a logistic equation, while the other terms represent the environmental impacts due to the supply of touristic services and to the presence of tourists.

Equations (6) and (7), which are linear, say very simply that the structures needed for producing services would become obsolete at an exponential rate δ if part of the profit (proportional to the number of visiting tourists) would not be reinvested in the service sector.

Assume that we are interested in the sustainability of the tourism industry and we want to discover the impact of the parameters ε (reinvestment of the touristic agents, for simplicity called *investment*) and a (competition of alternative touristic sites, called *competition*). Moreover, suppose that we also like to detect the effect on sustainability of λ and ν, which represent the preference of the tourists and of the agents for the first class of services. For this we must perform a bifurcation analysis of model (4-7) with respect to ε, a, λ, ν. The first step can be the study of the case $\lambda = \nu = 0.5$, in which the two classes of services are not distinguishable. Figure 1 shows the bifurcation diagram in the space (ε, a) for the reference parameter values given in the caption. The diagram has been obtained using specialized software for bifurcation analysis based on continuation techniques (Khibnik et al., 1993; Doedel et al., 1997). The attractors are either equilibria or limit cycles and there are only five types of bifurcations (Kuznetsov, 1995), that is

TC_{eq}	=	transcritical bifurcation of equilibria;
SN_{eq}	=	saddle-node bifurcation of equilibria;
$PLASN_{eq}$	=	saddle-node bifurcation of equilibria in the plane $y = 0, z_1 = z_2$;
$HOPF$	=	supercritical Hopf bifurcation;
HOM	=	homoclinic bifurcation.

Figure 1 shows that there are two codimension-2 bifurcation points: a Bogdanov-Takens (see point BT, where a saddle-node, a Hopf, and a homoclinic bifurcation curve merge) and a degenerate saddle-node (see, point $TCSN$, where a transcritical and a saddle-node bifurcation curve merge). The space (ε, a) is partitioned into 10 regions, each one characterized by a specific set of attractors. From Definition 1 it follows that the system is not profitable only in region 8, i.e. the tourists can be permanently present on the site if the competition of the alternative sites is not too high. Similarly, from Definition 3 it follows that the system is not compatible only in region 10, i.e. the environment is necessarily jeopardized by tourism activities if agents reinvest a lot (ε high) and alternative sites are not very competitive (a low). Finally, from Definition 5 it follows that the system is not sustainable in regions 8, 9, and 10, i.e. not only where it is not profitable (region 8) and where it is not compatible (region 10), but also in region 9 where the system is both profitable and compatible.

Figure 1 explicitly shows that not all bifurcations are boundaries of the profitability, compatibility, and sustainability regions. For example, the Hopf bifurcation curve is not a boundary of these regions.

The same bifurcation curves can be used to further partition the profitability, compatibility, and sustainability regions into safe and risky subregions. For example, Figure 2, which has been extracted from Figure 1, shows the subregions in which sustainability is safe and those in which it is at economic and/or environmental risk. From Figure 2 one can immediately conclude that the system can be sustainable and safe only if the alternative sites are not too competitive and the agents are not too aggressive in reinvesting their prof-

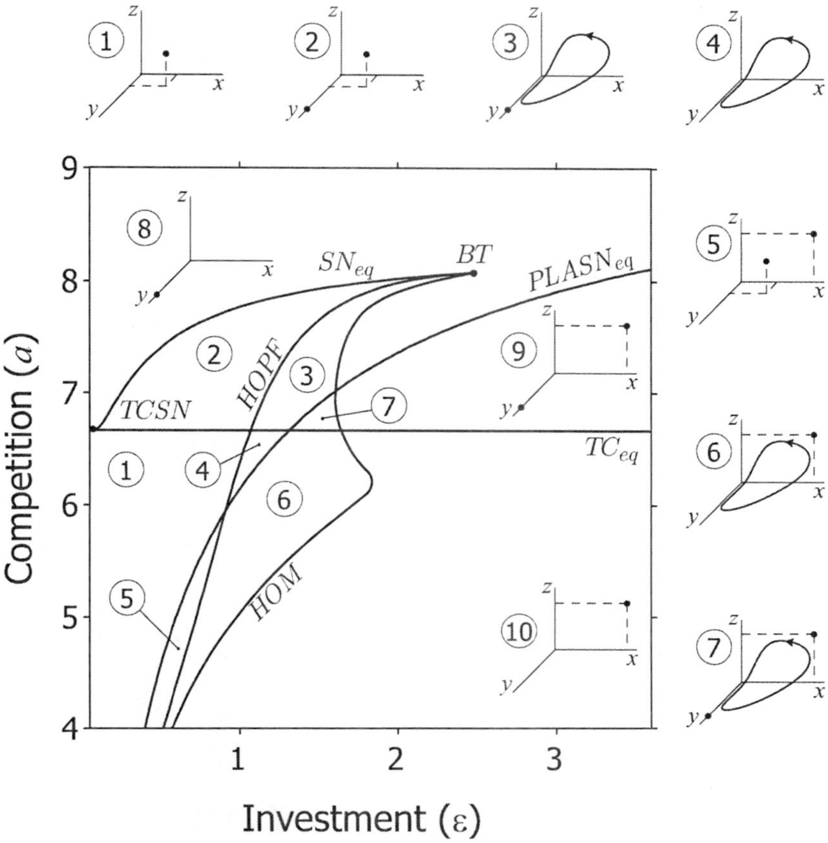

Figure 1. Bifurcation diagram of model (4-7) in the parameter space (ε, a) for the case $\lambda = \nu = 0.5$. The attractors in each region of the parameter space are sketched in the three dimensional space $(x, y, z = z_1 + z_2)$. Other parameter values are as follows: $r = K = \alpha = \gamma = \varphi_z = 1$, $\mu_y = \mu_z = 10$, $\varphi_y = \beta = 0.5$, $\delta = 0.1$.

its. An increase of competition first gives rise to an economic risk and then to the collapse of the tourism activities. Viceversa, an increase of the aggressivness ε of the agents first generates some environmental risk and finally jeopardizes the environment.

Once the analysis for the symmetric case $\lambda = \nu = 0.5$ has been performed, the parameters λ and ν can be relaxed and the same software can be used to complete the analysis through continuation, starting from Figs. 1 and 2. Two bifurcation diagrams produced in this way are shown in Fig. 3. The first refers to the case in which tourists are more interested in the first kind of services $(\lambda = 0.8)$, while agents invest primarily in the second class of services $(\nu = 0.2)$. The effect, with respect to the symmetric case, is an increase of the safe sustainability region and of that at environmental risk. In the opposite case, i.e. when agents adapt to the preferences of the tourists $(\lambda = \nu = 0.2)$, the above regions shrink and the system becomes more robust with respect to the competition of the alternative sites. All these properties, as well as others that could be easily obtained through continuation, are extremely useful for deriving qualitative but meaningful conclusions on tourism sustainability.

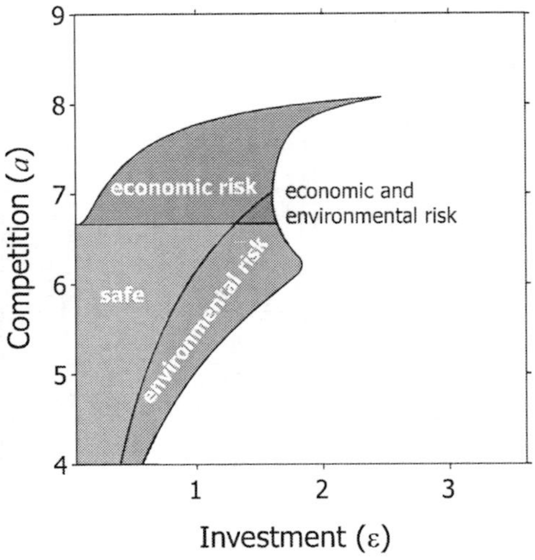

Figure 2. Sustainability diagram of model (4-7) with respect to investment and competition. Parameter values as in Figure 1.

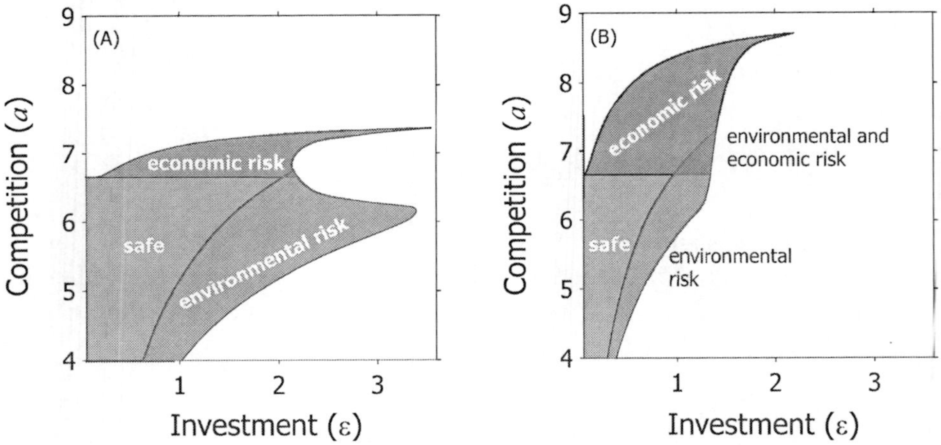

Figure 3. Sustainability diagram of model (4-7) with respect to investment and competition in the asymetric cases where $\lambda = 0.8$ and $\nu = 0.2$ (left panel) or $\lambda = \nu = 0.2$ (right panel). Other parameter values as in Figure 1.

5. Conclusions

Defining sustainability is a very difficult task, since everyone has different perspectives on what should be sustained. In this article we have tried to look at two of the most commonly visited viewpoints. According to the majority of socio-economical scientists, humans well-being and prosperity are issues of primary importance and should be saved with the highest priority. For them, Nature is often the lower trophic level at the expenses of which we can happily survive and grow. Other researchers, especially conservation biologists and philosophers, claim that we are but a very marvellous species that must coexist with other natural beauties, like animals and plants, clear lakes, pristine seas or green mountains with pure air. Many human activities often sounds to them like disturbances that interfere with the organic life of the biosphere. Mediation between these viewpoints is impossible. However, everyone of us has some economic interests in her/his everyday life *and* has the profound necessity of interacting with an alive environment, not simply for exploiting it in the present or in the near future.

We decided to be extreme here. Compatible is every policy that does not completely destroy the environment. Profitable is every policy that ensures some persistent income, no matter how little it can be. Evidently no conservation biologist or economist would easily accept these crude definitions, but surely they will agree that a non-compatible or non-profitable policy cannot be *realistically* sustained in the long run. If the policy is such that a situation which is simultaneously compatible and profitable does exist, thus we can hope to sustain a community. Of course, the income can be unaccettably too little or the environmental conditions can be too contaminated. This is matter of specific considerations or personal judgement and it is hard to see how a formal method applicable in general to any dynamical model can solve the issue. The method we have proposed, however, helps in selecting the set of sustainable policies in a rigourous way. Trashing the unsustainable can be a good starting point, indeed, in extremely complex situations.

Acknowledgments

The authors are grateful to Francesco Botti who has performed the numerical analysis for the proposed example.

References

Casagrandi R and Rinaldi S (2002). A theoretical approach to tourism sustainability. *Conservation Ecology* **6**(1):13. [online] URL: http//www.consecol.org/vol6/iss1/art13.

Clark WC and Munn RE, editors (1987). *Sustainable Development of the Biosphere*. Cambridge: Cambridge University Press.

Costanza R., editor (1991). *Ecological Economics: The Science and Management of Sustainability*. New York: Columbia University Press.

Dodds F, editor (2000). *Earth Summit 2002: A New Deal*. London: Earthscan.

Doedel EJ, Champneys AR, Fairgrieve TF, Kuznetsov YA, Sandstede B, and Wang XJ (1997). AUTO97: Continuation and bifurcation software for ordinary differential equations. Technical report, Department of Computer Science, Concordia University, Montreal, Canada. (Available by FTP from ftp.cs.concordia.ca in directory pub/doedel/auto).

Forrester JW (1971). *World Dynamics*. Cambridge, MA: MIT Press.

Hardin G (1968). The tragedy of the commons. *Science* **162**:1243–1248.

Khibnik AI., Kuznetsov YA, Levitin VV, and Nikolaev, EV (1993). Continuation techniques and interactive software for bifurcation analysis of odes and iterated maps. *Physica D* **62**:360–371.

Kuznetsov YA (1995). *Elements of Applied Bifurcation Theory*. New York: Springer Verlag.

Starke L, editor (2002). *State of the World 2002*. New York: Norton.

Wackernagel M and Rees W, editors (1995). *Our Ecological Footprint: Reducing Human Impact on the Earth*. NGabriola Island, Canada: New Society Publisher.

In: Chaos and Complexity Research Compendium
Editors: F.F. Orsucci and N. Sala, pp. 115-122
ISBN 978-1-60456-787-8

Chapter 9

Dynamical Prediction of Chaotic Time Series

Ulrich Parlitz * ***and Alexander Hornstein***
Third Physical Institute, University of Göttingen,
Bürgerstr. 42-44, 37073 Göttingen, Germany

Abstract

Externally driven dynamical networks are used for predicting signals from nonlinear (chaotic) dynamical systems. This dynamic approach for modelling and prediction is based on nonlinear functions emerging with generalized synchronization of driven dynamical systems. As an example we use random networks for prediction and cross prediction of signals from a chaotic Rössler system.

1. Introduction

Modelling and predicting the future evolution of a given time series from a (chaotic) dynamical systems is one of the main tasks of nonlinear time series analysis (Abarbanel, 1997; Kantz and Schreiber, 1997). Closely related is cross prediction of variables from bi-variate or multi-variate time series, where the evolution of some time series is predicted using samples from another signal that has been sampled simultaneously. In all these cases usually various regression methods are applied to approximate the underlying functional relation. The modelling consists basically in fitting some (static) function to data which are discretely sampled in time. This function is then used for predicting the observable of interest. In contrast to such a static and time discrete description we shall present in the following a *time continuous* and *dynamic* modelling method that exploits nonlinear functional relations emerging with *generalized synchronization*. Generalized synchronization (Rulkov et al., 1995, Kocarev and Parlitz, 1996) may occur when a dynamical system drives another (different) system. For proper coupling the state $\mathbf{y}$ of the response system is then asymptotically given by some nonlinear transformation h of the state $\mathbf{x}$ of the driving system, i.e. $\lim_{t\to\infty} \|\mathbf{y}(t) - h(\mathbf{x}(t))\| = 0$. The function h is apriori not known and can be smooth or

*E-mail address: parlitz@dpi.physik.uni-goettingen.de

very complicated depending on the dynamical features of the systems involved (Kocarev et al., 2000). Our goal is to exploit these emerging nonlinear functions as building blocks of a model or prediction scheme. In principle this could be done by feeding the given signal $u(t)$ into $m = 1, ..., M$ *different* response systems that are driven in a way such that they show generalized synchronization. In this case their state vectors $\mathbf{y}^m$ are asymptotically given by some functions $h_m(\mathbf{x})$ of the state of the driving system. Any output $f(\mathbf{y}(t))$ obtained from each of the response systems is thus also (asymptotically) given by some nonlinear function $g_m(\mathbf{x}(t))$ of the state $\mathbf{x}$ of the driving system. These functions g_m may be used as *basis functions* for approximating any other function F of the input

$$F(\mathbf{x}) = \sum_{m=1}^{M} c_m g_m(\mathbf{x}). \tag{1}$$

This approximation will be successful only when a sufficient variety of *different* basis functions g_m is available. Thus to provide a rich pool of functions one has to include many (very) different response systems. An alternative to such a set of individual systems are networks of dynamical systems. With generalized synchronization each dynamical variable of the network is (asymptotically) a nonlinear function of the input and again we may use for example a linear superposition of these variables for approximating the nonlinear function of interest.

Functional relations due to generalized synchronization have been exploited only recently in the context of (discrete time) recurrent neural networks by Jäger (2001) (*echo state networks*) and for spiking neurons by Maass et al. (2002) (*liquid state machines*), although these authors didn't point out the relation of their approaches to synchronization of dynamical systems. Inspired by their work we consider here time continuous networks with different local elements and different coupling topology. In particular we focus on the different roles of slow and fast dynamics in the network.

To illustrate our approach of dynamical prediction we shall use signals derived from the chaotic Rössler system

$$\begin{aligned} \dot{x}_1 &= -x_2 - x_3 \\ \dot{x}_2 &= x_1 + 0.25x_2 \\ \dot{x}_3 &= 0.4 + x_3(x_1 - 8.5) \end{aligned} \tag{2}$$

that are shown in Fig.1.

In Secs. 3 and 4 we shall demonstrate that one can predict the future evolution of x_1 as well as the time course of x_2 and x_3 using dynamical networks that are driven by x_1. The networks used will be introduced in the next section.

2. Dynamical Networks

The first network used for dynamical prediction consists of elements of the form

$$\dot{y}_m = -a_m(y_m + s_{i_m} s_{j_m}) \tag{3}$$

where $a_m \in [0, 2]$ is a randomly chosen parameter and s_{i_m} and s_{j_m} are signals that drive the m-th element. The signals s_{i_m} and s_{j_m} are randomly chosen from $\{1, u, y_1, ..., y_{m-1}\}$, i.e. they are constant, or the input u of the network, or from one of the preceeding elements. Due to the forward coupling structure the network is stable and shows generalized synchronization. The choice of the product $s_{i_m} s_{j_m}$ as nonlinearity is motivated by the fact that this function can easily be implemented electronically.

The *output* v of the network consists of a superposition of the signals y_m from all elements of the network

$$v = \sum_{m=1}^{M} c_m y_m \tag{4}$$

where the coefficients c_m are computed by solving a least squares problem to minimize the deviation from the desired output. For this purpose discretely sampled time series of the network variables $y_m(nt_s)$ and the signal to be approximated $z(nt_s)$ are used to formulate the least squares problem

$$\sum_{n=1}^{N} \left[z(nt_s) - \sum_{m=1}^{M} c_m y_m(nt_s) \right]^2 \stackrel{!}{=} \min \tag{5}$$

where N is the length of the time series and t_s denotes the sampling time. The least squares problem can be solved using singular value decomposition of the $N \times M$–matrix Y with elements $Y_{nm} = y_m(nt_s)$. Since the netsworks have a random structure and random parameters a_m their predictive power is also different. Therefore we repeated the random generation 100 times and choose the most efficient network for our task (see next section).

The second network has a layered structure. In each of the N layers there are M elements or nodes of the form

$$\dot{y}_{mn} = -a_{mn} y_{mn} + \frac{s_{mn}}{1.0 + |y_{mn}|}, \tag{6}$$

where $a_{mn} \in \mathbb{R}^+$ is the relaxation constant (free parameter) of the m-th element in the n-th layer and s_{mn} is the input signal which drives this element. The input signal s_{mn} is composed of the outputs from the elements in the preceding layer

$$s_{mn} = \sum_{i=1}^{M} c_{mi}^{(n)} y_{i(n-1)} \qquad \forall n = 2, \ldots, N, \quad \forall m = 1, \ldots, M,$$

where $C^{(n)} = \{c^{(n)}\}_{ij}$ is the $M \times M$ connectivity matrix between the $(n-1)$-th layer and the n-th layer. The elements in the first layer $n = 1$ receive their input $s_{m1} = c_m^{(1)} u$ directly from the input signal u. Assuming a one–dimensional (univariate) input $C^{(1)}$ is a vector whose coefficients $c_m^{(1)}$ are chosen randomly from the interval $[-1, 1]$. The extension to multi–dimensional (multivariate) inputs is straightforward.

The construction of the network is performed in two steps (see Fig. 2.). The first step consists of building a regular network. For this purpose the parameters a_{m1} of the first layer are set to some defined values from the interval $[a_{min}, a_{max}] \subset \mathbb{R}^+$. The same configuration

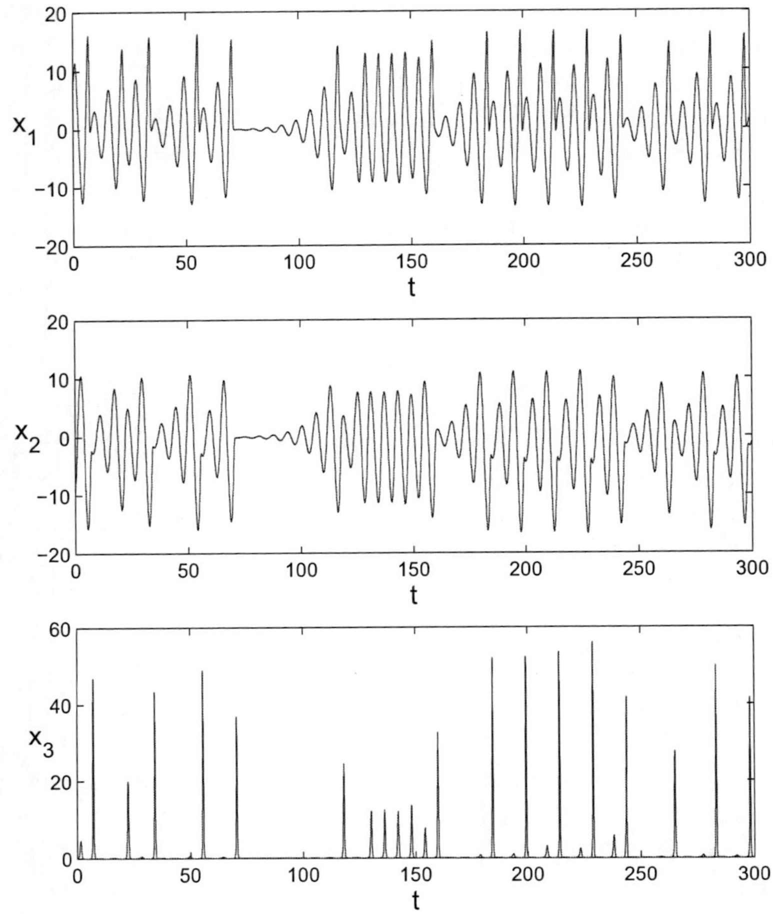

Figure 1. Chaotic dynamics of the Rössler system (2). (a) x_1 vs. time, (b) x_2 vs. time (c) x_3 vs. time.

is adopted for every other layer. In this way all elements from different layers but at the same position have the same relaxation constants

$$a_{mi} = a_{mj}, \qquad \forall i, j = 1, \ldots, N, \quad \forall m = 1, \ldots, M \, .$$

The regular network is completed by connecting only the elements with the same relaxation constant. This is simply done by choosing only the values of the diagonal elements $c_{ij}^{(n)}$ with $i = j$ and $n = 2, \ldots, N$ different from zero. Their value is determined randomly from the interval $[-1, 1]$. The relaxation constants a_{mn} affect the response time of an element to a stimulus in the input signal. Small values of a_{mn} produce a delay in the response while for large values the response follows almost immediately. Connecting nodes with the same response characteristics thus results in slow and fast propagation ways for the global input signal through the network as illustrated on the right hand side of Fig. 2..

To create a greater diversity the fast and the slow propagation ways can be intermingled. This is achieved in the second construction step by rewiring the elements of the regular network (Fig. 2.). Separately for each connection the coefficient $c_{ij}^{(n)}$ with $i = j$ is set

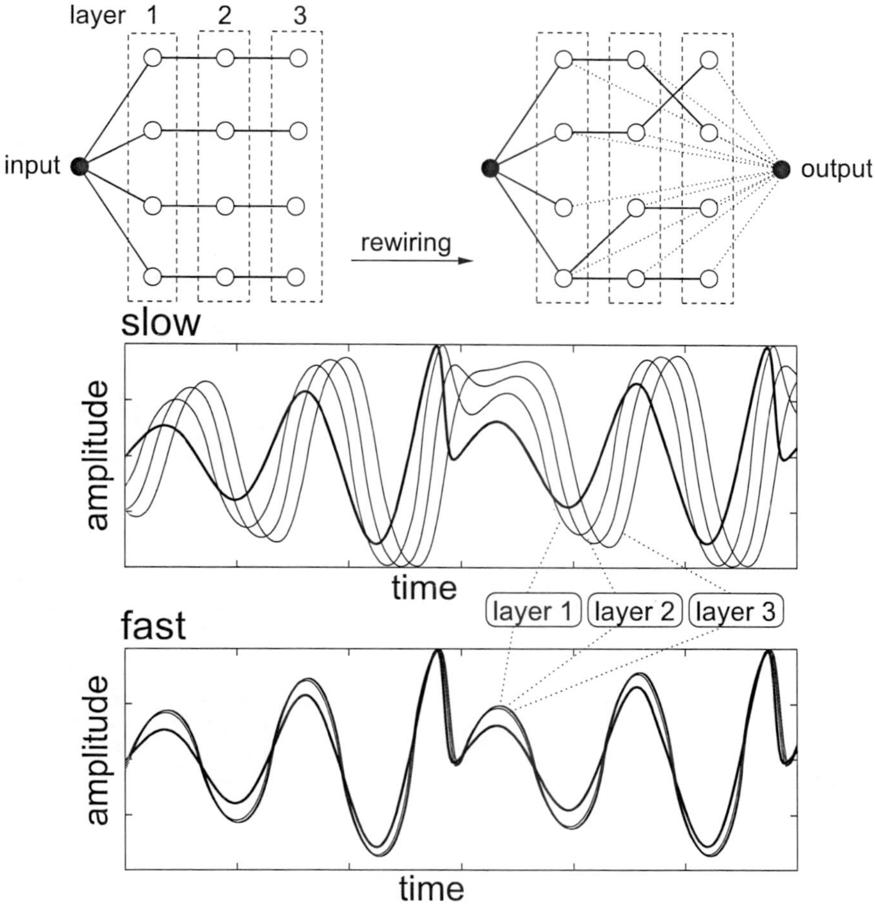

Figure 2. Left: Coupling topology of the second network before and after the rewiring step. Right: Input signal (fat line) propagating through the layers of the second network on a slow and a fast route with small and large values of the relaxation constants a_{mn}, respectively.

to zero with probability $p \in [0, 1]$. If a connection $c_{ij}^{(n)}$ is severed in such a way, a new connection is established by choosing k randomly from $1, \ldots, M$ and setting $c_{ik}^{(n)}$ to some random value in the interval $[-1, 1]$. The output of the second network is composed in analogy to the first network (see Eqs. (4) and (5)).

3. Prediction of Future Evolution

The two networks presented in the previous section will now be used to predict the future time course of the variable x_1 of the Rössler system (2). To achieve this goal the networks are driven by the signal $x_1(t)$. After transients decayed $N = 3000$ samples of the relevant variables of the network were taken with a sampling time $t_s = 0.1$. Simultaneously, the corresponding future values $x_1(nt_s + T)$ are sampled for approximating a prediction step of $T = 50t_s$ that corresponds to a typical cycle period of the chaotic oscillations of

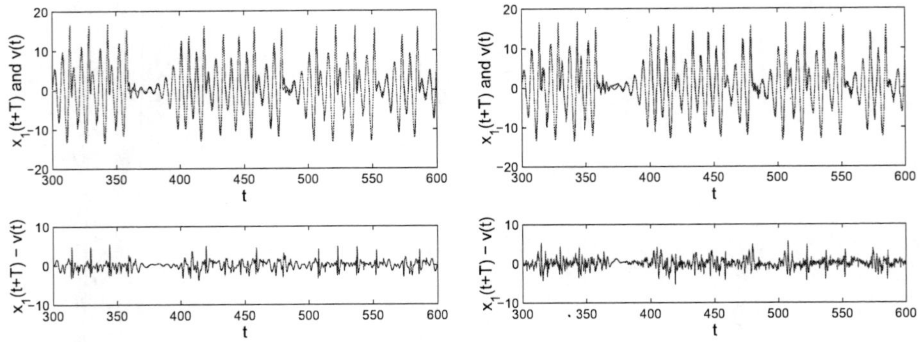

Figure 3. Prediction of the future values $x_1(t+T)$ of the x_1 variable of the Rössler system (2) from $x_1(t)$. Top: network output $v(t)$ and $x_1(t+T)$, bottom: error $x_1(t+T) - v(t)$; Left: network I, right: network II.

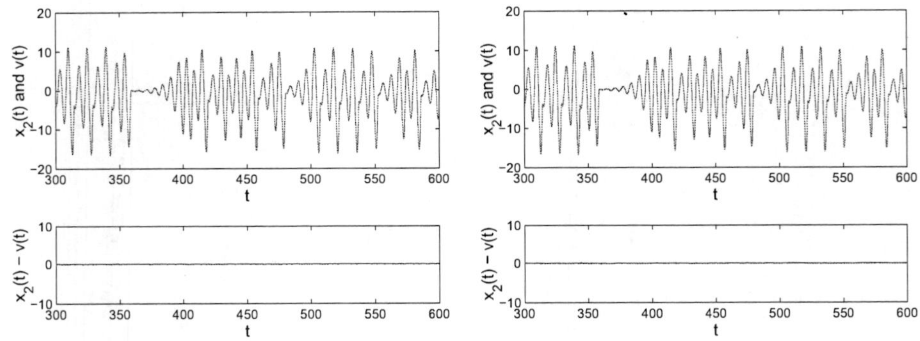

Figure 4. Cross prediction of the x_2 variable of the Rössler system (2) from x_1. Top: network output $v(t)$ and $x_2(t)$, bottom: error $x_2(t) - v(t)$; Left: network.I, right: network II.

the Rössler system. This data set of length N serves as *training set* and is used to compute the coefficients c_m by solving the least squares problem (5). Then another set of 3000 data points is sampled that constitutes an independent *test set* and is used to evaluate the predictive power of the networks on new data. The resulting prediction results for both types of networks are shown in Fig. 3. The plots in the first row show the future values $x_1(t+T)$ (red) and the prediction $v(t)$ (blue) from both networks (left and right), respectively. Since the true values and the predictions are hardly distinguishable we plotted in the second row the difference $x_1(t+T) - v(t)$. As can be seen both networks provide online predictions with some small fluctuating errors.

4. Cross Prediction

Instead of predicting a future value of a signal one may also be interested to predict some signal from another simultaneously sampled signal. This task is called cross prediction and can also be achieved by dynamical networks. As an example we predict the x_2 and the x_3 variable of the Rössler system (2) from x_1 that serves again as input of the networks.

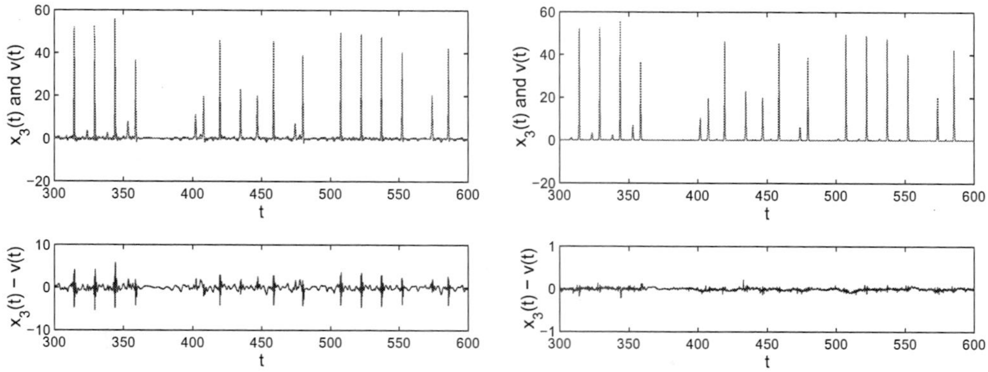

Figure 5. Cross prediction of the x_3 variable of the Rössler system (2) from x_1. Top: network output $v(t)$ and $x_3(t)$, bottom: error $x_3(t) - v(t)$; Left: network I, right: network II (note the different range of the vertical axis).

The same network structures as in the previous section are used. Only the coefficients c_m are recomputed by solving the least squares problem (5) for the desired output x_2 or x_3. Figures 4 and 5 show the results for x_2 and x_3, respectively. Again, prediction $v(t)$ and true values $x_2(t)$ or $x_3(t)$ are shown in the first row and their differences in the second row. Even the x_3 signal which is quite different in shape from the input x_1 can be predicted by the networks.

5. Conclusion

The success of dynamic prediction depends strongly on the right choice of the dynamical systems that are driven by the input signal. These systems not only have to respond with generalized synchronization but also have to provide the right nonlinear relations needed for approximating the (output) signal of interest. This problem is very similar to choosing the right structure or the right basis functions in any conventional (static) approximation scheme. Therefore, topics like overfitting or the so-called bias-variance-dilemma are of course also relevant for dynamic modelling. To improve the performance of the network term selection can be applied to generate better networks in a more systematic way. Random networks like those presented here are in this sense only the first step towards efficient dynamical prediction schemes. When implemented in analog hardware dynamic networks provide a continuous stream of real time predictions that may be used for monitoring or controlling technical devices or for online classification. The full potential of nonlinear dynamical systems for intelligent signal processing still has to be discovered.

Acknowledgment

We gratefully acknowledge financial support from the Volkswagen Foundation (grant I/76938).

References

Abarbanel HD (1997) *Analysis of Observed Chaotic Data*, New York: Springer-Verlag.

Jaeger H (2001) The "echo state approach to analysing and training recurrent neural networks: *GMD Report* **148**, German National Research Center for Information Technology.

Kantz H and Schreiber T (1997) *Nonlinear time series analysis*, Cambridge Nonlinear Science Series 7, Cambridge: Cambridge University Press.

Kocarev L and Parlitz U (1996) Generalized synchronization, predictability and equivalence of unidirectionally coupled systems *Phys. Rev. Lett.* **76**(11): 1816-1819.

Kocarev L, Parlitz U, and Brown R (2000) Robust synchronization of chaotic systems: *Phys. Rev. E* **61**(4): 3716-3720.

Maass W, Natschläger T, and Markram H (2002) Real-Time Computing Without Stable States: A New Framework for Neural Computation Based on Perturbations. *Neural Computation* **14**: 2531-2560.

Rulkov N F, Sushchik M M, Tsimring L S, and Abarbanel H D I (1995) Generalized synchronization of chaos in directionally coupled chaotic systems: *Phys. Rev. E* **51**: 980-994.

In: Chaos and Complexity Research Compendium
Editors: F.F. Orsucci and N. Sala, pp. 123-147
ISBN 978-1-60456-787-8

Chapter 10

DYNAMICS AS A HEURISTIC FRAMEWORK FOR PSYCHOPATHOLOGY

Jean-Louis Nandrino[a,b] **, *Fabrice Leroy***[a,b] ***and Laurent Pezard***[b,c,d*]
[a] UPRES "Temps, émotion et cognition",
Université Lille 3
[b] International Institute Erasme
[c] Neurosciences Cognitives et Imagerie cérébrale
LENA-CNRS UPR 640
[d] Institut de Psychologie,
Université Paris 5

Abstract

The development of the mathematics of dynamical systems now offers a rigourous framework to deal with complex phenomenon evolving with time. The possible euristic value of applying dynamical concepts to the field of psychopathology is investigated here. Three levels of applications found in the literature are reviewed: metaphoric, qualitative and quantitative. Psychopathology seems indeed a field where the concepts of dynamics can offer important tools, both theoretical and empirical. Nevetheless, specific problems should be emphasized to obtain a more profound insight in normal and pathological mental phenomenon.

1. Introduction

The science of the mind is usually fond of importing new concepts from other disciplines. In the last thirty years, the development of the scientific interest in the behavior of complex systems has led to the emergence of notions such as chaos, attractors, sensitivity to initial conditions, etc. and to related numerical methods. The goal of this article is to estimate, on the basis of a literature review[1], the possible heuristic value, for psychopathol-

*Address for correspondance: L. Pezard, LENA-CNRS UPR 640, 47 Bd de l'Hôpital, 75651 Paris cedex 13. France.

[1]The literature was scanned using two data bases: "pubmed" (url:) and "PsychInfo" (url:). Key words were: chaos, nonlinear dynamics, catastrophe theory, psychopathology, psychiatry, depression, schizophrenia, personality disorders, mood disorders, addiction.

ogy of the tools developed within the mathematical and physical framework of dynamical systems theory.

1.1. Explanation Levels in Psychopathology

Since mental diseases have been studied from biological to social level, psychopathology stands at the border between natural and human sciences. From the point of view of natural sciences, mental troubles are to be reduced to biological phenomena such as Korsakov syndrome or dementia in Alzheimer's disease. For the human sciences, mental disease are thought to be due to "mind" troubles or to be related to social factors such as relationships with close relatives (*i.e.* family) or to more general factors such as social frustrations. Nevertheless, the search for a linear causality from one level to another (from biology to social or backwards) has obviously failed. For example, no biological indicators are available yet to unambiguously decide for a specific mental trouble.

Such problems have led to emphasize the need for a multidisciplinary investigation of the bio-psycho-social nature of mental troubles (Engel, 1980; Freedman, 1995). These approaches usually explain the whole disease as the *sum* of each individual factor: biological, social and psychological. Nevertheless, a complex phenomenon, such as a mental disease, can hardly fit into a linear model and a co-determination of levels seems more probable. It is thus necessary to find tools to deal with circular causality and interactions between levels.

1.2. How Dynamic Are Mental Diseases?

The hallmark of mental troubles is the compulsive repetition of actions, fantasies or patterns of discourse which can be considered as successive conscious or unconscious acts. Mental diseases have an onset, evolve and can finally disappear. Moreover, specific temporal patterns appear in mental diseases whatever the observation scale: from milliseconds (response to stimuli, biochemical modulation or neuronal electrical activity) through minutes or hours (clinical interview) to years (time course of recurrence) or generations. During the acute period, changes in biological and behavioral rhythms are observed and during the whole life, specific alternations between disease and remission are also observed (Keller et al., 1986). The number of recurrences increases as a function of previous episodes and the illness patterns become more rhythmic with cycle acceleration finally resulting in rapid cycling or ultradian mood patterns (Kramlinger and Post, 1996; Huber et al., 2001a).

As an explanation for the occurrence and evolution of specific pathological patterns, several models have underlined the importance of initial conditions. In the psychoanalytic tradition, or even in cognitive psychotherapy, the possible influence of interactions and learning in infancy are assumed as important vulnerability factors for the development of mental disorders. Nevertheless, a longitudinal study, of more than one hundred subjects, from infancy to early adulthood, showed that the onset of behavioral disorder was highly variable (from 2 to 16 years). In most of the cases appearing during the adolescence, data revealed neither any prodromal or pathogenic symptoms nor excessive stress in earlier period (Thomas and Chess, 1984). The structural hypothesis of universal development stages and of early determinism of mental disorders is thus severely challenged. In fact, the evolution of mental troubles are highly contextualized and related to supports or constraints continuously acting on individuals.

1.3. From Dynamical Diseases to Psychopathology

The application of dynamical systems theory to the modeling of physiological systems led to the definition of "dynamical diseases" (Mackey and Glass, 1977; May, 1978; Mackey and Milton, 1987). The hallmark of a dynamical disease is a sudden qualitative change in the temporal pattern of physiological variables (Bélair et al., 1995). From a dynamical point of view, such changes are related to modifications in the control parameters that lead to abnormal dynamics. This kind of dynamical changes have been clearly observed in neurological diseases (Milton et al., 1989).

In a review of 32 neurological and psychiatric diseases, two main characteristics have been considered as landmarks for a "dynamical disease" (Milton and Black, 1995): the recurrence of symptoms (10/32) and the oscillations appearing in the functioning of nervous systems (22/32). Within these fields, epilepsy and affective disorders are the best candidates for the application of the "dynamical disease" concept.

Such a framework can be generalized so that psychopathology may fit into the framework of "dynamical disease". As in the case of physiological functioning, it may be hypothesized that a mental structure is an emergent property associated to an underlying dynamics. Clinical signs and symptoms, observed in psychiatry, would thus correspond to qualitative dynamical changes related to modifications in control parameters (Globus and Arpaia, 1994; Moran, 1991; Schmid, 1991). Within such a conception, mental disorders and changes in mental states (such as changes following psychotherapeutic activity) in emotional states or in developmental stages may be influenced by parameters acting at several levels from physiological to social one. The presence of a symptom would thus emphasize the stability of the system in a specific parameter domain and thus be seen as an attractor. The articulation between levels of observation would thus be defined on the basis of changes in dynamical observables.

We describe the psychopathological literature, dealing with time evolution of psychopathological phenomema, using mathematical and physical concepts from dynamical systems theory. We will distinguish three levels of application: firstly, the study of the dynamics of complex systems can offer a set of *metaphors* for the description of mental phenomena, secondly *qualitative* insights of the behavior of systems can be obtained with the study of various models (such as neural networks or catastrophe models) and then *quantitative* characteristics of dynamical behaviors can be infered using nonlinear modeling and time series analysis. At last, criticisms and interests are given in order to favor a *rigorous* development of the application of dynamical concepts to psychopathology.

2. Metaphors

On the basis of the similarities between *general properties* of nonlinear dynamical systems and temporal phenomena observed in mental life, metaphorical associations between concepts have been undertaken. We distinguish different attempts using dynamical system paradigm as a metaphor in psychopathology. It has been proposed to understand the Self as an emergent property issued from dynamics of multiple iterations of brain processes, perceptual and social experience. Moreover, psychotherapists have used terms from chaos theory as an analogy for phenomena emerging during the course of psychotherapies (psychoanalytical and systemic).

2.1. The 'Self' as a Dynamical System

Object-relation psychoanalysts (Mahler, 1968; Klein, 1948) have underlined how the extended system of personal relationship influence personality development throughout life. Intersubjectivity theory (Stolorow et al., 1994) examines how the interplay between the subjective worlds of the patients and the analyst gather into a new system. These two points of view led to conceptualize the 'Self' as adaptive and multi-stable state of consciousness about oneself and the 'world'. Thus, the 'Self' is able to adopt successively a set of discrete states evolving on the basis of contextual influences from microscopic level of physiology (Freeman, 1990) through macroscopic levels of psychology, social or cultural organization. This psychic structure could thus be conceptualized as an open, complex, dynamical system (Marks-Tarlow, 1999). Healthy selves self-organize and evolve to the edge of chaos, where they are capable of flexible reorganization in response to unpredictable social an environmental contingencies (Goldstein, 1997).

In these conditions, the 'Self' finds its origin in the continuous interactions between biological roots and the history of the subject. 'Self' is thus linked to preconscious and preverbal roots. Nevertheless, language is necessary to make the 'Self' conscious (Schwalbe, 1991). Consciousness, as a recursive process operating upon internal objects and external influences, does not precede acts but emerges out of it. An iterative loop of perception-action-reflection may lead to the emergence of a new level of complexity: a consciousness of consciousness.

2.2. Dynamical Metaphors for the Psychotherapeutic Processes

The course of psychotherapy is not a linear progression towards a new healthier mental state. Psychotherapy is a multidimensional process involving biological factors, psychological and social experiences leading the subjects towards a new state (Bütz, 1993). The course towards this change is an enchainment of stable and instable periods that could be described as a non linear dynamic phenomenon (Langs, 1986; Bütz, 1993; VanEenwyk, 1991; Spruiell, 1993; Levinson, 1994). Analysts perceive patients as different along the psychotherapy course; this change can be conceptualized as a qualitative shift in patients' state *i.e.* a bifurcation in the dynamical systems theory (Moran, 1991; Priel and Schreiber, 1994; Verhuslt, 1999).

Moreover, the process of interpretation during a psychotherapy can make the psychological system more sensitive to new perturbations (Bütz, 1993; Verhuslt, 1999). The psychotherapist's function, especially through his interventions, is to stabilize or destabilize patients' mental processes and their way of thinking or telling their narrative. The therapeutical situation can thus be viewed as a dynamical process where a common system is co-created in the interaction between therapist and patient (Elkaïm, 1990; Lonie, 1991).

The therapeutic frame (regular appointments and stable environment) is designed to allow the emergence of a sampling of the patient's inner world. This phenomenon has been interpreted through the concept of self-similarity. At any level of examination: within the whole case history, during a single session or a single dream, one can observe the patient own "signature", a recognizable pattern of his/her mental life (Lonie, 1991; Moran, 1991). Certain aspects of psychoanalytical situation such as unconscious fantasies have also been viewed as a form of strange attractor (Moran, 1991; Quinodoz, 1997; Galatzer-Levy, 1995) or the repetition of some themes in the course of the therapy as a limit cycle (Lonie, 1991).

The sensitivity to initial conditions and the unpredictability of complex phenomena is an important analogy between nonlinear dynamical systems and psychotherapeutic situations (Bütz, 1993; Lonie, 1991). Even if the individual's mental life and behaviour is powerfully affected and determined by precocious experiences, repetitions are not strictly identical and some small elements could make the evolution unpredictable. The evocation of the history of the patient or the focalization on certain events or feelings can have unpredictable effects. Therapy can thus be considered as an extended series of well-timed perturbations which serve gradually to disrupt the strange attractors characteristic of the patient's fantasy-behavioral coupling (Moran, 1991).

Systemic therapy has used the concepts from the general systems theory for a long time. The models from nonlinear dynamical systems are thus a kind of "natural" extension for this practice (Koopmans, 1998; Miller et al., 2001). The time evolution of a family system goes through ordered and disordered phases (Brabender, 2000) where the symptom signs the inability of the group to overcome crisis. Family therapist can be considered as a catalytic factor for changes in the family functioning leading the emergence of a new state (Ricci and Selvini-Palazzoli, 1984; Elkaïm, 1990).

2.3. Conclusion

The properties of nonlinear dynamical systems are obviously appealing for the description of complex mental phenomena. In fact, the metaphorical use of dynamical concepts might be a first movement to get away from strictly medical models based on a linear explanation of the onset and the evolution of mental disorders. In that sense, nonlinear dynamical analogies can offer new tools to deal with complex situations encountered in the clinical practice.

Nevertheless, several caveats need to be avoided. The distance between mathematical concepts and psychological (or psychoanalytical) theories needs to be questioned precisely (Denman, 1994; Kincanon and Powel, 1995). Does mathematics throw a light on psychology or does it darken it? What is exactly the nature of the explanation expected from such analogies (Gardner, 1994)? It is important to avoid errors due to superficial comprehension of precise scientific concepts (Sokal and Bricmont, 1999).

Finally, such analogies can be used as a starting point for a scientific enquiry into mental phenomena and should be tested on qualitative modeling or empirical quantitative studies.

3. Qualitative

Qualitative models are related to the introduction of explicit constraints on the definition of a *specific dynamical system* supposed to model the empirical system taken into account. Two types of dynamical systems have been taken as models in psychopathology: first, gradient systems related to "catastrophe theory" have been considered, then the development of neural network introduced another kind of modeling.

3.1. Gradient Systems

The state of a system at time t can be described by a set of variables $\psi(t) = \{\psi_i(t)\}$ (ψ_i are thus called *state variables*) and that a set of parameters, denoted c_α ($1 \leq \alpha \leq k$), controls the qualitative properties of the system's time evolution (c_α are thus called *control parameters*). The dynamics of the system is said to be described by a *dynamical system* when[2]:

$$\frac{d\psi}{dt} = \mathbf{f}\,(\psi, c_\alpha, t) \tag{1}$$

with $\mathbf{f} = \{f_i\}$. The general study of systems represented by equation (1) is a very difficult problem. It can be made more tractable when two assumptions are added (Gilmore, 1981):

1. If the functions f_i are considered as independent of time, the dynamical system is now an *autonomous dynamical system* and powerful statements can be made about such systems which depend on a small number of parameters ($k \leq 4$).

2. It can be noticed that in equation (1) the functions f_i look as the components of a force. With the assumption, inspired from mechanics, that all the functions f_i can be derived as the negative gradient (with respect to the ψ_i) of some potential function $V(\psi_j, c_\alpha)$:

$$f_i = -\frac{\partial V(\psi_j, c_\alpha)}{\partial \psi_i}$$

   the resulting system:

$$\frac{d\psi_i}{dt} + \frac{\partial V(\psi_j, c_\alpha)}{\partial \psi_i} = 0 \tag{2}$$

   is a *gradient system* ($\dot{\psi} = -\boldsymbol{\nabla}_\psi V$). This kind of system is much more tractable than the other systems described previously.

Dynamical systems theory deals with the solutions $\psi_1(t)$, $\psi_2(t)$, …, $\psi_n(t)$ of equation (1) which define trajectories (*i.e.* time evolution) of the system. Of particular interest are the equilibria ($d\psi_i/dt = 0$) of dynamical and gradient systems. They define the states where the system can settle in, either, a stable or unstable manner.

Elementary catastrophe theory is the study of how the equilibria $\psi_j^e(c_\alpha)$ of $V(\psi_j, c_\alpha)$ change as the control parameters c_α change for gradient systems. In that sense, elementary catastrophe theory is a quasi-static theory since it is only concerned by the equilibrium points of a dynamics and how they change when the control parameters are varied (Thom, 1977a; Arnol'd, 1992).

These models have been mainly used to model the emergence of discontinuous behaviors out of continuous parameter variations. The application of catastrophe theory to concrete phenomenon can be divided into the *'metaphysical'* way and the *'physical'* way (Thom, 1977b).

[2]For a more general statement about the time evolution of a system and the hypothesis that lead to the somehow reduced dynamical system description, see Gilmore (1981, p. 3–5).

The metaphysical way considers the generality of elementary catastrophe as justifying the use of archetype situations to describe phenomenon where the nature of the dynamical systems that produce them is unknown. This method lead to qualitative models that can be used analogically with real situations.

The dichotomy between anorexia and bulimia is an archetypic example (Zeeman, 1977). The starting point of the model was the observation that an anorexic loses access to normal attitudes toward food and that many sufferers develop bulimic phase. During theses cycles attitudes toward food switch catastrophically from one extreme to the other, and they never take on normal intermediate values. These are the hallmarks of the cusp catastrophe which was used to model this behavioral trouble. A more sophisticated model added the sleep/wake cycle to the preceding cusp model and thus develop a geometrical non-trivial double cusp model (Callahan, 1982).

Catastrophe theory has also been used in a set of other models in clinical psychology (Weiner, 1977; Galatzer-Levy, 1978; Scott, 1985). Catastrophe model based on the attention focus has been proposed to deal with manic/depressive illness (Johnson, 1986). Emotional numbing associated with post-traumatic stress disorder (Glover, 1992) and other emotional responses (Lanza, 1999) have also been modeled using cusp catastrophe such as the relationship between alcohol intoxication and suicidal behavior (Hufford, 2001). In the case of schizophrenia, catastrophe models provided ways in which neurochemical and environmental influences could interact so that very small changes in either variable may produce the rapid changes in intensity of psychosis (MacCulloch and Waddington, 1979) The dopaminergic hypothesis has also been investigated using this framework (King et al., 1981).

From a more general standpoint, the possible heuristic value of Thom's dynamical theory to the Freudian metapsychology has been evaluated (Porte, 1994). On the basis of a careful parallel between both authors, it can be estimated that positivists caveats of Freud's theory find a natural solution in modern dynamical theories.

The physical way applies when the dynamics is indeed described by a gradient system. It is the case for example in physical systems for phase transitions in thermodynamics or caustics in optics (Poston and Stewart, 1978; Gilmore, 1981).

An exemplary modeling of alcohol consumption follows such a perspective (an der Heiden et al., 1998). The model is based on the mathematical expressions relating general phenomenon supposed to drive alcohol consumption (denoted A). The authors reported several stage of a qualitative model which final expression is:

$$\frac{dA}{dt} = F - r.A + \sigma \frac{A^2}{1 + A^2} \tag{3}$$

where F (frustration) is considered as a constant force driving alcohol consumption (such as life conditions, habits, social environment...), r is related to the disagreement of alcohol intake (illness, social values...) and the last term with parameter σ is a nonlinear auto-catalytic model. The study of the equilibria of this model leads to describe the phenomenology of drinkers typology and a cusp catastrophe was found in the description of the bifurcations. The discussion of the model show how control parameters can be varied to change the drinking behavior and thus may be of interest in the clinical practice. Moreover, This study demonstrates that the interaction of very few "mechanisms" results in a large manifold of different kinds of behavior.

3.2. Neural Networks

The first use of neural networks has been devoted to provide *models of brain functioning*. Two major class of models can be differentiated: parallel distributed processing (PDP) models (McClelland et al., 1986a,b) and attractor neural network (ANN) models (Hopfield, 1982; Amit, 1989). We will only review here some models using ANN to deal with psychiatric syndromes (other models can be found in Rialle and Stip (1994), Aakerlund and Hemmingsen (1998) or Huberman (1987)). Neural networks have also been used as models of *symptoms dynamics*.

Attractor neural network models are based on systems such as (Hopfield, 1982):

$$S_i(t+1) = F(\sum_j w_{ij} S_j(t) - \theta_i) \qquad (4)$$

where $S_i(t)$ is the state of "neuron" i at time t, w_{ij} is the "synaptic weight" between neurons i and j and θ_i is the threshold. Such system has computational abilities since memories are stored as attractors of its dynamics; so that, as an content-addressable memory:

- memories (or patterns) are retrieved according to similarity to the input
- generalizations based on different memories are possible
- memories are distributed across all neurons, and are not localized

An alterations of these functions, related to changes in the control parameters, may thus simulate cognitive impairments in some mental disorders.

3.2.1. Models of Syndromes

Manic-depressive illness An interpretation of manic behavior has been proposed on the basis of a classical Hopfield network (Hoffman, 1987). The increase of noise (related to the steepness of the slope of the transition function) causes an increase of transitions between attractors. This behavior of the network has thus been related to the transitions between thoughts in manic patients.

Another model consider depression-like and manic-like behavior as attractors of a dynamical system (Globus and Arpaia, 1994). The same formalism is thus used at a higher level where attractors represent the overall behavior. It must be emphasized, that this model is clearly similar to a catastrophe model.

Schizophrenia On the basis of ANN, schizophrenia has been interpreted as the result of the overloading of the network memorization abilities (Hoffman, 1987). In fact, overloading causes the creation of spurious attractors from which the network cannot escape. Delirium has been associated with such a process. Troubles in cortical pruning, during development, lead to a decrease of cortical synaptic contacts and would thus decrease the memorization ability of the cortical network in schizophrenic patients (Hoffman and Dobscha, 1989; Hoffman and McGlashan, 1994, 2001). The presence of spurious attractors could be the analog of the three types of symptoms: strange outputs, independent submodules, and independence of modules functioning in front of inputs. This model has been discussed in David (1994). A network based on spreading activation was also proposed to model how an initial paranoid state becomes crystallized into a fixed delusion in schizophrenia (Vinogradov et al., 1992).

The defect of generalization and/or of taking into account the context in schizophrenic patients has been related to dysregulation of dopamine transmission (Cohen and Servan-Schreiber, 1992). Such changes in the interactions between cortical and sub-cortical structures could reduce the size of attractors in patients when compared to controls (Tassin, 1996). Nevertheless, the observed increased variability in behavior among schizophrenics, could also been related to chaotic dynamics in the central dopaminergic neuronal system (King et al., 1984).

3.2.2. Time-Course of Affective Disorders

Episodes of affective disorders have been analogically compared to firing in neuronal networks (Huber et al., 2000b,a, 1999). A mathematical model based on a nonlinear dynamical system influenced by noise has been proposed:

$$\tau_x \frac{dx}{dt} = -x - \sum_i a_i^\nu w_i(x - x_i) + S + gw \tag{5}$$

where τ_x is a relaxation time constant, a_i^ν represents the activation states ($\nu = 1$ or $\nu = 2$), i ranges over four different states, w_i are coupling constants and x_i describes different activation levels. S represents the control parameter (corresponding to an ongoing disease process), and gw represents a Gaussian white noise to take into account environmental or endogenous stochastic influences.

The dynamic behavior shows that, in the course of the illness, noise might amplify subclinical vulnerabilities into disease onset and could induce transitions to rapid-changing mood pattern. In this model, based on cooperative effects between deterministic and random dynamics, noise increases the spectrum of dynamic behaviors.

Furthers modifications of this model, based on a feedback mechanism for episode sensitization, permits to strongly support the importance of episode sensitization as fundamental mechanism for the disease's progression in affective disorders (Huber et al., 2001a,b).

3.3. Conclusion

The introduction of specific kind of dynamical systems as models in psychopathology provide a global framework for the description of changes in psychopathology. Based on the generality of the formalism it is thus possible to describe various levels of observations within the same model. Nevertheless, even if these models introduce more constraints than in analogical use of dynamical concepts, it is not always clear whereas they constitute real model or mere elaborated metaphors.

These models thus need development towards empirical empirical tests. The introduction of quantitative methods may fill the gap between qualitative modeling and empirically observed dynamics.

4. Quantitative

Empirical studies quantifying the characteristics of observed dynamics are needed to estimate the scientific and clinical value of dynamical paradigm in psychopathology.

4.1. Data Fitting

The presence of a "catastrophe" can be infered either on the basis of observation or from the study of a model. Empiricists would prefer that "catastrophe" could be proved and measured on the basis of experimental data.

The theoretical analysis of the behavior of systems in the neighborhood of singularities allow one to define critical phenomena that should be observed for a catastrophe model to apply. These phenomena have been called *catastrophe flags* (Gilmore, 1981). The first ones (modality, inaccessibility, sudden jump) have usually been taken as qualitative indices for the 'metaphysical' application of catastrophe theory. The other one (divergence, hysteresis, divergence of linear response, critical slowing down and mode softening, anomalous variance) are usually more difficult to observe or to describe. Such 'flags' have been infered in development stages (van der Maas and Molenaar, 1992).

Three quantitative approaches to the problem of testing the fit of behavioral data to catastrophe models have been developed. The first has taken stochastic difference equations as a basis and uses the methods of moment to estimate model parameters (Cobb and Watson, 1980). The second uses polytope search curve-fitting procedure to obtain maximum likelihood estimates of the model from the observed data (Oliva et al., 1987; Lange et al., 2001). The third approach is in the form of least-square regression (Guastello, 1982, 1987). This last method has been discussed in Alexander et al. (1992) and Guastello (1992).

The analysis of a cusp catastrophe used to model adolescent alcohol use have shown that dispositions should be viewed as the normal parameter and situation pressure as the splitting parameter of the cusp (Clair, 1998). Statistical analysis of empirical data using polynomial regression have shown that the cusp model better fit the data than the alternative linear models (Clair, 1998). Such procedure have also been used in the test of anxiety theory in the context of sport performance (Hardy, 1996).

4.2. Time Series Analysis

It is out of the scope of this review to develop a complete methodological overview. For complete references, see Kantz and Schreiber (1997); Grassberger et al. (1991); Abarbanel et al. (1993); Ott et al. (1994); Badii and Politi (1997).

Time series analysis deals with the quantification of the 'complexity' in the sequence of observed data. From the dynamical point of view, the first step is the reconstruction of the trajectory of the system within its phase space, then geometrical indices (such as dimensions) or dynamical indices (such as entropies) are computed. It has been shown that these indices should be statistically validated using surrogate data methods (for a review see: Schreiber and Schmitz, 2000). When data are discrete (or when the continuous dynamical system is 'properly' discretized), the characterization of the dynamics uses symbolic methods (Badii and Politi, 1997).

4.2.1. Brain Dynamics

The central nervous system can be considered as a complex system which can be modeled within the dynamical system theory. For example, nonlinear dynamics provides new methods for the investigation of EEG signals.

Depression Studies of brain dynamics in depression have mainly shown a decrease of the first Lyapunov exponent for sleep stage IV in depressed patients when compared either to controls (Roschke et al., 1995b) or schizophrenic patients (Roschke et al., 1994). Unipolar depression is characterized by specific brain dynamical patterns of low complexity which evolve during pharmacological treatments (Nandrino et al., 1994; Pezard et al., 1996). Nevertheless, the recovery of a healthy brain dynamics is dependent upon clinical history: in the case of patients with recurrent episodes, even after a clinical improvement similar to that of first episode patients, brain dynamics did not recover the complexity level of control subjects. Changes in brain dynamics have been correlated with clinical evaluation of depressive mood in three depressed patients (Thomasson et al., 2000). These results were confirmed in the case of a 48-hour rapid cycling patient (Thomasson et al., 2002).

Schizophrenia Brain dynamics was studied in schizophrenic patients both during sleep and awake states. REM sleep in schizophrenic patients is characterized by a lower Lyapunov exponent (Roschke et al., 1995a). This altered brain dynamics could correspond to an impairment of the safety function of dreams (Keshavan et al., 1990). In addition, it has been shown that EEG's dimensionality was reduced during sleep stages and REM in schizophrenic patients (Roschke and Aldenhoff, 1993).

During awake states, nonlinearity and correlation dimension computed with spatial embedding of EEG data are lower in schizophrenia (Lee et al., 2001b; Jeong et al., 1998). Moreover, Lyapunov exponents also decrease in schizophrenia (Kim et al., 2000). When time embedding is used, spatial heterogeneities are demonstrated by correlation dimension (Lee et al., 2001a).

Finally, using mutual cross prediction (Le Van Quyen et al., 1998), it has been shown that the driving system was shifted to the frontal channel after 4-week trial with clozapine in schizophrenia (Kang et al., 2001).

Other physiological indices Time series of heart period and respiratory rhythms obtained from normal controls and patients with panic disorder were analyzed (Yeragani et al., 2000, 2002). Results showed that approximate entropy and largest Lyapunov exponents were higher in patients in normal breathing condition (Yeragani et al., 2002).

4.2.2. Symptoms Dynamics and Therapies

Mood Disorders The alternation between depressed and manic episodes in bipolar troubles constitutes an important illustration of symptoms dynamics (e.g. Wehr and Goodwin, 1979; Wehr et al., 1982). In order to assess whether the time evolution of mood modifications in bipolar trouble are related to stochastic or deterministic dynamics, daily scores to analogical mood scales have been recorded from one to two years and a half (Gottschalk et al., 1995). Linear (autocorrelation function and power spectra) and nonlinear (phase space embedding, correlation dimension, recurrence plots and surrogate data testing) were performed on the data obtained from seven rapid-cyclers and twenty-eight control subjects. Six out of the seven patients depicted convergent estimates of the correlation dimension whereas none of the controls did. Together with the complex power spectra this result indicates that mood in patients with bipolar disorder is not really cyclic contrary to the current opinion. Nonetheless, self-rated mood in patients is more organized than in control subjects and can be characterized as a low-dimensional chaotic process.

In a similar study (Woyshville et al., 1999), patients and control generated time series data, using a visual analog scale to quantify their mood. The results showed that patients display more variability but less complexity (measured by fractal dimension) in their time series than controls.

Schizophrenia Time-course of schizophrenic episodes can be investigated as a non-linear phenomenon. Daily assessment of psychotic derealization in fourteen schizophrenics have been studied during a period lasting between 200 and 770 days. Phase space reconstruction, nonlinear forecasting methods and surrogate data testing were applied to these time series. Time evolution of psychotic symptoms were classified as non-linear dynamics (8 patients out of 14), linear dynamics (4/14), and stochastic evolution (2/14). These results show that schizophrenia can be considered as a nonlinear dynamical disease, controlled by a low dimensional attractor (Tschacher et al., 1997). More descriptive methods might also be valuable to the interpretation of symptoms trajectories in schizophrenia (Tschacher and Kupper, 2002; Kupper and Tschacher, 2002)

Addiction Single-case studies have shown that daily alcohol consumption assessed during a five-year period can be modeled using multi-scale nonlinear methods (Warren et al., 2003; Warren and Hawkins, 2002).

Psycho-social crisis intervention In a sample of 40 in-patients of a psychosocial crisis intervention unit, time series data were obtained by self-rated evaluation on mood, tension and cognitive orientation (Tschacher and Jacobshagen, 2002). In crisis intervention, outward cognitive orientation generally preceded improved mood so that cognitive orientation is responsible of the experienced affective effects of crisis intervention.

Psychotherapy courses To test empirically the proposal that psychotherapy can be viewed as a self-organized dynamical system, 28 psychotherapy courses have been evaluated (Tschacher et al., 1998). The course of the therapies was characterized by a decrease of degree of freedom and an increase of order. Moreover, these results were independent of the kind of therapy and increase of order was related to positive outcomes of therapy.

4.2.3. Dynamics of Cognitive Processes

Time series generated, in a simple binary choice task, by schizophrenics were more interdependent than that of controls, suggesting that their behavior is less complex (Paulus et al., 1996, 1999). Moreover, schizophrenic patients exhibited significantly less consistency in their response selection and ordering, characterized by a greater contribution of both highly perseverative and highly unpredictable subsequences of responses within a test session (Paulus et al., 1996). Schizophrenic patients also are significantly less influenced by external stimuli than are normal comparison subjects (Paulus et al., 1999). This dysregulation is stable over time and independent of psychosocial factors and symptomatic fluctuations (Paulus et al., 2001).

In motor and perceptual tasks, schizophrenic patients exhibit a higher instability in their movement's process (horizontal finger oscillations) and a higher reversal rate in the perception of an ambiguous figure (the Rubin vase) compared to matched controls. Moreover, motor and perceptual measures were unrelated. These results suggest that alterations observed in the motor and perceptual dynamics in schizophrenia are be supported by a common underlying mechanism (Keil et al., 1998).

Dynamical quantification of language in schizophrenia (Leroy et al., 2003) have shown that the probability transition between macro-clauses and micro-clauses is lower in schizophrenic patients than in controls. This result can be view as a deficit in the dynamical access to the context level in schizophrenia.

4.2.4. Clinical Interviews

During clinical interview, one can focus either on the patient himself, or on the patient-therapist interaction.

Brain dynamics In a pilot study (Rockstroh et al., 1997), time series were obtained from electroencephalographic records during clinical interviews with 10 schizophrenic (6 paranoid, 4 disorganized) and 2 depressive patients. The time sequence of thought disorders (unusual thought contents, sudden change in topic, thought stopping,...) were also assessed.

The paranoid subgroup has been characterized by a lower complexity but more critical transitions in the EEG when compared to disorganized and depressive patients. But, such results are hardly correlated with a particular symptom, or to an underlying cognitive process. Furthermore, sudden phase transitions in brain activity were significantly enhanced prior to expressions of thought disorders that were detected by the interviewer and an observer in the conversation, compared with time periods during the interview without such symptoms.

Cardiac dynamics Since cardiological markers are related to the emotional behavior, they might be of interest to assess the complexity of patient-therapist interaction (Redington and Reidbord, 1992; Reidbord and Redington, 1992, 1993). Patient's cardiac dynamics is less complex when talking about important topics than for more distant topics. In the case of the therapist, it has been shown that cardiac dynamics depict a higher complexity when the therapist feels something *with* the patient rather than *about* the patient. Similar results were found in a study of 20 patients where variation in the complexity of heart's dynamics was observed when topics changes (Pincus, 1991).

Patient-therapist interaction The communicative process between patient and therapist needs to be studied (Langs and Badalamenti, 1994). To contribute to the construction of research methodology, patient-therapist interactions were encoded by means a matrix, in which each column represent a time series obtained by responses at questions about the sequence of interactions (Rapp et al., 1991). By this method, time series were obtained and a complexity score was computed.

Psychotherapy is also viewed as a chaotic process, and tools of non-linear dynamics are used to quantify this qualitative hypothesis. A single case was analyzed, by means of a time series obtained from the patient-therapist interactions (Schiepek et al., 1997). It has been shown that the time series is non-periodic, and the technique of surrogate data demonstrates that this non-periodicity is caused by a chaotic dynamics, and not by a stochastic process (or by noise). Fractal dimension and largest Lyapunov exponents revealed the presence of an attractor, which characterized the chaotic process of the therapy. Nevertheless, from a clinical point of view, the goal of a therapy is to lead the patient toward change rather than to stability, thus the methods used to characterize stationary dynamical systems are hardly adapted. The same data were thus re-analyzed (Kowalik et al., 1997) and demonstrate that, critical transitions appear during the therapeutic process, so that a non-stationary approach of the phenomena is necessary.

4.2.5. Family System

Family systems may be described by a 5-R's model where the four components (rules, roles, relationships and realities) are determining the fifth R (response pattern). In order to test the basic assumption of this model a family discussion was video-taped and analyzed (Pincus, 2001) using the orbital decomposition procedure (Guastello et al., 1998). The author make the hypothesis that the family response patterns during the discussion will show evidence of both coherence and complexity.

The family conversation was transformed into a symbolic sequence. Entropy measurements demonstrate the existence of a local coherence for string lengths equal to 3 and provide evidence for low dimensional chaos within the global family discussion.

4.3. Conclusion

These studies demonstrate the importance of temporal evolution in psychopathology. Aside from methodological drawbacks, dynamical processes have been characterized at several levels from physiological to linguistic one. Moreover, several studies have shown correlation between dynamical processes at different levels: brain dynamics and mood assessment, cardiac dynamics and emotions induced during interviews.

5. Conclusion

We have explored three ways of using mathematical and numerical dynamical concepts in psychopathology. We can conclude that the metaphorical description of mental troubles and changes are beginning to be modeled and tested empirically. More efforts are still needed to introduce an adapted methodology to the field of psychopathology. In fact, empirical tests decribed here are usually either data fitting to models or time series analysis (either of continous or discrete data). These two approaches are mainly "data-driven" *i.e.* they do not rely on a "theoretical" model to be tested in the data exploration (even when they are based on a model such as a catastrophe model). This interaction between models and data exploration is certainly a promising perspective of the application of dynamical systems to psychopathology.

The application of dynamical methodology to the "human sciences" are, however, still in its infancy. Several problems are to be worked out:

1. The development of accurate quantitative tools on short time series are clearly needed since the numerical methods imported from physics are highly data demanding.

2. The emphasis has been mainly given to deterministic modelling because of the fascinating properties of deterministic chaos. Nevertheless, stochastic or deterministic description are only a problem of scale and choice (in physics, molecular dynamics are deterministic but stochastic and statistical description of a gas is usually prefered for macroscopic scale). Thus, the choice of a model should not be obscured by some 'fascination'.

3. Quantification has long being the ideal of science. However, carefully designed qualitative models might be more informative than the computation of (ill-founded) quantitative indices.

Psychopathology is an adapted field for dynamics since it deals with entities with clear time evolution. Nevertheless, it could be misleading to imagine that dynamics can be directly imported in the field of psychopathology without considering its specificity. Different scales usually means that different tools adapted are to each kind of measures.

The behavioral, biological and clinical data that are mostly used in the study of mental troubles are observed from one sample at a single time point. Those data are informative but lack sensitivity to the frequency of behavior and hence to its temporal organisation. Thus the measurements of dynamical complexity are complementary to the first kind of empirical data. These studies are an useful tool for the comprehension of mental and behavioral changes. They allow one the study of the interaction between several factors and thus avoid the reduction of mental trouble to the effect of one single factor.

Because several levels interact, it is important to focalize attention on break or changes of state. The ruptures or the dynamical changes are observable at the different observation levels. Clinical data are concordant with such a point of view since changes are simultaneously observed in neurophysiology, in the strategy of thinking, the kind of beliefs, the types of behavior or the transactional activities. The only common point susceptible to be study is these ruptures in dynamics.

Moreover, open systems are by definition coupled with their environment. Studying human being implies that researchers take into account the contexts in which a behavior is developed. We must not have only knowledge about the system itself but also about the way it uses to interact with its environment. Contexts are necessary the broadest possible and imply physiological parameters, ecological, familial, social and cultural elements. A last point must be underlined: the role of observers. An observer placed in an environment has necessarily an effect on the observed system.

The generalisation of the "dynamical disease" concept to mental troubles may open several clinical perspectives:

1. From the point of view of diagnostics, the possibility of defining dynamical characteristics specific of a disease (such as a specific rhythm in a biological functionning) would offer a tool for the biological side of psychopathology.

2. From the point of view of therapeutics, the isolation of factors that may influence the behavioral and/or mental changes would offer, to the clinicians, several paths of action. In that case, changes would be possible either on the basis of a changes in the control parameters or on the basis of a perturbation depending upon the level of the intervention (biological, psychological or social). It is thus possible to imagine new therapeutical ways based on valid models of the dynamics underlying the mental trouble.

3. From a theoretical point of view, the model of a "dynamical disease" underlying mental troubles seems more legitime than a linear "medical" point of view. Clinical signs or symptoms can be considered as discontinuous changes based on continuous changes in control parameters. Thus dynamical systems theory seems particularly well adapted to the study of mental troubles.

   It is thus important to develop the methodology of dynamical systems towards (rigorous) applications in the "human sciences" and then to integrate these tools into more classical psychopathological studies. It seems particularly important to emphasize the study of temporal dimension of psychopathological phenomena.

   Such a dynamical point of view decrease the ontological gap that has been hypothesized between normal and pathological mental activities: it favors an underlying continouous point of view even if the behavioral observables are clearly discontinuous.

References

Aakerlund, L., Hemmingsen, R., 1998. Neural networks as model of psychopathology. *Biological Psychiatry* **43**, 471–482.

Abarbanel, H. D. I., Brown, R., Sidorowich, J. J., Tsimring, L. S., 1993. The analysis of observed chaotic data in physical systems. *Reviews of Modern Physics* **65**, 1331–1392.

Alexander, R. A., Herbert, G. R., DeShon, R. P., Hanges, P. J., 1992. An examination of least-squares regression modeling of catastrophe theory. *Psychological Bulletin* **111** (2), 366–374.

Amit, D. J., 1989. *Modeling brain function: the world of attractor neural networks.* Cambridge University Press, Cambridge, UK.

an der Heiden, U., Schwegler, H., Tretter, F., 1998. Patterns of alcoholism: a mathematical model. *Mathematical Models and Methods in Applied Sciences* **8**, 521–541.

Arnol'd, V. I., 1992. *Catastrophe theory*, 3rd Edition. Springer-Verlag, Berlin.

Badii, R., Politi, A., 1997. *Complexity. Hierarchical structures and scaling in physics.* Cambridge University Press, Cambridge, UK.

Bélair, J., Glass, L., an der Heiden, U., Milton, J., 1995. Dynamical disease: identification, temporal aspects and treatment strategies of human illness. *Chaos* **5**, 1–7.

Brabender, V., 2000. Chaos, group psychotherapy, and the future of uncertainty and uniqueness. *Group* **24** (1), 23–32.

Bütz, M. R., 1993. Practical applications from chaos theory to the psychotherapeutic process. a basic consideration of dynamics. *Psychological Reports* **73**, 543–554.

Callahan, J., 1982. A geometric model of anorexia and its treatment. *Behavioral Science* **27**, 140–154.

Clair, S., 1998. A cusp catastrophe model for adolescent alcohol use: An empirical test. Nonlinear Dynamics, *Psychology and Life Sciences* **2** (3), 217–241.

Cobb, L., Watson, B., 1980. Statistical catastrophe theory: An overview. *Mathematical Modelling* **1** (4), 311–317.

Cohen, J. D., Servan-Schreiber, D., 1992. Context, cortex, and dopamine: a connectionist approach to behavior and biology in schizophrenia. *Psychol Rev* **99** (1), 45–77.

David, A. D., 1994. Dysmodularity: A neurocognitive model for schizophrenia. *Schizophrenia Bulletin* **20**, 249–255.

Denman, C., 1994. Strange attractors and dangerous liaisons: A reponse to Priel & Schreiber, 'On psychoanalysis and non-linear dynamics: The paradigm of bifurcation'. *British Journal of Medical Psychology* **67**, 219–222.

Elkaïm, M., 1990. *If you love me don't love me: constructions of reality and change in family therapy.* Jason Aroson, Northvale, NJ.

Engel, G., 1980. The clinical application of biopsychosocial model. American *Journal of Psychiatry* **137**, 535–544.

Freedman, A., 1995. The biopsychosocial paradigm and the future of psychiatry. *Comprehensive Psychiatry* **36**, 397–406.

Freeman, W., 1990. Consciousness as physiological self-organizing process. *Behavioral and Brain Sciences* **13**, 604.

Galatzer-Levy, R., 1978. Qualitative change from quantitative change: Mathematical catastrophe theory in relation to psychoanalysis. *Journal of American Psychoanalytic association* **26** (4), 921–935.

Galatzer-Levy, R., 1995. Psychoanalysis and dynamical systems theory: prediction and self-similarity. *Journal of American Psychoanalysis Association* **43**, 1085–1113.

Gardner, S., 1994. Commentary on Priel & Schreiber, 'On psychoanalysis and non-linear dynamics'. *British Journal of Medical Psychology* **67**, 223–225.

Gilmore, R., 1981. Catastrophe theory for scientists and engineers. John Wiley & Sons, New York, (republication in 1993 by Dover Publications, Mineola).

Globus, G. G., Arpaia, J. P., 1994. Psychiatry and the new dynamics. *Biological Psychiatry* **35**, 352–364.

Glover, H., 1992. Emotional numbing: A possible endorphin-mediated phenomenon associated with post-traumatic stress disorders and other allied psychopathologic states. *Journal of Traumatic Stress* **5**, 643–675.

Goldstein, J., 1997. Embracing the random in the self-organizing psyche. Nonlinear Dynamics, *Psychology and Life Sciences* **1** (3), 181–202.

Gottschalk, A., Bauer, M. S., Whybrow, P. C., 1995. Evidence of chaotic mood variation in bipolar disorder. *Archives of General Psychiatry* **52** (11), 947–959.

Grassberger, P., Schreiber, T., Schaffrath, C., 1991. Nonlinear time series analysis. *International Journal of Bifurcation and Chaos* **1,** 521–547.

Guastello, S., Hyde, T., Odak, M., 1998. Symbolic dynamic patterns of verbal exchange in a creative problem solving group. Nonlinear Dynamics, *Psychology and Life Sciences* **2**, 35–58.

Guastello, S. J., 1982. Moderator regression and the cusp catastrophe: Application of two-stage personnel selecction, training, therapy and policy evaluation. *Behavioral Science* **27**, 259–272.

Guastello, S. J., 1987. A butterfly catastrophe model of motivation in oragnizations: Academic performance. *Journal of Applied Psychology* **72**, 165–182.

Guastello, S. J., 1992. Clash of paradigms: A critique of an examination of the polynomial techique for evaluating catastrophe theory hypotheses. *Psychological Bulletin* **111** (2), 375–379.

Hardy, L., 1996. A test of catastrophe models of anxiety and sports performance against multidimensional anxiety theory using methods of dynamic differences. Anxiety, *Stress and Coping* **9**, 69–86.

Hoffman, R. E., 1987. Computer simulations of neural information processing and the schizophrenia-mania dichotomy. *Archives of General Psychiatry* **44** (2), 178–188.

Hoffman, R. E., Dobscha, S. K., 1989. Cortical pruning and the development of schizophrenia. *Schizophrenia Bulletin* **15** (3), 477–490.

Hoffman, R. E., McGlashan, T. H., 1994. Corticocortical connectivity, autonomous networks and schizophrenia. *Schizophrenia Bulletin* **20** (2), 257–261.

Hoffman, R. E., McGlashan, T. H., 2001. Neural network models of schizophrenia. *Neuroscientist* **7** (5), 441–454.

Hopfield, J. J., 1982. Neural networks and physical systems with emergent collective computational abilities. *Proceedings of the National Academy of Science USA* **79**, 2554–2558.

Huber, M., Braun, H., Krieg, J., 2000a. Effects of noise on different disease states of recurrent affective disorders. *Biological Psychiatry* **47** (7), 634–642.

Huber, M., Braun, H., Krieg, J., 2001a. On the impact of episode sensitization on the course of affective disorders. *Biological Psychiatry* **35** (1), 49–57.

Huber, M., Krieg, J., Braun, H., 2000b. Noise, nonlinear dynamics and the timecourse of affective disorders. *Chaos, Solitons and Fractals* **11**, 1923–1928.

Huber, M. T., Braun, H. A., Krieg, J. C., 1999. Consequences of deterministic and random dynamics for the course of affective disorders. *Biological Psychiatry* **46**, 256–262.

Huber, M. T., Braun, H. A., Voigt, K., Krieg, J. C., 2001b. Some computational aspects of the kindling model for neuropsychiatric disorders. *Neurocomputing* 38–40, 1297–1306.

Huberman, B. A., 1987. A model for dysfunctions in smooth pursuit eye movement. *Annals of The New York Academy of Sciences* **504**, 260–273.

Hufford, M. R., 2001. Alcohol and suicidal behavior. *Clinical Psychology Review* **21**, 797–811.

Jeong, J., Kim, D.-J., Chae, J.-H., Kim, S. Y., Ko, H.-J., Paik, I.-H., 1998. Nonlinear analysis of the eeg of schizophrenics with optimal embedding dimension. *Medical Engineering & Physics* **20**, 669–676.

Johnson, F. N., 1986. Different treatment modalities for recurrent bipolar affective disorders: An integrative approach. *Psychotherapy and Psychosomatics* **46**, 13–22.

Kang, U., Par, K., Ahn, Y., Koo, Y., Yoon, S., Yi, S., Kim, Y., 2001. Non-linear dynamics analysis of clozapine-induced electroencephalographic changes in schizophrenic patients – a preliminary study. *Progress in Neuro-Psychopharmacology & Biological Psychiatry* **25** (6), 1229–1239.

Kantz, H., Schreiber, T., 1997. *Nonlinear Time series analysis.* Cambridge University Press, Cambridge, UK.

Keil, A., Elbert, T., Rockstroh, B., Ray, W., 1998. Dynamical aspects of motor and perceptual processes in schizophrenic patients and healthy controls. *Schizophrenia Research* **33** (3), 169–178.

Keller, M. B., Lavori, P. W., Rice, J., Coryell, W., Hirschfeld, R. M., 1986. The persistent risk of chronicity in recurrent episodes of nonbipolar major depressive disorder: A prospective follow-up. *American Journal of Psychiatry* **143**, 24–28.

Keshavan, M., Reynolds, C., Kupfer, D., 1990. Electroencephalographic sleep in schizophrenia — a critical review. *Comprehensive Psychiatry* **30**, 34–47.

Kim, D. J., Jeong, J., Chae, J. H., Park, S., Yong Kim, S., Jin Go, H., Paik, I. H., Kim, K. S., Choi, B., 2000. An estimation of the first positive Lyapunov exponent of the EEG in patients with schizophrenia. *Psychiatry Research* **98**, 177–189.

Kincanon, E., Powel, W., 1995. chaotic analysis in psychology and psychoanalysis. *The Journal of Psychology* **129**, 495–505.

King, R., Barchas, J. D., Huberman, B. A., 1984. Chaotic behavior in dopamine neurodynamics. *Proceedings of the National Academy of Sciences U.S.A.* **81** (4), 1244–1247.

King, R., Raese, J., Barchas, J., 1981. Catastrophe theory of dopaminergic transmission: A revised dopamine hypothesis of schizophrenia. *Journal of Theoretical Biology* **92** (4), 373–400.

Klein, M., 1948. *Contributions to Psycho-analysis.* Hogarth Press, London, UK.

Koopmans, M., 1998. Chaos theory and the problem of change in family systems. Nonlinear Dynamics, *Psychology and Life Sciences* **2**, 133–148.

Kowalik, Z. J., Schiepek, G., Kumpf, K., Roberts, L. E., Elbert, T., 1997. Psychotherapy as a chaotic process II. The application of nonlinear analysis methods on quasi time series of the client-therapist interaction: a nonstationary approach. *Psychotherapy Research* **7**, 197–218.

Kramlinger, K., Post, R., 1996. Ultra-rapid and ultradian cycling in bipolar affective illness. *British Journal of Psychiatry* **168**, 314–323.

Kupper, Z., Tschacher, W., 2002. Symptom trajectories in psychotic episodes. *Comprehensive Psychiatry* **43**, 311–318.

Lange, R., Oliva, T. A., McDade, S. R., 2001. An algorithm for estimating multivariate catastrophe models: GEMCAT II. *Studies in Nonlinear Dynamics and Econometrics 4 (3), 137–168.*

Langs, R., 1986. Clinical issues arising from a new model of the mind. *Contemporary Psychoanalysis* **22**, 418–444.

Langs, R., Badalamenti, A., 1994. Psychotherapy: the search for chaos and the discovery of determinism. *Australian and New Zealand Journal of Psychiatry* **28** (1), 68–81.

Lanza, M. L., 1999. Catastrophe theory: Application of nonlinear dynamics to assault victim responses. *Journal of the American Psychiatric Nurses Association* **5**, 117–121.

Le Van Quyen, M., Adam, C., Baulac, M., Martinerie, J., Varela, F., 1998. Nonlinear interdependencies of eeg signals in human intracranially recorded temporal lobe seizures. *Brain Research* **792**, 24–40.

Lee, Y., Zhu, Y., Xu, Y., Shen, M., Tong, S., Thakor, N., 2001a. The nonlinear dynamical analysis of the eeg in schizophrenia with temporal and spatial embedding dimension. *Journal of Medical Engineering & Technology* **25** (2), 79–83.

Lee, Y., Zhu, Y., Xu, Y., Shen, M., Zhang, H., Thakor, N., 2001b. Detection of non-linearity in the eeg of schizophrenic patients. *Clinical Neurophysiology* **112** (7), 1288–1294.

Leroy, F., Pezard, L., Nandrino, J.-L., Beaune, D., 2003. *Dynamical quantification of schizophrenic speech.* Submitted .

Levinson, E. A., 1994. The uses of disorders: Chaos theory and psychoanalysis. *Contemporary Psychoanalysis* **30** (1), 5–24.

Lonie, I., 1991. Chaos theory: a new paradigm for psychotherapy? *Australian and New Zealand Journal of Psychiatry* **25**, 548–560.

MacCulloch, M. J., Waddington, J. L., 1979. Catastrophe theory: A model interaction between neurochemical and environmental influences in the control of schizophrenia. *Neuropsychobiology* **5,** 87–93.

Mackey, M. C., Glass, L., 1977. Oscillation and chaos in physiological control systems. *Science* **197**, 287–289.

Mackey, M. C., Milton, J. G., 1987. Dynamical diseases. *Annals of the New-York Academy of Science* **504**, 16–32.

Mahler, M., 1968. *On symbiosis and the vicissitudes of individuation.* International Universities Press, New York.

Marks-Tarlow, T., 1999. The self as a dynamical system. *Nonlinear Dynamics, Psychology, and Life Sciences* **3**, 311–345.

May, R. M., 1978. Dynamical diseases. *Nature* **272**, 673.

McClelland, James, L., Rumelhart, D. E., the PDP research group, 1986a. *Parallel distributed processing. Explorations into the microstructure of cognition.* Vol. 1: Foundations. The MIT Press, Cambridge, MA.

McClelland, James, L., Rumelhart, D. E., the PDP research group, 1986b. *Parallel distributed processing. Explorations into the microstructure of cognition.* Vol. 2: Psychological and biological models. The MIT Press, Cambridge, MA.

Miller, W. L., McDaniel, R. R., Crabtree, B. F., Stange, K. C., 2001. Practice jazz: understanding variation in familly practices using complexity science. *Journal of Family Practice* **50** (10), 872–878.

Milton, J., Longtin, A., Beuter, A., C., M. M., Glass, L., 1989. Complex dynamics and bifurcations in neurology. *Journal of Theoretical Biology* **138**, 129–47.

Milton, J. G., Black, D., 1995. Dynamic diseases in neurology and psychiatry. *Chaos* **5**, 8–13.

Moran, M., 1991. Chaos theory and psychoanalysis: the fluidic nature of the mind. *International Review of Psychoanalysis* **18**, 211–221.

Nandrino, J.-L., Pezard, L., Martinerie, J., El Massioui, F., Renault, B., Jouvent, R., Allilaire, J.-F., Widlöcher, D., 1994. Decrease of complexity in EEG as a symptom of depression. *NeuroReport* **5** (4), 528–530.

Oliva, T. A., Desarbo, W. S., Day, D. L., Jedidi, K., 1987. GEMCAT: A general multivariate methodology for estimating catastrophe models. *Behavioral Science* **32**, 121–137.

Ott, E., Sauer, T., Yorke, J. A. (Eds.), 1994. Coping with chaos. *Analysis of chaotic data and the exploitation of chaotic systems.* Wiley & sons, New-York.

Paulus, M. P., Geyer, M. A., Braff, D. L., 1996. Use of methods from chaos theory to quantify a fundamental dysfunction in the behavioral organization of schizophrenic patients. *American Journal of Psychiatry* **153** (5), 714–717.

Paulus, M. P., Geyer, M. A., Braff, D. L., 1999. Long-range correlations in choice sequences of schizophrenic patients. *Schizophrenia Research* **35** (1), 69–75.

Paulus, M. P., Rapaport, M. H., Braff, D. L., 2001. Trait contributions of complex dysregulated behavioral organization in schizophrenic patients. *Biological Psychiatry* **49** (1), 71–77.

Pezard, L., Nandrino, J.-L., Renault, B., El Massioui, F., Allilaire, J.-F., Müller, J., Varela, F. J., Martinerie, J., 1996. Depression as a dynamical disease. *Biological Psychiatry* **39** (12), 991–999.

Pincus, D., 2001. A framework and methodology for the study of nonlinear, self-organizing family dynamics. *Nonlinear Dynamics, Psychology, and Life Sciences* **5** (2), 139–173.

Pincus, S., 1991. Approximate entropy as a measure of system complexity. *Proceedings of the National Academy of Sciences* **88**, 2297–2301.

Porte, M., 1994. *La dynamique qualitative en psychanalyse.* Presses Universitaires de France, Paris.

Poston, T., Stewart, I., 1978. *Catastrophe theory and its applications.* Pitamn Publishing Ltd, London.

Priel, B., Schreiber, G., 1994. On psychoanalysis and non-linear dynamics: The pardigm of bifurcation. *British Journal of Medical Psychology* **67**, 209–218.

Quinodoz, J., 1997. Transitions in psychic structures in the light of deterministic chaos theory. *International Journal of Psychoanalysis* **78**, 699–718.

Rapp, P. E., Jimenez-Montano, M. A., Langs, R. J., Thomson, L., Mees, A. I., 1991. Toward a quantitative characterization of patient-therapist communication. *Mathematical Biosciences* **105** (2), 207–227.

Redington, D., Reidbord, S., 1992. Chaotic dynamics in autonomic nervous system activity of a patient during psychotherapy. *Biological Psychiatry* **31** (10), 993–1007.

Reidbord, S. P., Redington, D. J., 1992. Psychophysiological processes during insight-oriented therapy. further investigations into nonlinear psychodynamics. *Journal of Nervous and Mental Disease* **180** (10), 649–657.

Reidbord, S. P., Redington, D. J., 1993. Nonlinear analysis of autonomic responses in a therapist during psychotherapy. *Journal of Nervous and Mental Disease* **181** (7), 428–435.

Rialle, V., Stip, E., 1994. Modélisation cognitive en psychiatie : des modèles symboliques aux modèles paralèles et distribués. *Journal of Psychiatry and Neuroscience* **19** (3), 178–192.

Ricci, C., Selvini-Palazzoli, M., 1984. interactional complexity and communication. *Family Process* **23** (2), 169–176.

Rockstroh, B., Waltz, H., Kowalik, Z. J., Cohen, R., Sterr, A., Müller, M., Elbert, T., 1997. Dynamical aspects of the EEG in different psychopathological states in an interview situation: a pilot study. *Schizophrenia Research* **28**, 77–85.

Roschke, J., Aldenhoff, J., 1993. Estimation of the dimensionality of sleep-eeg data in schizophrenics. *European Archives of Psychiatry and Clinical Neuroscience* **242** (4), 191–196.

Roschke, J., Fell, J., Beckmann, P., 1995a. Nonlinear analysis of sleep EEG data in schizophrenia: calculation of the principal Lyapunov exponent. *Psychiatry Research* **56**, 257–269.

Roschke, J., Fell, J., Beckmann, P., 1995b. Nonlinear analysis of sleep EEG in depression: calculation of the largest Lyapunov exponent. *European Archives of Psychiatry and Clinical Neuroscience* **245**, 27–35.

Roschke, J., Mann, K., J., F., 1994. Nonlinear EEG dynamics during sleep in depression and schizophrenia. *International Journal of Neuroscience* **75**, 271–284.

Schiepek, G., Kowalik, Z. J., Schütz, A., Köhler, M., Richter, K., Strunk, G., Mühlnickel, W., Elbert, T., 1997. Psychotherapy as a chaotic process I. Coding the client-therapist interaction by means of sequential plan analysis and the search for chaos: a stationary approach. *Psychotherapy Research* **7**, 173–194.

Schmid, G., 1991. Chaos theory and schizophrenia: Elementary aspects. *Psychopathology* **24** (4), 185–198.

Schreiber, T., Schmitz, A., 2000. Surrogate time series. *Physica D* **142** (3–4), 346–382.

Schwalbe, M., 1991. The autogenesis of self. *Journal for the Theory of Social Behavior* **21**, 269–295.

Scott, D. W., 1985. Catastrophe theory applications in clinical psychology: A review. *Current Psychological Research and Reviews* **4**, 69–86.

Sokal, A., Bricmont, J., 1999. *Fashionable nonsense: Postmodern intellectuals' abuse of science.* Picador, New York, NY.

Spruiell, V., 1993. Deterministic chaos and the sciences of complexity: psychoanalysis in the midst of a general scientific revolution. *Journal of American Psychoanalysis Association* **41**, 3–44.

Stolorow, R., Brandchaft, B., Atwood, G., 1994. *Psychoanalytic treatment: An intersubjective approach.* The Analytic Press, Hillsdale NJ.

Tassin, J.-P., 1996. Schizophrénie et neurotransmission : un excès de traitement analogique. *L'Encéphale* **22** (Spec. 3), 91–98.

Thom, R., 1977a. Stabilité structurelle et morphogenèse. Essai sur une théorie générale des modèles. InterÉditions, Paris.

Thom, R., 1977b. Structural stability, catastrophe theory and applied mathematics. *SIAM Review* **19**, 189–201.

Thomas, A., Chess, S., 1984. Genesis and evolution of behavioral disorders: from infancy to early adult life. *American Journal of Psychiatry* **14**, 1–9.

Thomasson, N., Pezard, L., Allilaire, J.-F., Renault, B., Martinerie, J., 2000. Nonlinear EEG changes associated with clinical improvement in depressed patients. Nonlinear Dynamics, *Psychology and Life Sciences* **4** (3), 203–218.

Thomasson, N., Pezard, L., Boyer, P., B.Renault, Martinerie, J., 2002. Nonlinear EEG changes in a 48-hour cyclic manic-depressive patient. Nonlinear Dynamics, *Psychology and Life Sciences* **6** (3), 259–267.

Tschacher, W., Jacobshagen, N., 2002. Analysis of crisis intervention processes. *Crisis* **23** (2), 59–67.

Tschacher, W., Kupper, Z., 2002. Time series models of symptoms in schizophrenia. *Psychiatry Research* **113**, 127–137.

Tschacher, W., Scheier, C., Grawe, K., 1998. Order and pattern formation in psychotherapy. *Nonlinear Dynamics, Psychology, and Life Sciences* **2** (3), 195–215.

Tschacher, W., Scheier, C., Hashimoto, Y., 1997. Dynamical analysis of schizophrenia courses. *Biological Psychiatry* **41** (4), 428–437.

van der Maas, H. L., Molenaar, P. C., 1992. Stagewise cognitive development: an application of catastrophe theory. *Psychological Review* **99** (3), 395–417.

VanEenwyk, J., 1991. Archetypes: the strange attractors of the psyche. *Journal of Analytical Psychology* **36**, 1–25.

Verhuslt, F., 1999. Psyoanalysis and chaos theory. *International Journal of Psychoanalysis* **80**, 623–625.

Vinogradov, S., King, R., Huberman, B., 1992. An associationist model of the paranoid process: application of phase transitions in spreading activation networks. *Psychiatry* **55** (1), 79–94.

Warren, K., Hawkins, R., 2002. Multiscale nonlinearity in a time series of weekly alcohol intake. *Psychological Reports* **90** (3), 957–967.

Warren, K., Hawkins, R., Sprott, J. C., 2003. Substance abuse as a dynamical disease. evidence and clinical implications of nonlinearity in a time series of daily alcohol consumption. *Addictive Behavior* **28** (2), 369–374.

Wehr, T. A., Goodwin, F. K., 1979. Rapid cycling in manic-depressives induced by tricyclics antidepressants. *Archives of General Psychiatry* **36**, 555–559.

Wehr, T. A., Goodwin, F. K., Wirz-Justice, A., Breitmaier, J., Craig, C., 1982. 48-hour sleep-wake cycles in manic-depressive illness. *Archives of General Psychiatry* **39**, 559–565.

Weiner, P., 1977. Les applications de la théorie des catastrophes en psychopathologie. *L'Évolution Psychiatrique* **42**, 955–974.

Woyshville, M. J., Lackamp, J. M., Eisengart, J. A., Gilliland, J. A. M., 1999. On the meaning and measurement of affective instability: Clues from chaos theory. *Biological Psychiatry* **45** (3), 261–269.

Yeragani, V., Nadella, R., Hinze, B., Yeragani, S., Jampala, V., 2000. Nonlinear measures of heart period variability: decreased measures of symbolic dynamics in patients with panic disorder. *Depression and Anxiety* **12** (2), 67–77.

Yeragani, V. K., Radhakrishna, R. K., Tancer, M., Uhde, T., 2002. Nonlinear measures of respiration: Respiratory irregularity and increased chaos of resporation in patients with panic disorder. *Neuropsychobiology* **46** (3), 111–120.

Zeeman, E. C., 1977. *Catastrophe theory: Selected papers 1972-1977.* Addison-Wesley, Reading Mass.

In: Chaos and Complexity Research Compendium
Editors: F. Orsucci and N. Sala, pp. 149-157
ISBN: 978-1-60456-787-8

Chapter 11

COLLECTIVE PHENOMENA IN LIVING SYSTEMS AND IN SOCIAL ORGANIZATIONS

***Eliano Pessa*[1,a], *Maria Petronilla Penna*[2,b] and *Gianfranco Minati*[3,c]**

[1]Dipartimento di Psicologia, Universita' degli Studi di Pavia,
Piazza Botta 6, 27100 Pavia, Italy
[2]Dipartimento di Psicologia, Università degli Studi di Cagliari,
Via Is Mirrionis, 09100 Cagliari, Italy
[3]AIRS, Italian Association for Systems Research,
Via P.Rossi 42, 20161 Milano, Italy

Abstract

In this paper we introduce a distinction between two kinds of collective behaviours: the ones in which cognitive interactions between elementary units (i.e. individuals) are not essential to understand the behaviour of whole system, and the ones in which these interactions are at the basis of the birth of collective behaviours themselves. We argue that the methods so far introduced by theoretical physics are able to describe only the first kind of collective behaviours. As regards the second type, we claim that new concepts have to be introduced. To this regard, we propose a new prototypical model of interaction between individuals endowed with a cognitive system. We show that the model is able to exhibit processes of flock formation, characterized by suitable order parameters. However, the behaviours of single individuals belonging to a flock differ very much, in this model, from the one of decaying fluctuations within ordered structures occurring in physico-chemical systems. We conjecture that such a difference can be considered as the main cue for evidencing the operation of individual cognitive activities underlying collective behaviours within socio-economical systems.

[a] E-mail address: eliano.pessa@unipv.it
[b] E-mail address: maria.Pietronilla@unica.it
[c] E-mail address: gianfranco.minati@airs.it

1. Introduction

The emergence of collective phenomena is one of the most intriguing and fascinating behavioural features of complex systems whose components are living beings. Some particular cases, in recent times, the subject of intensive investigations, both theoretical and experimental: we will quote, to this regard, the processes of formation of flocks, swarms, herds, schools, and of nest building by insects such as ants or termites.

There is a number of different reasons for studying collective phenomena, here defined, in a broad sense, as macroscopic behaviours occurring as a consequence of interactions between microscopic units. We will list some of these reasons as follows:

(1) *technological reasons*, connected to the need for implementing complex functions in future nanocomputers through suitably controlled collective behaviours taking place on atomic scale (such attempts have a long history, cfr. [5]);

(2) *reasons routed in distributed artificial intelligence*, connected to the hope of solving complex problems through collective behaviours occurring in a set of interconnected *agents* (see, e. g., [17] [18]);

(3) *socio-economical reasons*, connected to the hope of improving the productivity of particular socio-economical systems (such as firms) through the implementation, within them, of suitable rules underlying collective behaviours;

(4) *philosophical reasons*, connected to the need for understanding whether the study of collective phenomena will require or not a generalization or a modification of conventional scientific approach.

We remind that the whole body of theoretical knowledge so far existing about collective phenomena is deeply rooted in theoretical physics. As a matter of fact the only methods which can be used in a successful way to gain some insight about collective behaviours are the ones of statistical mechanics and of quantum physics (including Quantum Field Theory). Without insisting on technical details, we will limit ourselves to remark that, within this context, collective behaviours have been described by resorting to a number of key concepts, such as *stable equilibrium states*, *collective variables*, *order parameters*, *Bose-Einstein condensation*, and so on. The technical apparatus underlying these concepts led us describe, forecast, and control a number of interesting collective phenomena, such as the ones related to superconductivity, superfluidity, laser effect, ferromagnetism, vibrations in crystals, and many others.

Unfortunately such an approach appears as unsuited to model collective behaviours taking place in biological and socio-economical systems (cfr. [10]). The main difficulties so far encountered can be thus summarized:

(a) observed collective behaviours evidence a lack of coherence on spatial scales which are small which respect to overall system's spatial scale; such deviations from coherence follow evolutionary histories very different from the ones of decaying fluctuations within coherent systems, such as superfluids, or ferromagnets;

(b) observed collective behaviours are *metastable* and have a very limited lifetime;

(c) observed collective behaviours appear to be very sensitive to context, boundary conditions, and environmental influences;
(d) observed collective behaviours involve beings endowed with a *cognitive system* (even if rudimentary), and whence with goals, strategies, i.e. with features which are not represented in usual physical models, based, in an ultimate way, on the concept of point particle;
(e) observed collective behaviours sometime appear to violate fundamental constraints on formation of macroscopic ordered structured, imposed by physical laws; to this regard a well known example is given by the celebrated Mermin-Wagner theorem [8], asserting the impossibility of formation of stable macroscopic ordered 2-dimensional structures; as it is well known, most flocks are, on the contrary, just typically 2-dimensional structures.

Such a situation favoured the search for *phenomenological models* able to exhibit (in a qualitative sense) some features of observed collective behaviours. Most models of formation of flocks, swarms, herds, schools, or of problem solving by insect colonies just belong to this latter category. A common assumption underlying all these models is that observed collective behaviours must not depend on cognitive abilities of single living beings which the systems under study are constituted by. According to this assumption, which is strongly connected with the fundamental assumptions of the so-called "reactive school" in robotics (cfr, e.g., [1] [4]), every system component is able only to react to immediate local stimulations.

In this paper we claim that such an assumption prevents from a correct description, and understanding, of collective behaviours observed in socio-economical systems. On the contrary, we hold that every model of these latter should include a more o less complicated description of cognitive systems of single individuals. We will present, in the following, a simple prototypical example of such a description relative to the process of flock formation, and amenable, to some extent, to a mathematical analysis. We obtained that our model, when its parameters lie in a suitable range, exhibits a flocking behaviour. Besides, in it individual motion behaviours differ somewhat from individual behaviours taking place in a physical system in which coherent collective behaviours occur as a consequence of a phase transition. Namely, individuals belonging to flocks, in our model, evidence irregular fluctuations (not decaying with time) with respect to average motion behaviour. Such a circumstance appears to be in agreement with individual behaviours in flocks observed in Nature. On the other hand, statistical properties of such fluctuations seem to offer a way to characterize in a quantitative fashion the degree of influence of single cognitive systems on a collective behaviour. Such a circumstance favours the application of ideas underlying our model to the analysis of experimental data relative to flocking behaviours. Before starting the exposition of main principles underlying our model, however, we will shortly review some other phenomenological model of collective behaviours in biological systems, in order to stress the differences and the analogies with our proposals.

2. Some Models of Collective Behaviours in Biological Systems

Between the first models of this type we will quote *Boids*, a computer program created by Craig Reynolds [13] to do realistic graphic simulations of swarm formation and motion in a

3-D environment. In order to obtain such a result Reynolds introduced three main kinds of rules of steering behaviours each individual must follow:

2a) *rules of separation*: the individual must steer to avoid crowding of locally neighbouring individuals;
2b) *rules of alignment* : the individual must steer towards the average heading of locally neighbouring individuals;
2c) *rules of cohesion* : the individual must steer to move towards the average position of locally neighbouring individuals.

Even if programs based on suitable choices of these rules are very effective in producing swarm-like behaviours, however their operation cannot be analyzed, so far, by resorting to conventional methods of statistical mechanics. Besides, they cannot be considered as realistic models of swarm formation in the case of concrete living beings, such as bees. Namely, how could an individual recognize what individuals are his immediate neighbours, without being endowed with a cognitive system? And the description of such a system should constitute the essential part of the model of swarm-like behaviour we want to build.

Some other models try to overcome a fundamental shortcoming of Reynolds model, by introducing mathematical structures amenable to a treatment by statistical mechanics. Between these models we will quote *Fluid Neural Networks* [6] [14] [15], and the *Fixed Threshold Model* [2] [3]. Both are based on discretized lattices of units whose state is determined by local evolutionary rules. A model belonging to this category, but described through a nonequilibrium statistical field theory, is constituted by *Swarm Networks* [9] [12]. All these models exhibit phase transitions from disordered to ordered behaviours (of collective nature), typically consisting in collective problem solving by insect colonies (such as ants).

A model of bird flock formation based on a very different approach has been, instead, built by Tu and Toner [16]. This model could be called a "fluid" model because flock dynamics is described through a continuous nonequilibrium dynamical model based on Navier-Stokes equations of fluid motion. Such a framework lets the authors introduce in a explicit way a mechanism of structure formation based on a balance between short-range activation (in this case due to model nonlinearity) and long-range inhibition (in this case due to convection), in conformity with ideas first proposed by Gierer and Meinhardt (cfr. [7] [11]). Between the results obtained by the authors there is the description of formation of 2-dimensional flocks (circumstance which constitutes a violation of Mermin-Wagner theorem).

3. A New Approach to Collective Behaviour Description and a Prototypical Example

As already stressed in the Introduction, we claim that, contrarily to what assumed in all models quoted in the previous paragraph, in order to model collective behaviours in socio-economical systems, we should include a description of cognitive systems of constituent individuals. However, instead of building a detailed model of a particular type of cognitive system, we choose to build only a very single prototypical model of interaction between

individuals endowed with a cognitive system. The reason of our choice is that such a kind of model is amenable to a mathematical analysis, owing to its simplicity, and such a circumstance lets us gain some insight, useful to build more detailed models.

The main idea underlying our prototypical model can be formulated as follows:

3a) the operation of every individual cognitive system requires a processing of information coming from external stimuli, which, in a first approximation, come from spatially neighbouring sources of stimulation;
3b) such a processing requires a suitable amount of *energy*; it, can whence, be considered as equivalent to the effect produced on the individual by a contrary *force*, which opposes to individual motion;
3c) such a force, in a first approximation, depends only on individual immediate neighbours, and can therefore be represented as a *local inhibition*.

We propose, as a consequence, that collective behaviours in socio-economical systems arise from a suitable combination of three kinds of influence acting on every individual: a *local inhibition* (due to the operation of his cognitive system), a *middle-range activation*, and a *long-range inhibition*. This scheme can be viewed as a generalization of the one already proposed by Gierer and Meinhardt.

In order to implement our idea, we introduced a system of N *point individuals*, living in a 2-dimensional world, interacting through forces influencing their instantaneous velocities. More precisely, let us denote by $x(i, t)$, $y(i, t)$ ($i = 1, \ldots, N$) the instantaneous coordinates of the i-th individual at time t, and by $v_x(i, t)$, $v_y(i, t)$ the two components of its velocity vector. Then, the equations of motion for the i-th individual are:

$$dv_x(i,t) / dt = \Sigma_{j\neq i} F(r_{ij}) \{[x(j, t) - x(i, t)] / r_{ij}\} - Kv_x(i, t)$$

$$dv_y(i,t) / dt = \Sigma_{j\neq i} F(r_{ij}) \{[y(j, t) - y(i, t)] / r_{ij}\} - Kv_y(i, t) \qquad (1)$$

where:

$$r_{ij} = \{[x(j, t) - x(i, t)]^2 + [y(j, t) - y(i, t)]^2\}^{1/2} \qquad (2)$$

and:

$$F(r) = -A(r^2 - a^2)(r^2 - b^2) \quad \text{if } r \leq r_c \qquad (3)$$

$$F(r) = 0 \text{ if } r > r_c$$

The symbols A, a, b, r_c, K denote suitable model parameters, obeying the following constraint:

$$r_c > b > a > 0. \qquad (4)$$

It is easy to see that the three types of influence previously introduced are described through the form itself of the function $F(r)$. Namely $F(r) < 0$ for $r < a$ (local inhibition),

$F(r) > 0$ for $a < r < b$ (middle-range activation), and $F(r) < 0$ for $r_c \geq r > b$ (long-range inhibition).

Owing to the impossibility of finding an analytical solution of (1), we were forced to study them through numerical integration methods. As we were interested in values of parameters granting for flock formation, we were forced to introduce a suitable *order parameter*, so as to characterize flock formation as a phase transition. To this regard, we made resort to spatial autocorrelation functions of point individual velocities. These latter can be computed through the following formulae:

$$S_x(\rho, t) = \Sigma_i \, \Sigma_{\xi(i)} v_x(i, t) \, v_x(\xi(i), t) \, / \, N_F \tag{5.a}$$

$$S_y(\rho, t) = \Sigma_i \, \Sigma_{\xi(i)} v_y(i, t) \, v_y(\xi(i), t) \, / \, N_F, \tag{5.b}$$

where N_F is a normalizing factor and $\xi(i)$ denotes the set of all point individuals such that, at time t, $x(i, t) = x(\xi(i), t)$ and $/y(i, t) - y(\xi(i), t)/ = \rho$, or $/x(i, t) - x(\xi(i), t)/ = \rho$ and $y(i, t) = y(\xi(i), t)$.

Now we must take into account that a flock can be considered as a entity which is coherent with itself in time. This means, in turn, that the structure of the field of velocities of point individuals, described by the autocorrelation functions (5.a) and (5.b), should not change (or change very little) with time. So we should expect that, within a flock, values of $S_x(\rho, t)$ and $S_y(\rho, t)$ should last, more or less, constant. Such a constancy can be checked through the time autocorrelation functions of S_x and S_y, defined by:

$$T_x(\tau) = \Sigma_k \, \Sigma_i \, S_x(\rho_k, t_i) \, S_x(\rho_k, t_i+\tau) \, / \, Q \tag{6.a}$$

$$T_y(\tau) = \Sigma_k \, \Sigma_i \, S_y(\rho_k, t_i) \, S_y(\rho_k, t_i+\tau) \, / \, Q, \tag{6.b}$$

where Q is a suitable normalization factor. In the case of nearly constant values of S_x, S_y we should expect $T_x(\tau)$, $T_y(\tau)$ to assume very high values for τ great enough. This can be considered as criterion for characterizing the existence of a flock, and, e.g., the average value of $T(\tau)$ could be used as an order parameter in order to describe flock formation as a phase transition. By using such a criterion, we were able to find, though numerical experiments, model parameter values grating for flock formation. The best ones we found are:

$$r_c = 20, \; b = 16, \; a = 2, \; A = 0.0001, \; K = 0.2.$$

The goodness of this criterion was tested against phenomenological observations of flock formation. In all cases the presence of a flock was associated to high values of $T_x(\tau)$, $T_y(\tau)$, and the absence of a flock to small values of these quantities. Two particular examples are shown in figures 1a), 1b).

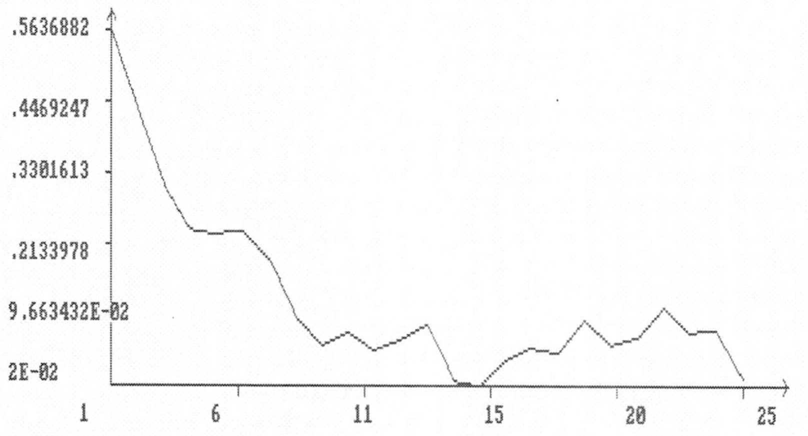

Figure 1a). Time autocorrelation of v_x ; vanishing initial velocity common to 12 point individuals; random fluctuations around initial common velocity with amplitude 0.05; 100 steps of time evolution; phenomenological observation of a flock formation; as it is possible to see, time autocorrelation , at least up to $\tau = 6$, is characterized by high values.

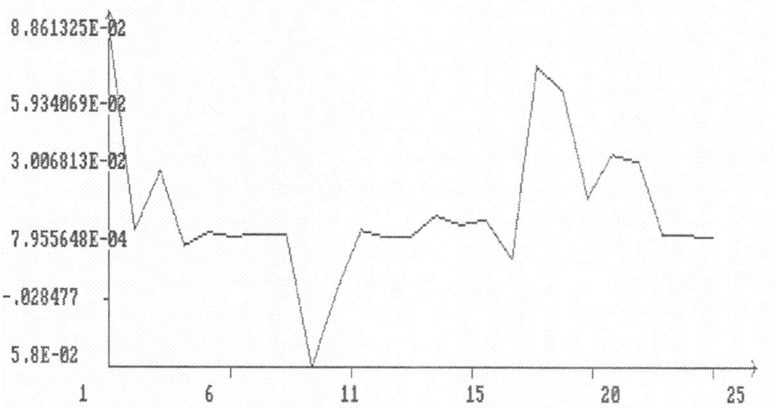

Figure 1b). Time autocorrelation of v_x ; vanishing initial velocity common to 12 point individuals; random fluctuations around initial common velocity with amplitude 1; 100 steps of time evolution; absence of flock formation; as it is possible to see, time autocorrelation is characterized by values which are mch smaller than in the case of Figure 1a).

From the phenomenological point of view we observed a number of features, such as:

f1) within a flock point individuals execute individual motions (more or less oscillatory) around flock momentaneous barycentre;

f2) starting from initial random velocities and positions of point individuals, not all point individuals will aggregate within a flock, every in the case of flock formation;

f3) flock formation is not favoured by the fact that all point individuals hare a common velocity, but the existence of such a velocity makes easier phenomenological observations; individual random fluctuations of initial velocity around a common value are allowed, but when such fluctuations become too great (of the order of unity) flock formation doesn't happen;

f4) an increase of *A* or the vanishing of *K* prevents from flock formation.

f5) We can therefore assert that our model allows for flock formation, once satisfied suitable conditions. The flocks thus generated, however, are associated to features partly different from the ones characterizing ordered structures as observed in most physical or chemical systems. Namely these latter depend, as regards their formation, only on parameter values and not on initial conditions. Besides, in physico-chemical systems individual fluctuations with respect to overall structure tend to decay with time, whereas in our case we have a persistence of individual motions. We hypothesize that the mathematical structure of such motions constitutes a sort of *signature* denoting the operation of cognitive systems underlying collective behaviours in socio-economical systems. In order to test such an hypothesis, however, further studies and simulations will be needed.

Conclusion

We proposed a prototype model of collective behaviours in socio-economical systems, based on the introduction of a parameter designed to represent in a explicit way the operations of cognitive systems of individuals belonging to the system under study. Numerical simulations showed that our model allows for flock formations, once chosen suitable parameter values and initial conditions. However, the observed collective behaviour is associated to features very different from the ones characterizing collective behaviours in physico-chemical systems. The most important difference is the occurrence, in our model of typical patterns of individual motions, not decaying with time. We conjecture that such a circumstance could be used to characterize collective behaviours in socio-economical systems.

References

[1] Beer R. D., A dynamical system perspective on agent-environment interaction. *Artificial Intelligence* **72** (1995) pp.173-215.

[2] Bonabeau E., From Classical Models of Morphogenesis to Agent-Based Models of Pattern Formation. *Artificial Life* **3** (1997) pp.191-211.

[3] Bonabeau E., Theraulaz G., Deneubourg J.L., Quantum study of the fixed threshold model for the regulation of division of labour in insect societies. *Proceedings of the Royal Society of London*. **B 263** (1996) pp.1565-1569.

[4] Brooks R., A layered intelligent control system for a mobile robot. *IEEE J. Robotics Automat*. **RA-2** (1986) pp.14-23.

[5] Conrad M., Speedup of Self-Organization Trough Quantum Mechanical Parallelism. In R. K. MISHRA, D. MAAZ, E. ZWIERLEIN (Eds.) *Self-organization: An Interdisciplinary Search for a Unifying Principle* (Springer , Berlin 1994) pp. 92-108.
[6] Delgado J., Solé R. V., Mean Field Theory of Fluid Neural Networks. *Physical Rewiew E* **57** (1998) pp.2204-2211.
[7] Gierer A., Meinhardt H., A Theory of Biological Pattern Formation. *Kybernetik* **12** (1972) pp.30-39.
[8] Mermin N. D., Wagner H., Absence of ferromagnetism or antiferromagnetism in one- or two-dimensional isotropic Heisenberg models. *Physical Rewiew Letters* **17** (1966) pp.1133-116.
[9] Millonas M. M., A Connectionist Type Model of Self-Organized Foraging and Emergent Behaviour in Ant Swarms. *Journal of Theoretical Biology* **159** (1992) pp.529-542.
[10] Minati G., Pessa E. (Eds.), *Emergence in Complex, Cognitive, Social, and Biological Systems* (Kluwer, New York 2002).
[11] Oster G. F., Lateral inibition models of developmental processes. *Mathematical Biosciences* **90** (1988) pp.265-286.
[12] Rauch E. M., Millonas M. M., Chialvo D. R., Pattern Formation and Functionality in Swarm Models. *Physics Letters A* **207** (1995) pp.185-193.
[13] Reynolds C. W., Flocks, Herds, and Schools: A Distributed Behavioural Model. *Computer Graphics* **21** (1987) pp.25-34.
[14] Solé R. V., Miramontes O., Information at the edge of chaos in Fluid Neural Networks. *Physica D* **80** (1995) pp.171-180.
[15] Solé R. V., Miramontes O., Goodwin B. C., Oscillations and Chaos in ant societies. *Journal of Theoretical Biology* **161** (1993) pp.343-357.
[16] Toner J., Tu Y., Long-Range Order in a Two-Dimentional Dynamical XY Model: How Birds Fly Together. *Physical Review Letters* **75** (1995) pp.4326-4329.
[17] Wolpert D., Tumer K., Frank J., Using Collective Intelligence To Route Internet Traffic. In *Advances in Neural Information Processing Systems, vol. 11* (Morgan Kauffman, Denver 1998) pp. 952-958.
[18] Wolpert D., Wheeler K., Tumer K., General Principles of Learning-Based Multi-Agent Systems. In *Proceeding of the Third International Conference on Autonomous Agents* (Seattle, WA 1999) pp. 77-83).

In: Chaos and Complexity Research Compendium
Editors: F. Orsucci and N. Sala, pp. 159-170
ISBN: 978-1-60456-787-8

Chapter 12

CONTRIBUTION TO THE DEBATE ON LINEAR AND NONLINEAR ANALYSIS OF THE ELECTROENCEPHALOGRAM

***F. Ferro Milone*[1,a], *A. Leon Cananzi*[2,b], *T.A. Minelli*[3,c], *V. Nofrate*[4,d] *and D. Pascoli*[3,e]**

Università di Verona I, Verona, Italy
Research & Innovation, Padova, Italy
Dipartimento di Fisica dell'Università e Sez. INFN, Padova, Italy
Research & Innovation-Dipartimento di Fisica dell'Università di Padova, Italy

Abstract

In the last ten years many papers have been devoted to a description of the electroencephalographic (EEG) activity in terms of chaos, namely in terms of a hypothetical underlying low dimensional nonlinear dynamics (attributed to neuron synchronization). However, the imperfect scaling of correlation sums to a power law and the incomplete saturation of the same for increasing embedding dimension have strongly reduced the expectation of an exhaustive EEG description in terms of low dimensional deterministic chaos. Further evaluations of embedding dimension, like the false nearest neighbors method or the singular value decomposition, confirm the limits of this approach. However, such a result would not look surprising. In fact, the EEG activity is only approximately stationary, as required by dimension evaluation; furthermore, the EEG time series are contaminated by intrinsic dynamical noise (sensory stimulation and membrane fluctuations) and by minor electrical noise of measurement apparatus. These characteristics of the EEG activity have been pointed out by the analysis, performed by linear and nonlinear methods, in some experiments of periodic photo-stimulation. This is a significant result since measurement of noise can put severe restrictions on the

[a] E-mail address: f.ferro@libero.it
[b] E-mail address: albertaleon@researchinnovation.com
[c] E-mail address: minelli@pd.infn.it
[d] E-mail address: nofrate@pd.infn.it
[e] E-mail address: pascoli@pd.infn.it

classification of the tracings and, on the other side, the noise reduction methods may severely disturb the structure of the signal in the time and phase-like space domain.

Introduction

From a general point of view, the electroencephalogram (EEG) may be looked on as a "global phenomenon" addressed by a global wave theory [17]. It may be considered, in the clinical practice, as a continuous shuffling of slow and fast rhythms (brain waves) that correspond to different behavioral states as, in healthy subjects, alert (eyes open), rest (eyes closed), drowsy, sleepy and, in pathological conditions, anxiety states, depression, dissociative behavior, dementia, paresis and paralysis of sensory or motor function, epilepsy etc. In an oversimplified view many of these states, in physiological as well as in pathological conditions, correspond in the EEG to slow and fast waves and/or rhythms (asymmetric and/or symmetric in times), that are, from the neurophysiological point of view, to desynchronization and synchronization of the neuronal population: this means that desynchronization and synchronization have temporal as well as spatial configuration.

The physiological mechanism we use to induce synchronization is the periodic photo-stimulation. A typical synchronizing activity in pathological conditions is that recorded in epileptic patients. With the aim to contribute to clarify the applicability of linear and nonlinear analysis in clinical electroencephalography, we have studied some EEG time series in healthy subjects without and with synchronizing stimuli (photo-stimulation driving).

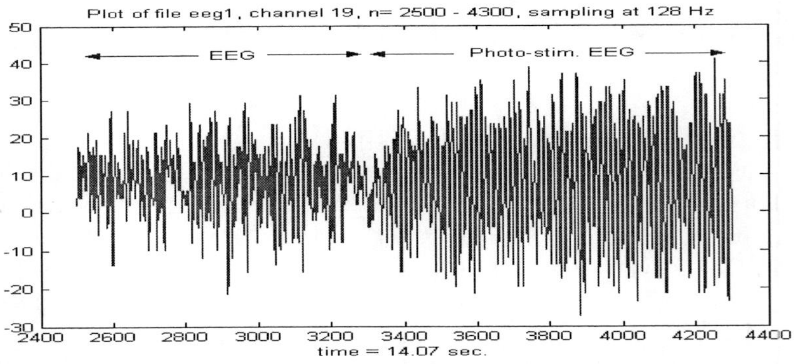

Figure 1. Plot of channel Pz of an EEG (recorded with the International Electrode System 10-20) without photo-stimulation (about from 1 to 6.25 sec., corresponding to points 2500 - 3300) and a with photo-stimulation (about from 7.81 to 14.06 sec., corresponding to points 3500 – 4300). The sampling is at 128 Hz.

Spectral Analysis

In the clinical practice slow and fast waves/rhythms are detected by means of computed spectral analysis (absolute and relative power and spectral coherence), that is by means of linear analysis.

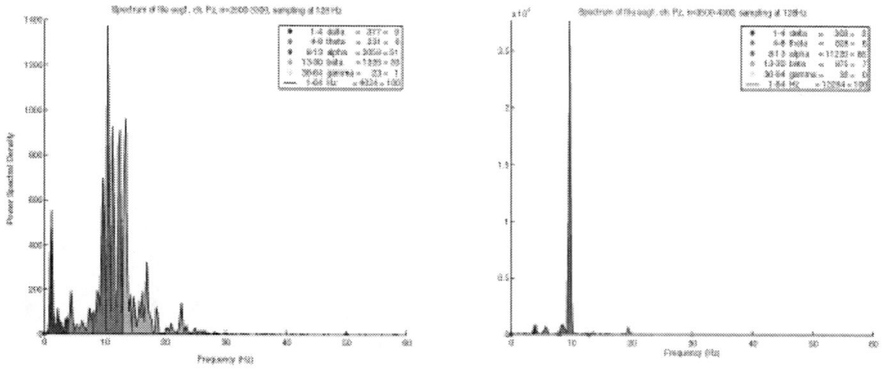

a) b)

Figure 2. Power spectral density of the two epochs of Fig. 1: in b) the resonance induced by periodic photo-stimulation at 10 Hz is evident (notice the different scales).

Besides the spectral power, measuring the amount of the spectral components of the EEG, the mainly used measure for the one-channel synchronism is the spectral coherence [2, 21]

$$C_{xy}^2(\omega) = \frac{[G_{xy}(\omega)]^2}{G_{xx}(\omega)G_{yy}(\omega)}$$

where $G_{xy}(\omega)$ is the cross-power spectral density and $G_{xx}(\omega)$, $G_{yy}(\omega)$ are the respective auto-power spectral densities, largely used in the EEG measure for the two-channel synchronism [3, 10, 22, 24]. In clinical practice the spectral coherence is used as an index, ranging from 0 to 1, of the channel synergy.

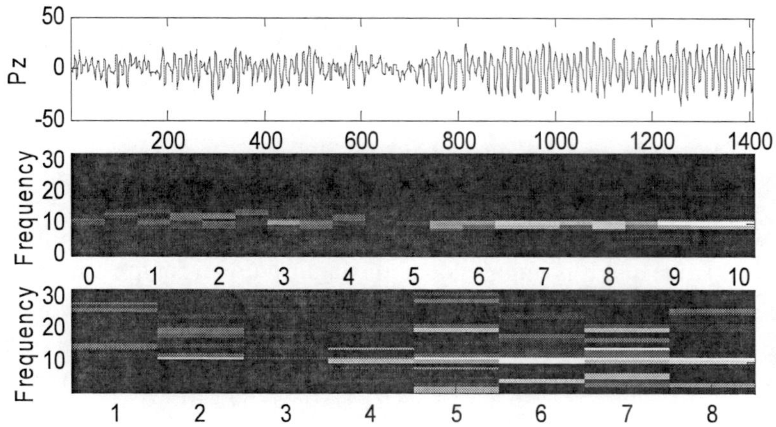

Figure 3. The figure exhibits an enhanced activity in a narrow band centered at 10 Hz both for the windowed spectrogram of channel Pz (middle) and for the windowed spectral averaged coherence in the region around electrode Pz (bottom), during the photo-stimulation (in the right part). Notice the evidence of harmonics generation.The spectral power and the coherence are measured with levels of gray and tones increasing from black (value zero) to white (value one) and the scansion scale has been chosen to mark the window time progression.

A windowed extension of the spectral coherence [9] has been tested to analyze the photo-stimulation experiment (Fig. 3 top). In the same figure, respectively in the middle and at the bottom, the windowed spectrogram of channel Pz and the windowed averaged coherence, mediated with the channel neighbors of Pz [3], are reported for the alpha rhythm enhanced by periodic stimulation. Both the spectral pictures exhibit, with levels of gray from black to white, an enhanced activity in a tight band centered at 10 Hz during the stimulation; while the first of the two patterns confirms the essentially periodic nature of this epoch, in agreement with the recurrence plot (see later), the second one reveals an underlying co-operation, justified by a neuron hyper-synchronization.

Chaos and Noise in the Electroencephalogram

During the last decade many Authors developed time series analysis in EEG that takes into account the hypothesis that the generating process of the time series may be reproduced in terms of a hypothetical underlying low dimensional nonlinear dynamics [15], an assumption justified by the neuron synchronization.

Nonlinear time series analysis involves complex computational procedures and the application to clinical diagnostics is still under evaluation. The strong unpredictability of the EEG time series suggests that the underlying dynamics may be characterized by a chaotic behavior, as a possible paradigm for the neuron dynamics. Chaos quantification is usually measured by attractor's dimension and by Lyapunov exponents, while the second index, calculating the exponential divergence of near trajectories (the attractor is one of the structures characterizing the asymptotic behavior of the dynamics), is inversely proportional to the time in which the trajectories distance remains smaller than a fixed quantity [16].

Packard and Ruelle introduced the time-delay coordinates in order to reconstruct the phase-space of the observed dynamical system and Takens' theorem guarantees that the reconstructed dynamics (with appropriate values of the time-delay τ and of the embedding dimension d_E) is equivalent to the original one. Having the original scalar time series $\{s_i, i=1,2,\ldots ndata\}$ the reconstructed trajectory ($\mathbf{x}_1, \mathbf{x}_2, \ldots, \mathbf{x}_N$) of the phase-space are obtained by the delayed vectors $\mathbf{x}_i = (s_i, s_{i+\tau}, s_{i+2\tau}, \ldots, s_{i+(dE-1)\tau})$, where the time lag τ and the embedding dimension d_E must to be found, where $N = (ndata - d_E * \tau)$.

Being the dynamics time constants unrelated to the sampling time, the time delay τ must be evaluated with proper criteria, because if it is too small the coordinates s_i and $s_{i+\tau}$ have too similar numerical value and it is impossible to distinguish one from the other, while if it is too large s_i and $s_{i+\tau}$ are, in statistical terms, completely independent. If one denotes the attractor's dimension d_A, one has to find the embedding dimension d_E large enough to be able to unfold the points $\{\mathbf{x}_j\}$ without ambiguity, that is with no superposition or self-crossings due to projections in a phase-space too small. The sufficient condition for this is $d_E > 2d_A$ and is due to Mané and Takens [1, 18].

In order to choose the delay τ there are at least two methods: one chooses for τ the first zero of the linear autocorrelation function [1] while the second one identifies τ as the first minimum of the average mutual information, as proposed by Fraser and Swinney [8]. The equivalence of the values obtained with both methods, as we report in Fig. 4, is evident.

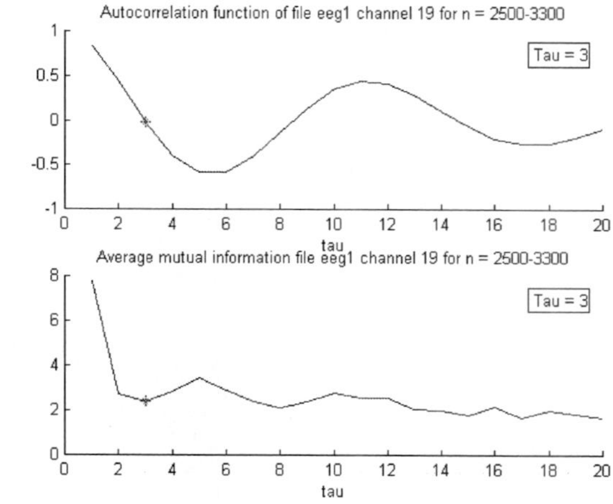

a)

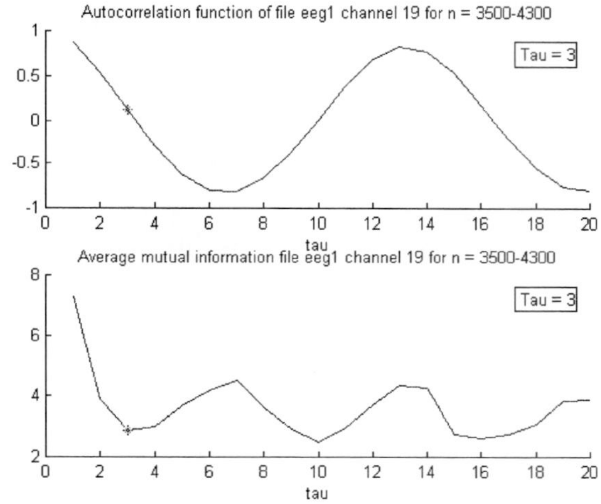

b)

Figure 4. Autocorrelation function and average mutual information for the channel Pz in EEG, sampled at 128 Hz, before (a) and during the photo-stimulation (b).

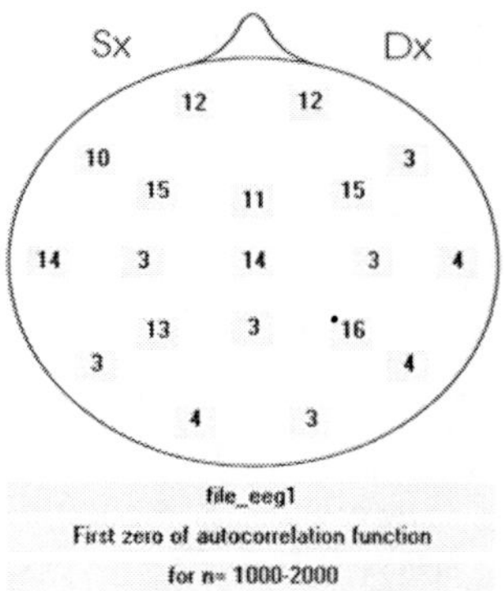

a)

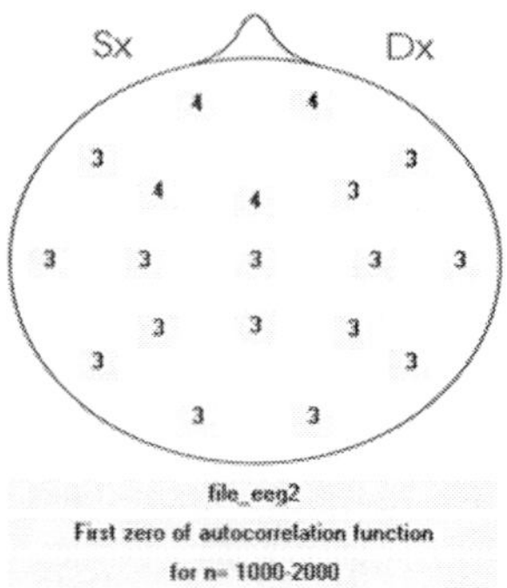

b)

Figure 5. An example of the values of the first zeroes of autocorrelation function for each electrode position of not stimulated (a) and stimulated (b) EEG.

In Fig. 5 the dependence of the values of τ (calculated as the first zero of the correlation function) on the electrode position is reported. The values in the anterior electrodes are clearly diminished and leveled by the stimulation.

In Fig. 6 the attractors, two-dimensional reconstruction of the analyzed epochs of the time series of Fig. 1, are reproduced.

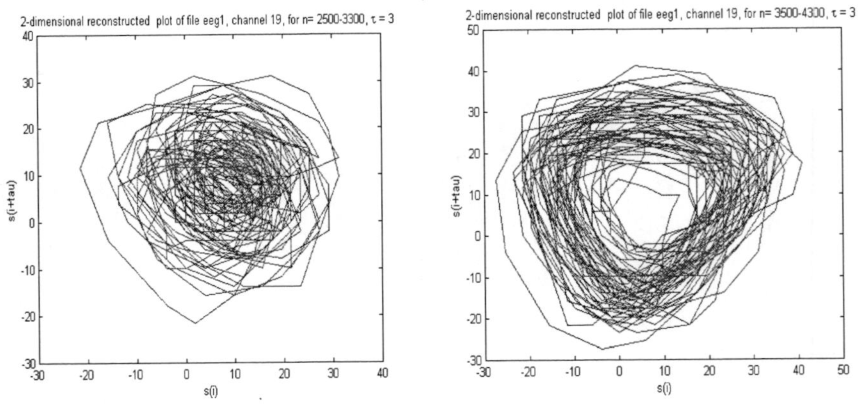

a) b)

Figure 6. Two-dimensional reconstruction of s_i against $s_{(i+\tau)}$ for not stimulated (a) and stimulated (b) epochs of Fig.1 obtained by using the time lags previously evaluated. Notice the ring structure of the right attractor signaling the essentially periodic nature of the photo-stimulated activity.

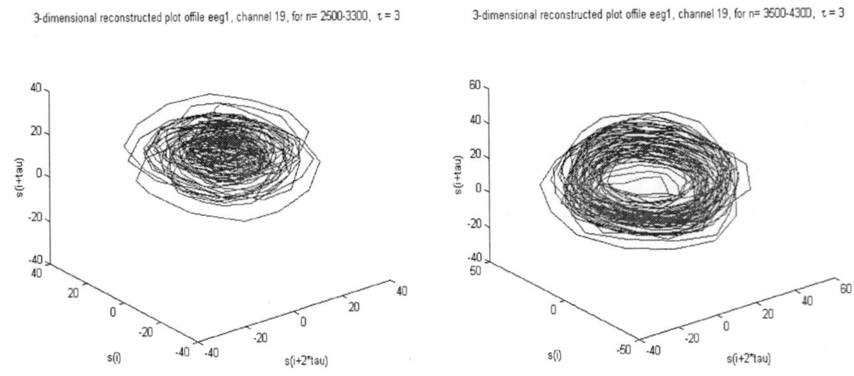

a) b)

Figure 7. Three-dimensional reconstruction of s_i against $s_{(i+\tau)}$ and $s_{(i+2\tau)}$.

The relationship between Lyapunov exponents and attractor dimension has been quantified by Kaplan and Yorke [16], and possesses a heuristic explanation in terms of recurrence plot and correlation sum. The recurrence plot exhibits, by white dots on black background, the points of the plane $(\mathbf{x}_i,\mathbf{x}_j)$ for which the distance is smaller than a fixed quantity ε [14]. In case of periodic motions, it reveals zones covered by line segments parallel to the diagonal: segments length is proportional to the time in which the trajectories distance remain smaller than ε. Recurrence plot can be used, for instance, to reveal the resonance induced by periodic photo-stimulation in the time series of Fig. 1. In the patterns of recurrence plot of Fig. 8b (corresponding to 2 seconds of the stimulated epoch) a covering by white line segments parallel to the diagonal is a sign of regularity: this structure is less evident in plot 8a, corresponding to the unstimulated alpha activity.

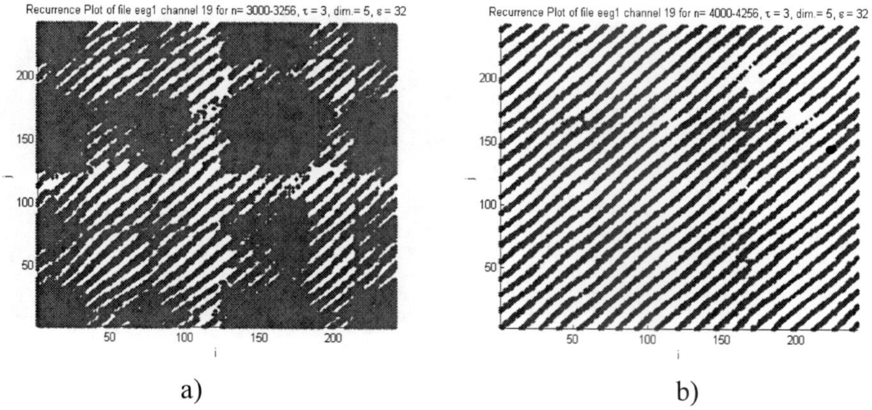

Figure 8. Recurrence plot in Pz channel of EEG for free alpha activity (a) and in the stimulated one (b) with the cut-off at ε = 32, with the distance calculated with the Euclidean Norm, for 2 seconds of the original time series.

The Grassberger and Procaccia correlation sums is founded on an account of the fraction of pair of points i and j that are closer than ε:

$$C(\varepsilon) = \frac{2}{N(N-1)} \sum_{i=1}^{N} \left[\sum_{j \neq i}^{N} \Theta(\varepsilon - \|\mathbf{x}_i - \mathbf{x}_j\|) \right]$$

where the norm can be the Euclidean one or the Max Norm.

In the case of an existing underlying low dimensional dynamics a power law like

$$C(\varepsilon) \propto \varepsilon^{\nu}, \varepsilon \to 0$$

is obtained from the correlation sum. While in case of high dimensions (for example noise) the exponent ν increases without bound with the embedding dimension d_E, in the case of low dimensional dynamics one attains saturation and the corresponding asymptotic value D defines the correlation dimension itself [16, 20].

In Fig. 9 the correlation sums versus ε for different values of d_E are reported (in log-log scale) for both the free and the photo-stimulated segments of the time series of Fig. 1. The incertitude of the slope of these curves and the incomplete saturation with the increasing of d_E have strongly reduced the early assumption on an exhaustive EEG description as low dimensional deterministic chaos [13]. This warning has been confirmed by further checks, for instance by the surrogate data test on the resistance to the phase disruption, and is a property characteristic of the noise [4, 20]. However, the false nearest neighbors test, which accounts the nearest points of the time series in terms of the embedding dimension and, for d_E large enough, select only points which are contiguous because of the dynamics and not because of self-crossings due to projection into low spaces, allows to estimate the embedding dimension and to signalize the presence of noise [13, 20]; also the Singular Value Decomposition (SVD) [4, 16, 20] can be used for this purpose.

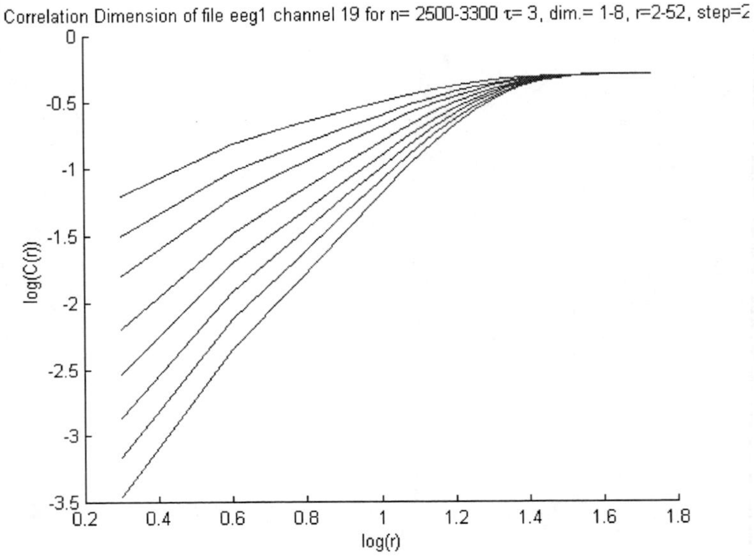

a)

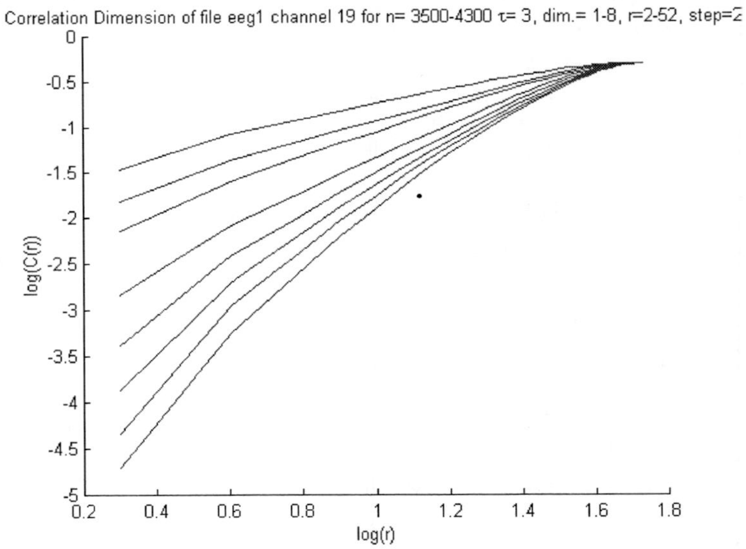

b)

Figure 9. The logarithm of the correlation sums plotted against the logarithm of the distance ε respectively for not stimulated and stimulated EEG.

The first method analyzes for each reconstructed time series (using an enough large embedding dimension) how many points which are nearest in a given dimension d become not near in dimension (d+1), that is how many are false nearest neighbors. When this number drops to zero the embedding dimension d_E is found and the system can be unfolded in a d_E - dimensional Euclidean space.

The second method used to determine d_E, the Singular Value Decomposition, calculate the eigenvalues of the Covariance Matrix.

According to recent surrogated and reversibility tests only a little fraction of the EEG records can be explained in terms of an underlying dynamics, while the largest set of the tracings behaves more as a random signal then as a chaotic trajectory, so that cannot be distinguished from linearly filtered Gaussian noise [6, 23]. This different behavior can be seen in the two epochs of the rhythm of Fig.1 respectively during and before periodic photo-stimulation.

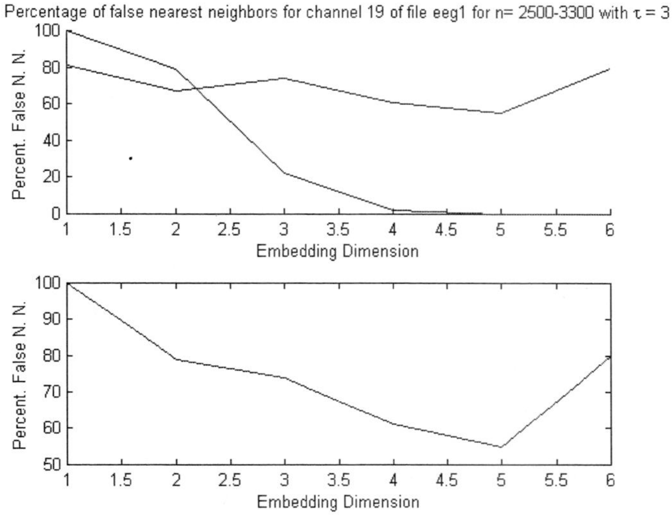

a)

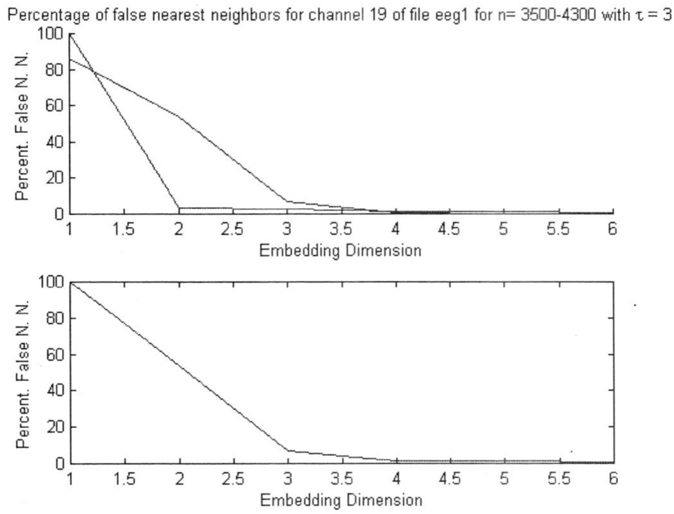

b)

Figure 10. Percentage of false nearest neighbors for the channel Pz of EEG sampled at 128 Hz. For not stimulated (a) and stimulated (b) time series. At the top are plotted both tests proposed by Abarbanel et al. and at the bottom there is the resultant curve.

Following the suggestion of Abarbanel et al. [1, 18] for the false nearest neighbors we find $d_E = 5$ and $d_E = 4$ respectively for the not photo-stimulated and the stimulated time series (Fig.10 at the bottom). We can notice the presence of noise especially in the first time series because of the high "tail" of the plot.

The computing of the Singular Value Decomposition [4] allows to find $d_E = 6$ for the not stimulated time series and $d_E = 4$ for the stimulated one (Fig.11).

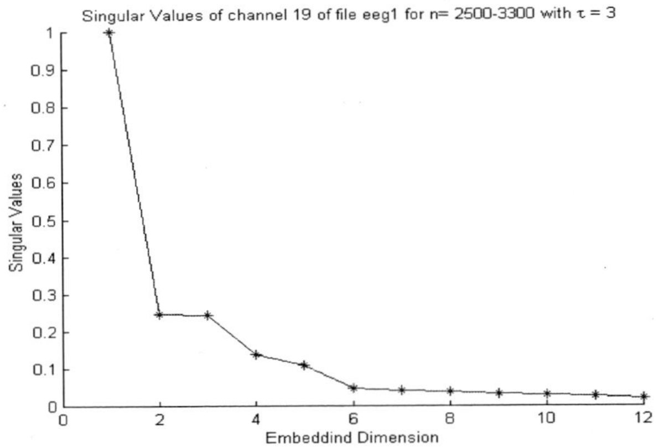

a)

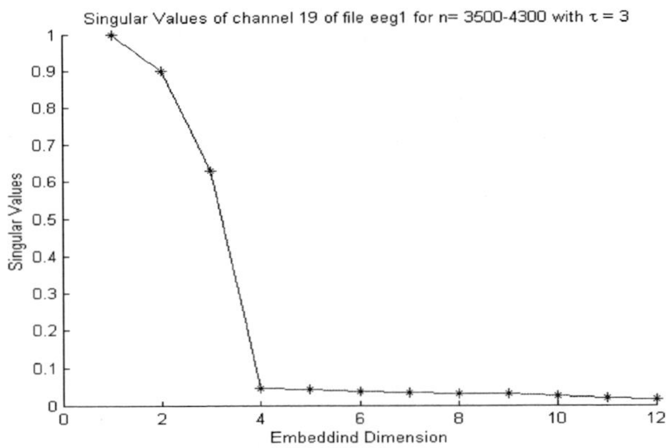

b)

Figure 11. Spectrum of the singular value decomposition of the not photo-stimulated (a) and stimulated (b) EEG.

Therefore the values of the time delay and of the embedding dimension are $\tau = 3$ and $d_E = 5\text{-}6$ for the not stimulated time series and $\tau = 3$ and $d_E = 4$ for the stimulated one. The dominance of the internal noise on the external one, illustrated by the previous figures, only partially clarifies the doubts on the approach in terms of chaos; also the signal fluctuations, peculiar to sensory stimulation and membrane noise, disturb the neuron synchronization and contribute to the inadequacy of the correlation dimension as an exclusive index for the phenomenological classification of the rhythms [5, 7, 11, 12, 19].

Concluding Remarks

A contribution to the debate on the possible utility of EEG in diagnosis founded on the dimension of a hypothetical underlying nonlinear dynamics has been presented.

The role of the time delay and of the embedding dimensions used for the reconstruction in a phase-like space, for segments of approximatly 6-7 seconds (for which the times series can be treated as a near stationary signal) is stressed. The dependence of the time-delay on the electrode position, and therefore on the cortex area, has been observed and discussed. An outline of the standard linear and nonlinear diagnostic methods is furthermore presented and applied to the EEG analysis during basal conditions and after photo-stimulation at 10 Hz. This latter phenomenology may be considered as an experimental condition inducing a surplus of synchronization (and in extreme pathological conditions may give rise to real hyper-synchronization, as in photo-stimulated epilepsy).

Since the correlation dimension seems to be inadequate to discriminate basic and stimulated EEG, limits to the embedding dimension have been estimated by using the false nearest neighbors test and the singular value decomposition method.

Acknowledgments

This work has been supported by the MURST project (DM 2125/98) Patologie immunoinfiammatorie e degenerative del sistema nervoso: aspetti patofisiologici e sviluppo diagnostico e terapeutico. The authors are indebted to L. Turicchia for his help in signal processing and to V. Aricò for his help in graphic processing.

References

[1] Abarbanel, H. D. I., Brown, R., Sidorowich, J. J., Tsimring, L. S., The analysis of observed chaotic data in physical systems, *Review of Modern Physics*, vol. 65, N°4 (1993) pp.1331-1392.

[2] Bendat J.J. and Piersol A. G., Random Data, Wiley, New York (1986).

[3] Besthorn, C., Förstl, H., Geiger-Kabisch, C., Sattel, H., Gasser, T. and Schreiter-Grasser, U., EEG Coherence in Alzheimer Disease, *Electroencephalography and Clinical Neurophysiology* , **90** (1994) pp. 242-245.

[4] Broomhead, D. S., King, G. P., Extracting qualitative dynamics from experimental data, *Physica* **20 D** (1986) pp.217-236.

[5] Deutsch, S., Deutsch, A., *Understanding the Nervous System*, New York: IEEE, (1993).

[6] Diks, C., Nonlinear Time Series Analysis, World Scientific, Singapore (1999).

[7] Faure, P., Korn, H., A nonrandom dynamic component in the synaptic noise of a central neuron, *Proc. Natl. Sci*, USA, **94**, (1997) pp. 6506-6511.

[8] Fraser, A. M., Swinney, H. L., Independent coordinates for strange attractor from mutual information, *Physical Review A,* vol. 33, N° 2 (1986) pp.1134-1140.

[9] Gabrieli, C., Ferro Milone, F., Ferro Milone, G., Minelli, T. A., Turicchia, L., From the mathematical anatomy to the mathematical physiology of brain co-operative

phenomena, in *Chaos, Fractals & Models* Ed. by F. Marsella & G. Salvadori, Italian University Press, Pavia (1998) pp. 406-410.

[10] Glass, A., Zappulla, R. A., Nieves, J., Diamond, S. P., EEG Coherence as a Predictor of Spike propagation, *Electroencephalography and Clinical Neurophysiology* , **82** (1992) pp. 10-16.

[11] Grzywna, Z.J., Siwy, Z., Chaos in Ionic transport through membranes, from H. T. Moon, S. Kim, R. P. Behringer, Y. Kuramoto (Eds.), *Nonlinear Dynamics and Chaos*, Singapore, World Scientific (1997) pp. 329-337.

[12] Han, S. K., Park, S. H., Yim, T. G., Kim, S., Chaotic bursting behavior of coupled neural oscillators, from H. T. Moon, S. Kim, R. P. Behringer, Y. Kuramoto (Eds.), *Nonlinear Dynamics and Chaos*, Singapore, World Scientific, (1997) pp. 43-50.

[13] Ivanov, D. K., Posh, H. A., Stumpf, Ch., Statistical measures derived from the correlation integrals of physiological time series, *Chaos*, **6** (1996) pp. 243-253.

[14] Ivanski, J. S., Bradley, E., Recurrence plot of experimental data: To embed or not to embed?, *Chaos*, vol. 8, N° 4 (1998) pp.861-871.

[15] Kantz, H., Schreiber, T., Dimension Estimates and Physiological Data, *Chaos*, **5** (1995) pp. 143-154.

[16] Kantz, H., Schreiber, T., Nonlinear time series analysis, Cambridge University Press, (1997).

[17] Katznelson, R. D., Normal modes of the brain: Neuroanatomical basis and a physiological theoretical model, from Nunez, P. L., *Electric Fields of the brain: The Neurophysics of EEG*, New York: Oxford University Press, 1981.

[18] Kennel, M. B., Brown, R., Abarbanel, H. D. I., Determining embedding dimension for phase-space reconstruction using a geometrical construction, *Physical Review* **A**, vol.45, N° 6 (1992) pp.3403-3411.

[19] Lee, S. G., Kim, S., Synchrony and clustering in two and three synaptically coupled Hodkin-Huxley Neurons, from Moon, H. T., Kim, S., Behringer, R. P. and Kuramoto, Y. (Eds.), *Nonlinear Dynamics and Chaos* , Singapore, World Scientific (1997) pp. 63-69.

[20] Nayfeh, A.H., Balachandran, B., Applied nonlinear dynamics, New York: Wiley, (1995).

[21] Newland, D. E., An introduction to Random Vibrations, Essex, Longman (1993).

[22] Nunez P. L., Neocortical Dynamics and Human EEG Rhythms, New York, Oxford University Press (1995).

[23] Stam C. J., Pijn, J. P. M., Suffczynski, P., Lopes da Silva, F. H., Dynamics of the human alpha rhythm: evidence for non-linearity?, *Clinical Neurophysiology,* **110** (1999) pp. 1801-1813.

[24] Thatcher, R. W., Krause, P. J., Hrybyk, M., Cortico-Cortical Associations and EEG Coherence: A two-Compartimental model, *Electroencephalography and Clinical Neurophysiology,* **64** (1986) pp. 123-143.

In: Chaos and Complexity Research Compendium
Editors: F. Orsucci and N. Sala, pp. 171-193
ISBN: 978-1-60456-787-8

Chapter 13

COMPLEX DYNAMICS OF VISUAL ARTS

***Ljubiša M. Kocić*[1] *and Liljana Stefanovska*[2]**
University of Niš, 18000 Niš, Serbia and Montenegro
University »Sv. Kiril i Metodij«, Skopje, Macedonia

Abstract

History of art is a very complex entity. Our purpose is to show that (pitchfork) bifurcation is the basic phenomenon that creates this complexity. Based on the very complex structure of human mind, the aesthetic values suffers from local unstability, i.e. they often undergo revisions, turn stable criteria into unstable, and are being replaced by two new stable aesthetic values. It causes bifurcations that make a certain artistic style split into two new ones. This generates the period doubling (or Figenbaum) route to chaos. Consequently, the body of art-history characterizes by all typical features: self-similarity to different scales and hierarchy of forms: from constancy and linearity to periodicity, complexity and even chaos. Although we conjecture that bifurcations are common in all art movements, we are forced to narrow our examples to visual arts, mostly to painting, otherwise the paper might be too extensive.

1. Introduction

There is no doubt that the structure of the Universe is much closer to fractal than to regular geometry. This is the consequence of nonlinearity and permanent dynamics that take part in many different aspects and scales. The human being, as a specific mirror, reflects the fabulous complexity of the Universe and its psychology makes the smaller individual copy of it inside his mind by the mental process that we call *▱-prism*. Layer by layer, the selected stuff from the outside world is being placed in the subconsciousness of the individual during his life-time, in the process that is similar to producing fractal attractors by baker's transformation. Then, all the collective or individual actions in human history bear the seal of this fractal-like "archive" of concentrated experience. Therefore, one expects that the art

[1] E-mail address: kocic@elfak.ni.ac.yu
[2] E-mail address: liljana@ereb1.mf.ukim.edu.mk

history possesses a specific dynamics that reflects such a complex structure of both individual and collective mind.

In this paper, we study bifurcation phenomena in art history that seem to be present in all times of the human creativity although these are not so explicitly noticeable. These bifurcations relate to splitting of some dominant art movement into two new directions, each based on new locally stable aesthetic criteria being accepted by some avant-garde group of artists. Since the history of art is too vast, we limited ourselves to visual arts and, among them, predominantly paintings. Nevertheless, music, literature, architecture, theatre etc., follow very similar patterns. Examples that we study are simple but we hope clear enough to illustrate how bifurcation in aesthetics causes bifurcation of art styles through history.

The second section considers the baker's transformation and its action in Ψ–space (space of psychology of an individual). The third one discusses the aesthetic unstability and appearance of bifurcations. The fourth section is devoted to self-similarity in different cases, and the final, the fifth section deals with hierarchy of complexity, which is something that characterizes the complex dynamics.

2. Baker's Transform in Ψ-Space

From the point of view of an individual, the Universe roughly divides into two parts: Outer region (objective space) and Inner region (psychic space of individual or Ψ–space). There is an extensive and complex interaction between these two spaces. The main topic of art is to explore this relationship. On the other hand, the Universe is permeated with hierarchy, and this hierarchy is present to different measure scales [12], [1], [2]. The same property applies to the Ψ–space. This hierarchy in Ψ–space can be followed starting by the "pyramidal algorithm" of seeing [13], over the mental mechanism of selection and transfer of information to the organ of intelligence up to the psychic activity of deposition of used information in the subconscious domains of Ψ–space. In each stage of this information processing, one can see that the output information is smaller in amount than the input information, i.e., that there is a whole bundle of different *contractive mappings* that act inside our minds both in sequential or in parallel. This continuous compression and condensation of information is the only way to extract some facts from the Outer space that are relevant for the living process of an individual. However, once being deposited the process of rejecting of part of information will continue. All pictures and other stimuli that are being stored are still objects of the subconscious selection process. Some of them remain other go to oblivion and are gradually completely forgotten.

As a conclusion, one can state that the dynamics of such processes in Ψ–space is characterized by:

(1) Being iterative;
(2) A kind of contractive mapping applies in each iteration.

In the Theory of Dynamic Systems it is known that activities 1. and 2. are components of the so called *baker's transformation* [14] (also: *horseshoe map*, although sometimes not with identical meaning, or *Hénon map*), which is essential in complex motions. The set of points that is invariant under this transformation is known as *strange attractor* [14] and (usually) has *fractal* structure [14].

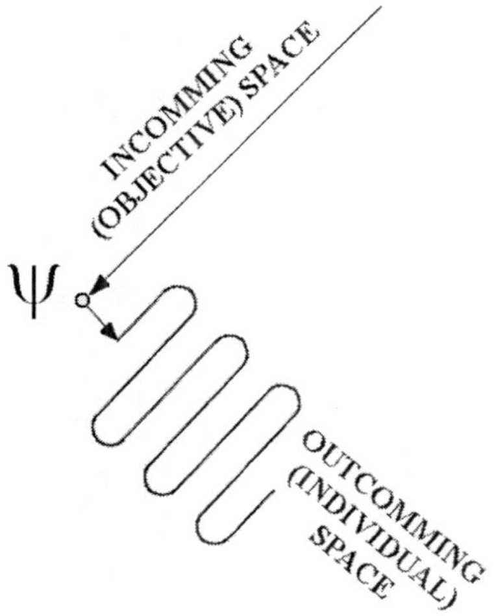

Figure 1. ▱-prism.

So, the nature of processing of data from their input (objective space stimuli) to their output (deposit in individual Ψ–space) is regarded as a special baker's transformation that applies during the lifetime of the subject. The subconscious deposit itself is a fractal-like attractor for this transformation. Like a baker's dough after many hours of kneading, this deposit has a vast number of very thin layers, containing tremendous amount of data mainly in pictorial form. This transformation, just for the purpose of this paper will be named ▱–*prism*. This term comes from the similarity with the Newton's prism that turns a single light ray into a fan of rainbow colored rays of light. In a similar manner, the ▱–prism analyses the incoming space, classifies the perceived data and stores them after selection. Kandinsky, in his famous essay *Concerning the Spiritual in Art* describes it by the following words: "Eye is a hammer. Soul is a piano with many strings" [9]. Our consciousness makes a special storage archive made out of wrinkled, many times overlapped layers of data being received through the life-time. The older layers are more compressed, and the more compressed they are the more they resemble to a fractal (or multi-fractal) set. Therefore, the ▱–prism can be understand as a generator of fractal-like archive in the subconscious mind of some individual. It completes the action of *perception*.

According to Herbert Read [17] the creativity process of an artist starts with *perception* and finishes with *expression*. In this scope, ▱–prism supplies an artist with the data, necessary for his creations.

Example 2.1. Consider the William Blake's (1757-1827) illustration *Urizen as the Creator of the Material World*, from 1794 (outlined in Figure 2).

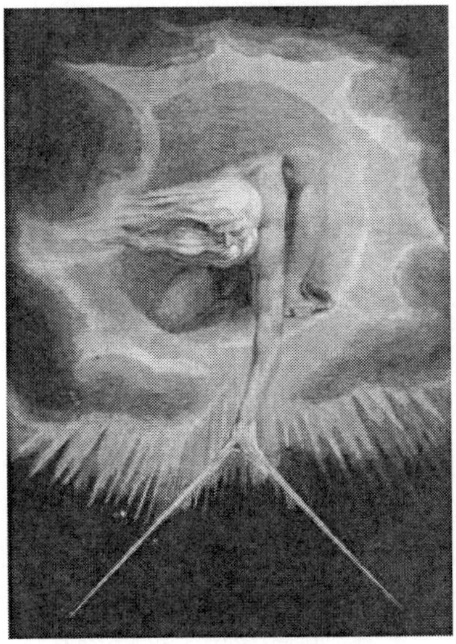

Figure 2. William Blake's Urizen as the Creator of the Material World.

A line can schematically represent the individual experience of William Blake necessary in creation of *Urizen*, which more or less coincides with the time-line (Figure 3). Some important impressions from his experience that was relevant for *Urizen* are shown as points on this line. These are moments of perception of key notions such as *Old man*, *Rays of light*, *Circle*, *Compass*, *Creator* and so on. In his further career, Blake learned more about every of these key concepts and even modified or upgraded them. For example, a compass, the school device for drawing circles, is formed in early youth as a basic notion. In further experience, some variants of the compass, such is a set of huge calipers (a measuring instrument that *Urizen* holds) was also added to Blake's, fractal-like subconscious archive. His own mythological being, *Urizen* replaced the idea of the Creator. All of this belongs to the process performed by the Blake's ▱–prism. At the moment of creation, his mind connected these basic data (1-2-3-4-5 in Figure 3) and combined them into a nice, expressive picture.

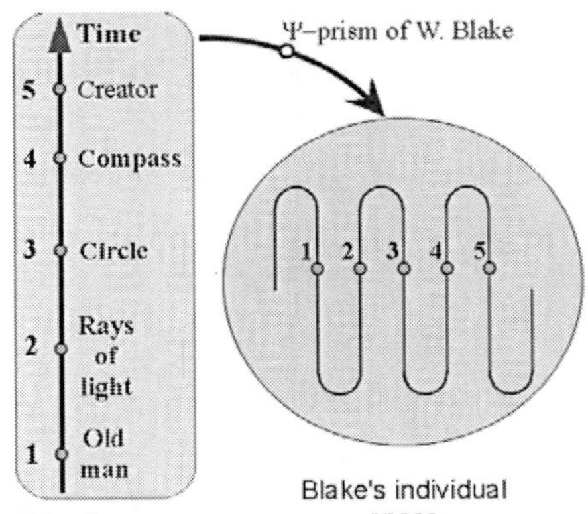

Figure 3. William Blake's ▱–prism.

In the same way as Blake, who found in his "memory storage" key ideas and associated them into a compact whole, needed for *Urizen*, Hiram and Fidias had all elements important for building Temple of Jerusalem or Parthenon in their own Ψ–spaces. Nevertheless, none of them had *all* the elements needed. In the very process of materialization of initial ideas, they supplemented their visions up to the final form. This process resembles to spanning the vector space as the linear combination of basic elements.

Two conclusions are possible. First, that the deposited experience in the Ψ–space, made by continuous baker's transform activity of $\square$–prism has very complex structure. This deposit contains a rich basis of ideas stored in the cortex of the cerebrum, and probably has fractal or multi-fractal structure, even in the sense of organization of energy and bio-substance.

The second one is that existence and phenomenology of the $\square$–prism is necessary but not sufficient condition for art creation. Namely, the mechanism of the $\square$–prism is responsible for creating rich layers of experience. If the individual has the capability to combine, through associative process, the necessary basic elements, and to construct a vision of the future creation, this individual is creative. If there is a will to materialize this vision and to step forward in public with it, one can say that the artist is born. The famous example of Picasso, who found elements such are handlebars and seat of a bicycle in his "fractal storage", and by combining them created his famous "Bull's head", is illustrative [8].

3. Aesthetic Unstability Initiates Bifurcation

Unstability is one of the characteristics of the nonlinear Universe. Fortunately, what appears is not *global* but *local* unstability. It is familiar from Theory of Dynamical Systems that local unstability may create a fairly complex dynamics (see [14] for *multiple potential energy wells* problem). What is said in the previous section about the structure of our subconscious archive of deposited data being acquired through our senses during the life-period stands in favor of complex rather than of a simple dynamics of human psychology. This dynamics, in turn, results in having complex dynamic in aesthetics. This agrees with Collingwood who states, "Either in big or in small, the equilibrium of aesthetic life is permanently unstable" [3] (see also [17]). The phenomena of unstability lead to typical pre-chaotic dynamics in which *bifurcation* phenomenon has important role.

Dynamics of art movements in the history of civilization exhibits bifurcations, caused by changing of aesthetic criteria. In fact, what is stable aesthetics in one period is replaced by a short period of unstability that leads to period doubling or *pitchfork bifurcation* of aesthetic criteria in two opposite directions. So, the old prong, once stable, becomes unstable and plays the role of *repellor*, while new prongs are stable and act as *attractors* [19, p. 272]. The classic example is dynamics of *logistic map* $x \# f(x) = \lambda x(1\text{-}x)$, where λ is a positive parameter, and behavior of solutions of *logistic equation* $x = f(x)$. The orbits $\{x_k\}_{k\in N}$, where $x_k = \lambda x_{k-1}(1\text{-}x_{k-1})$, $k = 1, 2, 3,\ldots, x_0 \in [0, 1]$ is fixed, converge to the unique solution of logistic equation, provided $\lambda < 3$. The first bifurcation occurs for $\lambda_1 = 3$, the next ones for $\lambda_2 = 3.449489...$, then for $\lambda_3 = 3.544090...$, $\lambda_4 = 3.564407...$ etc., which is known as Figenbaum scenario of passing to chaos.

The aesthetic value is the main "metric" for measuring intensity of feelings induced in a human being that is faced with nice and pleasant objects. This intensity, caused by different aesthetic values like *beauty* (harmonic reconciliation of Ideal and Real), *sublimity* (Edmund Burke, Kant), *simplicity* (Kant), *tragic*, *comic* (Hegel), *cute* (unconscious reconciliation of Ideal and Real) etc., is sometimes referred to as *aesthetic tension*. Suppose that aesthetic tension can be measured, and let it be T. Obviously, $T > 0$ and since it depends on many parameters (different aesthetic criteria), it is a multi-variable function. This function will be called (aesthetic) *tension function*. Let x be one of these variables (criteria), and let the others be fixed. Let this restriction of T be denoted by $T(x)$. Then, by definition, $T(x)$ must have a local maximum at the point where the criterion x is optimally reached. Consequently, the reciprocal of the aesthetic tension, $1/T(x)$, will have local minima at the point where the tension is maximal.

Example 3.1. (Golden rectangle) Consider the aesthetic problem of finding the most pleasant rectangular form with sides a and b ($a \leq b$). Then, x can be defined as the ratio b/a. The graph of the function $T(x)$ is constructed by changing b, providing fixed a. In fact, Gustav Theodor Fechner made a kind of such graph [11]. He varies x from 1 (square case) to 2.7 (long rectangle) and the graph of $T(x)$ exhibits the local maximum around the famous Golden Mean value, $x_{max} \approx 1.618$. Moreover, this maximum is parabolic-like. So, the reciprocal, $1/T$ (x) has the local minima at the same point.

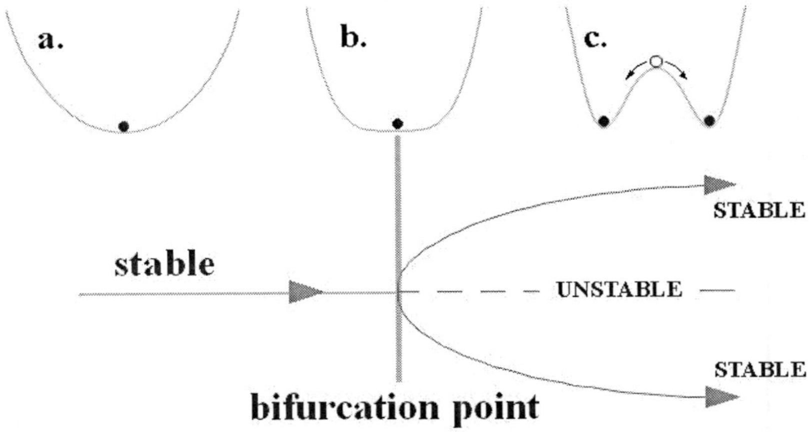

Figure 4. Stable (one-well), indifferent and unstable (two-well) equilibrium and apparition of bifurcation.

In the light of Example 3.1, the aesthetic stability/unstability has a strong mathematical description. Graphical representation is given in Figure 4. The three reciprocal tension functions with graphs *a.*, *b.* and *c.*, represent consequently *stable*, *indifferent* and *unstable* aesthetic situation. The first, leftmost parabolic-like curve has one single minimum i.e., it generates *one-well dynamics* [14]. Therefore, in this minimum, the small imaginary ball will stay still in the stable equilibrium. The little ball occupies the position of the most beautiful object according to the aesthetic criteria, accepted by a local historic and social frame. However, at a certain moment, some aesthetic element, previously ignored, becomes gradually accepted. This causes changes in the average observer's criteria and now two opposite tensions replace the old leading element. It makes two maximums in the observer's aesthetic tension T or minima on the graph of $1/T$, like in the rightmost curve in the upper part of Fig. 4. This is a framework for *two-well dynamics* [14]. Then, our ball will go to one or another side and the bifurcation are born. Note that bifurcation takes place in the *Art-space*. The Art-space is a collection of pairs $(\boldsymbol{p}, \boldsymbol{q})$, where $\boldsymbol{p} = (p_1, p_2,\ldots, p_n)$ is the vector of aesthetic local parameters (proportion, rhythm, color) and $\boldsymbol{q} = (q_1, q_2,\ldots, q_m)$ is the vector of aesthetic values or criterions (beauty, simplicity, tragic). The dimensions of $\boldsymbol{p}$ and $\boldsymbol{q}$ depend on the definition of aesthetics applied, but both are certainly multi-dimensional.

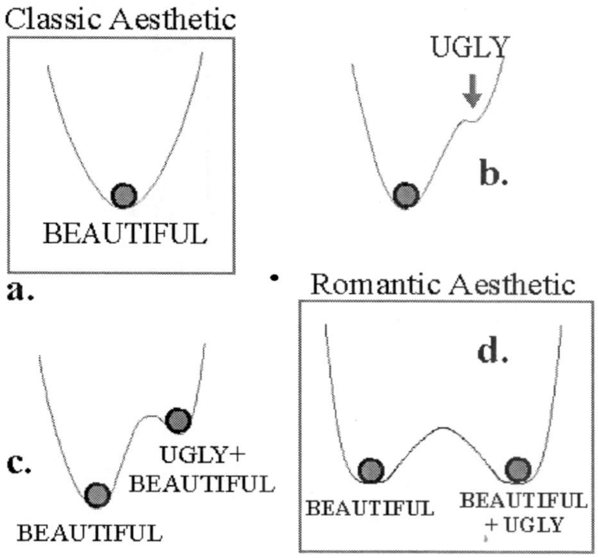

Figure 5. Unstability in classic aesthetics causes Romantic art.

Bifurcations considered in this paper are period-doubling (as it is said above) and these represent qualitative changes on attractors (collection of attracting points) caused by a (one-dimensional) local parameter p_i (bifurcations of co-dimension 1). The value of local parameter at which bifurcation occurs is known as bifurcation value $(p_i)_B$.

Example 3.2. Let the art period in which *beauty* was the main quality of classic aesthetics be call "Classicism". The graph of aesthetic tension reciprocal $1/T$ is shown in Figure 5*a*. At a certain historical moment, people became convinced of the ugliness being present in reality and promoted it as a new, more truthful quality. This quantity, in contrast with beauty may look even more attractive than pure beauty itself (Fig. 5 *b*. and *c*.). Therefore, the ugly things become increasingly popular, in addition to the beautiful ones and the new function $1/T$ acquired two local minima (two-well dynamics). In fact, the "Classic" aesthetics bifurcates into two branches (Figure 6), to some revisited "Classicism" that keeps beauty and "Romanticism" that accepts mixture of beauty and ugliness as a supreme aesthetic value.

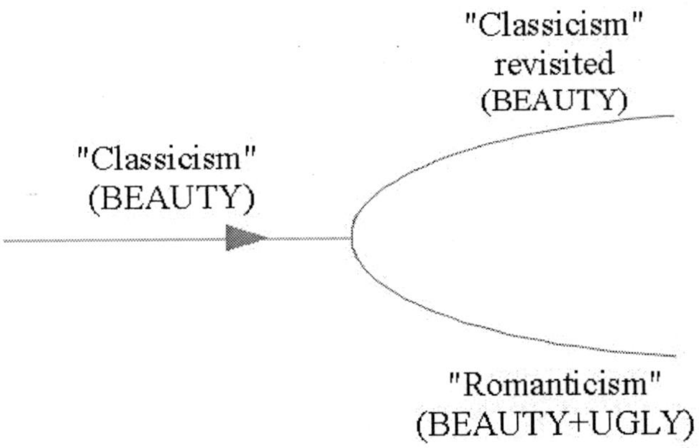

Figure 6. Bifurcation of Classicism.

It is important to note that the "horizontal axis" in Figure 6 does not necessarily coincide with the time-axis. It is rather homeomorphic to it. This means that the "certain historical moment" mentioned above does not need to be a single point on the abscissa-line. It may be a time-interval instead.

Why bi-furcations? Why not three-, four-, or multi-furcations? The answer seems to be closely connected to the architecture of the Ψ–space and the fractal-like archive of the individual. The development of every individual's Ψ–space is based on the answers on the whole series of antinomies: Good↔Bad, Light↔Dark, True↔False, Life↔Death, Active↔Passive and so on.

In modern art, bifurcations are even more frequently present. It is enough to see the taxonomic chart of evolution of modern art (made by Alfred Barr [18], [5]). In addition, these bifurcations become more complicated due to the influence of more than one aesthetic criterion (vector $\boldsymbol{q}$). Thus, sometimes there arises the illusion that one art movement splits into more than two branches. Nevertheless, really nothing but bifurcations occur although they may be in different dimensions, i.e. with respect to different variables (aesthetic criteria). Writing about fauvist painters, sir Herbert Read [17] says that it was a close parallel between contemporary developments of fauvism in Paris and Munich and continues "But...parallels have certain beginning point and never meet", which is a perfect description of bifurcation.

The next example belongs to the European painting scene from the end of the nineteenth and the beginning of the twentieth century.

Example 3.3. The movements of (French) *Impressionism* and *Post-Impressionism* were among the most influential in modern art. The theory and practice of Impressionism/Post-Impressionism contain seeds of many later movements. Further branching of impressionists' ideas results in many bifurcations in different aesthetic planes.

One and the most important bifurcation of Post-Impressionism seem to be to cubism and expressionism. This is shown in Figure 7. The post-impressionists: Cezanne, Gauguin and van Gogh preserved and further developed the main ideas of impressionism. Paul Cezanne, the great admirer of Manet, used to establish a peculiar way of definition of volumes and masses by using light and color. Thereby he had shown a new direction in understanding the relationship between the physical world and the world of painting. He strongly influenced young Braque and Picasso who further developed Cezanne's ideas by usage of simplified geometric forms in the legendary 1908-1909 winter when Cubism was born (group *du Beatu-Lavoir*). On the other hand, Gauguin and van Gogh's vision of impressionism moved in the opposite direction. They used color to underline the inner state of soul rather than to define volumes and physical forms. Fauvists (Matisse, Marquet, Derain, Vlaminck), who in fact were expressionists [17], elaborated this function of color. The movement of Expressionism was mainly spread out in Germany through the groups like *Die Brücke*, *Der Blaue Reiter*, *Die Neue Sachlichkeit* and *Bauhaus*.

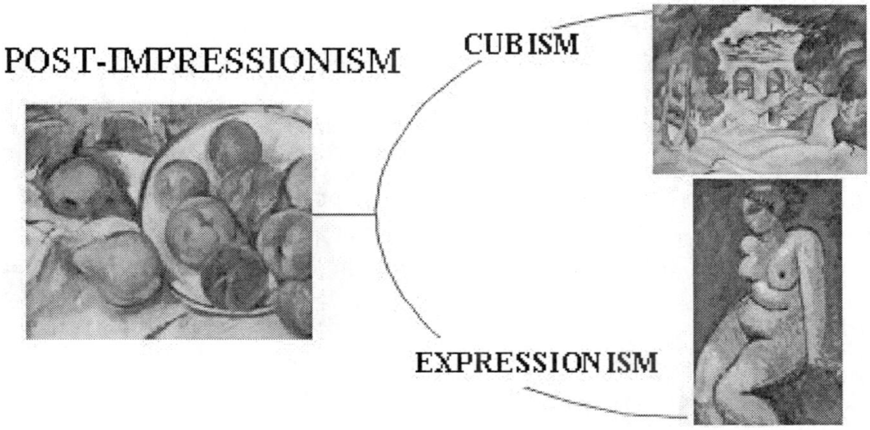

Figure 7. Bifurcation of the Post-Impressionism. Left: Paul Cézanne, *Table, Napkin, and Fruit (Un coin de table)*, 1895-1900; Right above: Georges Braque, *Le viaduct de L'Estaque*, 1907; Right below: Alexei von Jawlensky, *Seated Female Nude*, 1910.

Replacing of impressionistic aesthetic criteria by two opposite new directions: *formal* and *emotional*, causes bifurcation (Figure 7).

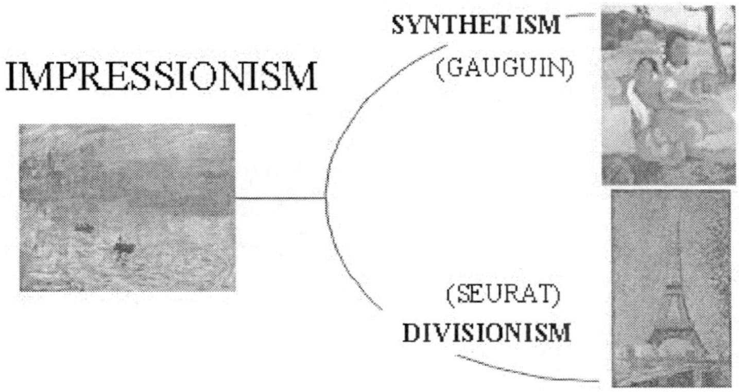

Figure 8. Another bifurcation of Impressionism. Left: Claude Monet, *Impression: soleil levant*, 1872; Right above: Paul Gauguin, *Nafea Fanipoipo? ("When Will You Marry?")*, 1892; Right below: Georges Seurat, *The Eiffel Tower*, 1889;

Another bifurcation, displayed in Figure 8 occurs in a different aesthetic plane. Again, Impressionism initializes two new movements heading two opposite directions: Synthetism and Divisionism.

Divisionism is the term used by Georges Seurat for his sophisticated scientific approach to painting (*Bathing at Asnières*, 1883, National Gallery, London). In contrast to the Impressionists, he uses logic rather then intuition in making his paintings. He used very small brush strokes right next to each other. When viewed from a distance, an observer's eye does integration of paint's particles into a homogenous color. This phenomenon is known as *spatial summation*. Divisionism is very similar to *Pointillism*, a form of painting in which tiny primary color dots are used to generate secondary colors. Among the relatively few artists following this style were Camille Pissarro, Paul Signac and Henri-Edmond Cross. The term, coined in 1886 by the art critic Félix Fénéon to describe this new offshoot of Impressionism was *Neoimpressionism*.

Paul Gauguin chose the opposite direction. About this, du Colombier and Muller in [4] say "In times when Neoimpressionists intended to open a new way in painting by offering opportunities to use scientific truths, another painter started looking for salvation in a totally opposite direction, in primitive, originating beauty." They are speaking about Gauguin, who was the main figure of a group of artists who worked in and around the town of Pont-Aven in Brittany. When Gauguin met in 1888 Émile Bernard, the Synthetistic style was established.

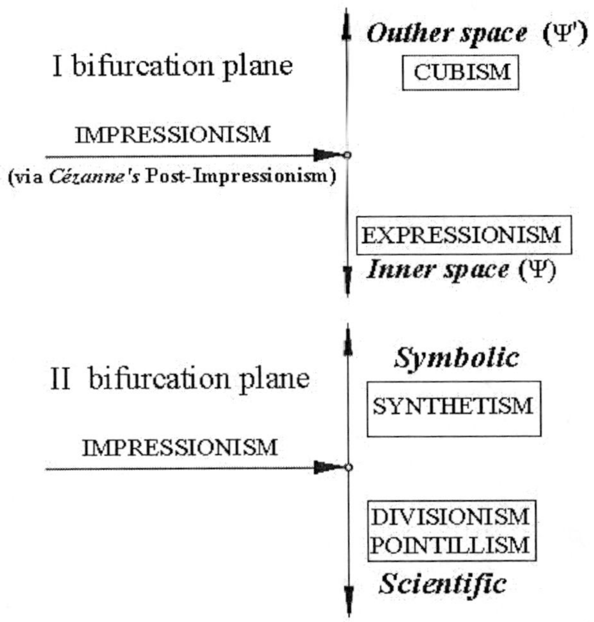

Figure 9. Two different bifurcation planes of impressionism.

Figure 9 summarizes these two independent bifurcations of Impressionism/Post-Impressionism from Figures 7 and 8. In Cubism-Expressionism bifurcation that emerges from Post-Impressionistic heritage of Impressionism, the aesthetic criteria embrace the difference between emotions from the Inner space (Ψ–space) and formality of Outher space (Ψ'–complement of Ψ–space). The second bifurcation is based on contrariety of Symbolic

(Synthetism) and Scientific (Divisionism) function of color. The irrational origin of Synthetism makes it evolve into a new variant of symbolism. In fact, Gauguin introduced pure color, stated the criteria of its use and established standards of its symbolic function [6].

Example 3.4. (Suzon's case) The evolution of a female portrait is considered from Impressionism and further on. The perception of light and function of color, summarized in the famous Suzon's portrait from the *Bar at the Folies Bergére* painting from 1882 by Edouard Manet, was masterly condensed in Post-Impressionistic Paul Cezanne's *Madame Cézanne in blue* (1886). Then, at the bifurcation point **A** in Figure 10 it bifurcates into Synthehism and Divisionism. Representatives of Synthetism in Fig. 10 are Émile Bernard with the *Woman and Haystacks, Brittany*, (c.1888) and Paul Gauguin with the *Ancestors of Tehmana* made in1893. Georges Seurat and Paul Signac represent Divisionists by their portraits in *The Models* (1887-8) *Women at the Well*, from 1892. Synthehism in turn, splits into Nabis (as a specific wing of Symbolism, point **B** in Fig. 10) and a part of Expressionism. The first movement, illustrated by Maurice Denis's portrait from his painting *The Muses in the Sacred Wood* (1893) further influenced the development of Surrealism and here is a characteristic portrait of Milena Pavlovic-Barilli (*Self-portrait with Brush* from 1936 [16]). Edvard Munch and his *Madonna* (1894-95), Franz Marc in *Girl with Cat II*, from 1912 and Emil Nolde's *St. Simeon and Women*, from 1915 represent the expressionistic branch. Divisionism bifurcates at point **C** into Fauvism and Futurism. The Fauvism is represented by Matisse *Green Stripe (Madame Matisse)* from 1905 and Maurice de Vlaminck with *The dancer in "Rat Mort"* from 1906. Fauvism further evolves into Cubism (Picasso's oil *Dora Maar Seated* from 1937). Giacomo Balla's Portrait of Benedetta, c.1924, illustrates the Futirist's style. The Futurist line combined with a specific mixture of expressionism, cubism and surrealism results in "hard-to-classify" style of Paul Klee (Winter's Dream, 1938).

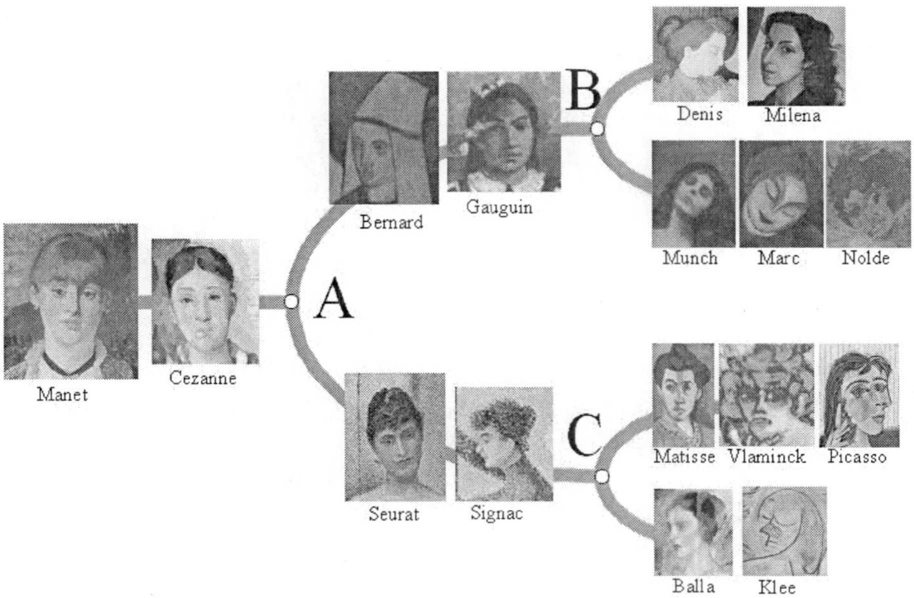

Figure 10. From Suzon to Dora Maar and Milena.

The opposite aesthetic forces that cause bifurcation of Impressionism to Synthetism and Divisionism are discussed in the previous example. What causes bifurcations at points **B** and **C** (Fig. 10) in this example? The point **B** is the node at which two opposite aesthetic valuations of colored fields are active: The upper one is symbolic, with the tendency towards magic and *onirique*. The lower one inclines towards deep emotions and inner dispositions. Bifurcation at point **C** also follows two choices, where the color is dominant over the form: The upper one has more figurative meaning, trying to define volumes with the language of colors whilst the lower one is more intuitive and instinctive.

What is to be stressed is that the examples of paintings in Figure 10 are illustrative. For ex. Paul Gauguin participates Nabis as well and Klee's opus does not completely lean on the Futurists' heritage, but was influenced by many other precursors. Therefore, these examples should be seen as just accents and guidelines of some dominant stream of the time.

One conclusion might be that the specific artistic styles or movements are, in fact, *attractors* in Art-space. They attract prosperous artists of the specific times. For ex. Mannerism, Baroque, Rococo, Neo-Classical, Romanticism etc. Individuals, staying out of the group rarely survive as artists. They are or attracted towards or repelled out of the group. Bunching-up the movements in modern times (more than hundred in twentieth century) clearly show that bifurcations are the main phenomena in art history.

4. Self-Similarity in Art Space

So far, three objects have been mentioned as highly complex: The Universe (all physical and social entities and relationships), the Individual space (Ψ–space) and the Art space (collection of all human activities and products connected with art). It has been shown that the Ψ–space is a specific "projection" of the Universe, and a part of it. On the other hand, this space is a main creator of human actions that lead to embodied artifacts. In this way, the Art space bears the seal of the complexity of the Ψ–space and therefore the complexity of the Universe. It is familiar that artistic attempts of a subject can tell to an expert much about his subconscious mind, even about its eventual psychical irregularities or diseases. Accordingly, all these three spaces share a similar degree of complexity as well as the characteristic features of complexity: self-similarity and hierarchy, which is expected.

Both can be found both in the Art space. The evidence of self-similarity at different levels is more than striking. There is coincidence between development of different art movements and styles in different periods. For instance, there were two opposite tendencies in antique paintings: one favors surface another three-dimensional paintings. This tendency awakes in early renaissance and then many times after, especially in the twentieth century art. An example of similar bifurcations is given in Figures 11 and 12. The first shows splitting of the so called *Byzantine style* [8] (or *maniera greca*) of the late thirteenth century, represented by the famous Cimabue (Cenni di Pepo, c.1240-c.1302) on Giotto and Duccio schools. Linearity, the main characteristic of the Byzantine style, bifurcated in a voluminous manner by the Florentine school, led by Giotto di Bondone (c.1267-1337) and the more decorative "colored surfaces interplay" cultivated by the Siena school, with Duccio de Buoninsegna (c. 1255-1319). The first one favored three-dimensional sculpturality in order to gain spatial illusion of reality. The latter uses flat and geometrically simplified forms to accentuate predomination of spiritual values over material ones (see Figure 11).

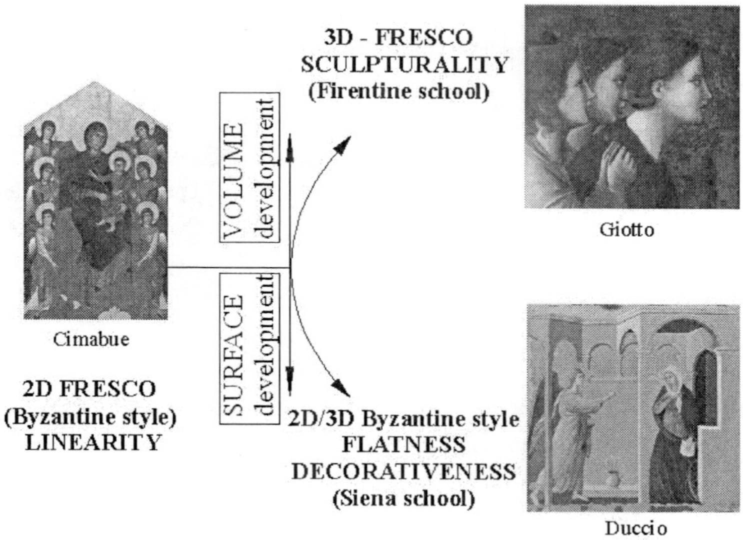

Figure 11. Bifurcation of fresco painting at the end of 13. and beginning of 14. century.

Similar bifurcation occurred in the beginning of XX century when some elements of Cubism developed in two opposite directions similar as 700 years before (Figure 12). One, known as *Purism*, was launched 1918 with a book *Après le Cubisme*, by Amédée Ozenfant and Charles Edouard Jeanneret (Le Corbusier). They espoused for clarity of forms and objectivity by restoring the representational nature of art based on precision and mathematical order. The contemporary "machine aesthetics" used by Fernand Léger, and the 3D picture of the world influenced them. Le Corbusier even rejected ornamentation in architecture. Instead, they liked forms like cubes, cones, spheres, cylinders or pyramids, which are great primary forms whose light reveals an advantage.

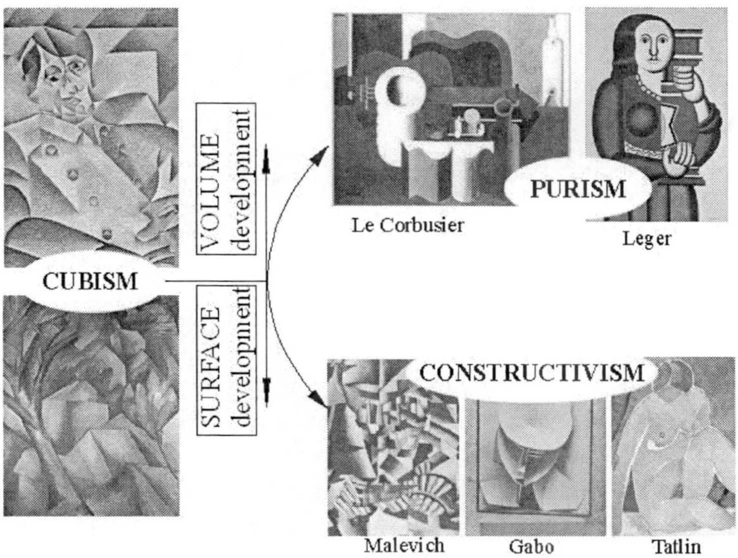

Figure 12. Bifurcation at the beginning of the 20-th. century.

The other prong of the bifurcation fork produced *Constructivism*. Although it was mainly a movement in sculptural art and architecture, founded in about 1913 by the Russian artist Vladimir Tatlin, later joined by Antoine Pevsner and Naum Gabo, Constructivism was, in fact, a deep, scientific study of certain abstract properties of picture such as surface, construction, lines and colors. In paintings, Constructivism uses surface and geometric elements as well as collages (Rodchenko, Lissitsky), or crossing of reflected rays from various objects as in *Rayonism* of Mikhail Larionov and Natalia Goncharova. Thus, it may be a profound parallel to the Siena school.

There are also other similarities. The development of an individual artist often resembles the history of art itself. There are many nice examples among the so-called "modern" artists. The typical development of a twentieth century painter includes: Realism, Impressionism, Fauvism, Cubism, etc. Illustrate is the career of the Dutch painter, Piet Mondriaan (1872-1944). Some of his key works are shown in Figure 13 (see Appendix).

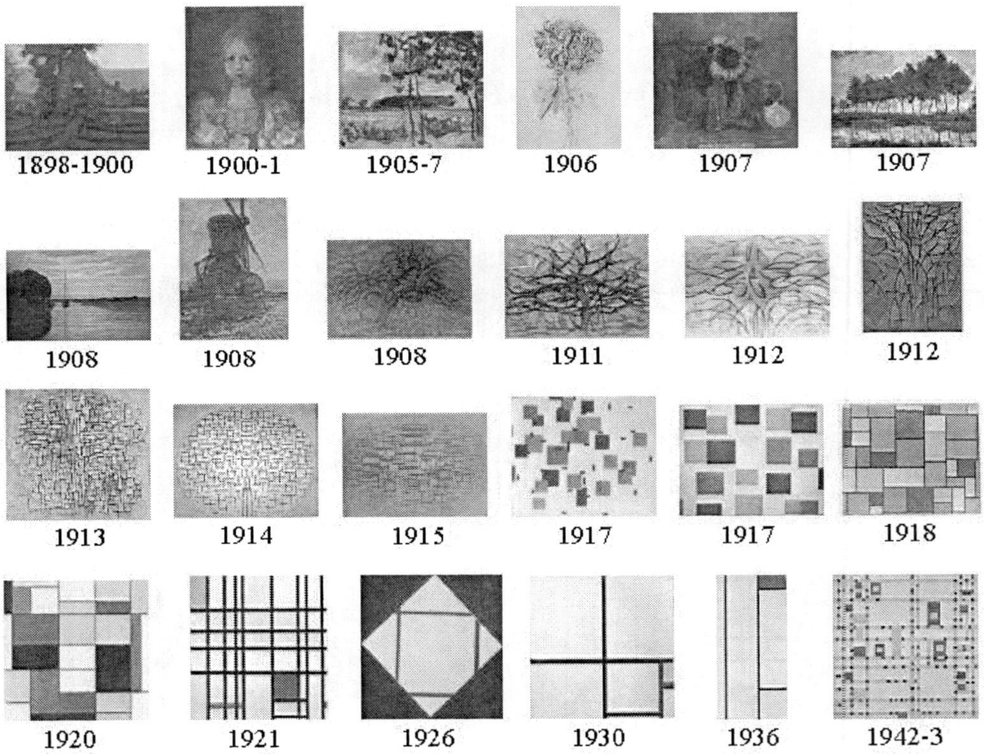

Figure 13. Piet Mondriaan's developing line: From realism to neo-plasticism.

From his early drawing and painting experiments up to about 1908, he had experimented within realistic and naturalistic manner to all scales up to Impressionism. From 1908 to 1910, young Mondriaan accepted the symbolist style after which he was under the influence of Pointillism and Fauvism. Then, about 1911, he began to work in a cubist mode. The magic of Braque and Picasso attracted Mondriaan to move to Paris (end of 1911). Here, he undertook a profound and systematic study of analytic cubism. In 1914 he moved back to Holland and in 1916 he joined the new artists alliance *De Stijl* (The Style) founded by Theo van Doesburg. In 1917 he experimented with more clean geometric elements. The link he had missed was found, and from 1918 on he gained his own, independent abstract style, that he called *Nieuve Bleeding* (Neo-plasticism). His paintings became subtle and harmonic compositions made out of vertical and horizontal lines, rectangles and squares. From 1920, he started reducing his palette to a very few colors and minimized the number of geometric elements in his paintings. After moving to New York 1940, a new, dynamic element was added to his compositions as the artist's response to the dynamism of the big city.

The inductive way of thinking has its antipode in deduction. These opposites reflect on the modern science and technology development, as well as arts, making bifurcations the all the time. Two similar bifurcations are present in Figure 14. The first one refers to Cubism. Being the outcome of intellectualized rather than impressionistic vision, the first wave of cubistic efforts (known as Proto-cubism or Facet Cubism, 1907-1909) split in two opposite sub-styles, Analytic and Synthetic Cubism. Following the inductive thinking patterns, artists analyze the cubic and fractured forms into predominantly geometrical structures with overlapping planes making a shallow relief-like space of some depersonalized pictorial style referred to as Analytic Cubism (1910-1912). The inductively accumulated experience of the Analytic Cubism led to its deductive counterpart – Synthetic Cubism (1913-1915). This is a more decorative phase in which objects were constructed (or synthesized) from flat fragments in brighter colors or ornamental patterns. What is important and worthy of stressing again is that the horizontal axis on the bifurcation diagram is not the time-axis. In fact, if considered in the time domain, the Analytic Cubism is a precursor to the Synthetic one. However, here the element of stylistic advancement is the only relevant variable, and it increases in the usual sense of x-axis.

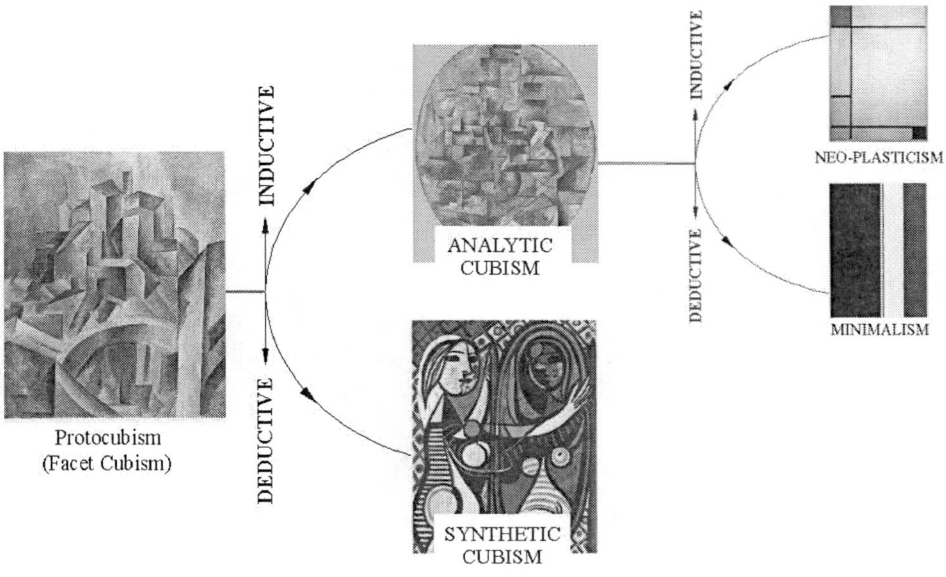

Figure 14. Two similar bifurcations in modern art (see Appendix).

The second bifurcation that Figure 14 shows is similar to the first one and performs in a smaller scale. It illustrates one of the outcome influences of the Analytic Cubism. The depersonalized mode of it attracted some artists that were looking for the essence behind the form. They believed they found the essence in the magic interplay of sharply defined and regular geometry. In addition, the inductive-deductive bi-pole split this stream in two offspring again. The first one represents the geometric abstraction of Mondriaan that relates to the Analytic Cubism in the similar way the Analytic Cubism relates to the Facet Cubism. As it is said above, Mondriaan used the Analytic Cubism to dissolve natural forms profoundly enough to gain (possible influenced by Kandinsky) the *De Stijl* ideal: a non-objective pictorial language that describes changeableness of nature by plastic expression of certain, pure geometric relations. On the other hand, Analytic Cubism, through the influence of Kazimir Malevich Suprematism paved the way for a new level of synthesis. After Malevich's *Black Cross* from 1915, artists have more refined and more impersonal elements for their anti-expressionistic movements. Similarly, as one uses elements of the Analytic Cubism for making a Synthetic one, the combinations of Supremacist elements make a new geometric abstraction. This new style uses a highly reduced geometric/color arsenal to produce pure self-referential compositions, emptied of all external references. Centered in New York in 1960s, under different names such as Minimalism, ABC art, Primary Structure art etc, it gathered sculptors as well as painters, such as Frank Stella, Ellsworth Kelly, Barnet Newman, and so on.

Figure 15. Linearity in ancient and modern art: Egyptian wall painting (left) and *Succession*, the oil by Kandinsky from 1935.

Finally, one should note that there are "repetitions" and influences of old to new art. The influence of Japanese "estampes" and African masks on Postimpressionism and Cubism are famous. In Figure 15 one sees two similar solutions in two artworks that are distanced almost three millennia.

5. Hierarchy of Complexity

In the end, it may be interesting to point out that Art space possesses a real structure of a high complexity that has chaotic elements too. It is enough to have an insight into its hierarchy. Namely, very complex structures have all levels of complexity: Constancy, Linearity, Periodicity, Complexity and Chaos. In all the periods of art, there are such wholes. Some have briefly been selected from the art history.

<u>*Constancy*</u>. It is, maybe, the most spread-out element present in all periods of visual (and other) arts as well. The primal artistic tendency is to isolate some aesthetic absolutes and

constants that should reflect the eternity of Beauty. The earliest expression of the sensation of such an absolute is symmetry (Figure 16, left). Another important human answer is module and proportion. Module (the unit of measure) and proportion is the essence of all art works, architectural, sculptural, musical or literary. Let us recall some of the most important proportion systems in visual arts:

(1) The Golden mean system, based on the *golden number* $\phi = (1+\sqrt{5})/2 \approx 1.618$, connected with the regular pentagon. Incorporated in ancient Greek temples, such as Parthenon (Fig. 16, middle), where the module was 30,86 cm long [15], and used by Le Corbusier in creating his *Modulor*;

(2) The Roman system, based on the number $\theta = 1+\sqrt{2} \approx 2.4142$, known as *holly section*. It is connected with the regular octagon;

(3) The Paladio system, based on the number $\psi = 1+\sqrt{3} \approx 2.732$, connected with the regular dodecagon;

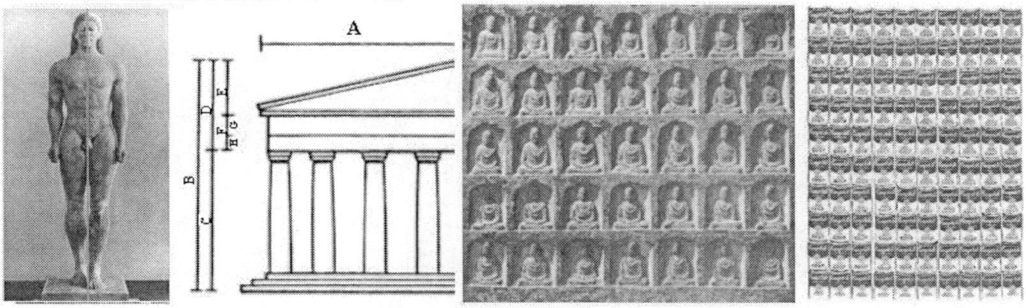

Figure 16. Constancy expresses: symmetry (Kriosos, 525 BC), proportion (Parthenon, 437-438 BC), important universal principles (stela with 96 Buddha's reincarnations, China VI century) or impersonal status as in *Campbells Soup* by Andy Warhol.

Except symmetry and proportion, calm, still or massive buildings, temples, sculpture or relief represent the feeling of intransitive and immanent. To stress stable and unchanging nature of God or laws of the Universe, ancient artists often repeat the figure or the picture of some still, well balanced object or icon. Multiplications of Buddha, Fig. 16 (right) extend the law of reincarnation to cosmic relations. According to Theology, the only real constant in the Universe is God, and contribution to His glory was the (almost) only objective of the mediaeval artists.

Linearity. The next more complicated form is linearity. It is always present in logical thinking as a mean of approximation to the real Universe. Therefore, linearity brings just a partial and local truth. In visual arts, linearity has multiple functions. The still ornaments can become "alive" by introducing the sense of movement, like in the stone relief, Figure 17(left). Movement may also be described by separating the line of movement from the usual vertical-horizontal framework like in the fresco by Giotto (Fig. 17, right).

Figure 17. Horizontal or diagonal movement by linearity: Left. *Masons at work*. Stone relief, Kajuraho, North India, X century; Right. Giotto di Bondone, *Legend of St. Francis –8. Vision of the Flaming Chariot*, 1297-9, fresco (detail).

The linear gesture is used by Etruscan art in metaphorical function – to show the magnitude of a god or a goddess (Figure 18a). The linear perspective that was introduced in the thirteenth and fourteenth centuries uses linearity as a geometric mean used to produce illusion of spatial depth (Fig. 18b). Modern art reveals decorative function (Fig. 18c) and expressive strength of linear forms (Fig. 18d)

Figure 18. Linearity in different contexts: a. *Aphrodite* (?), Etruscan art; b. *The Annunciation* by Fra Angelico, (1437); c. Charles Mackintosh, design of chair from 1904; d. Design by Hans Hartung.

<u>*Periodicity.*</u> The next dynamic form beyond linearity is the periodic one. The essence of periodic movement is the circle. It is the simplest geometric figure that, decomposed into two orthogonal spaces gives sine waves.

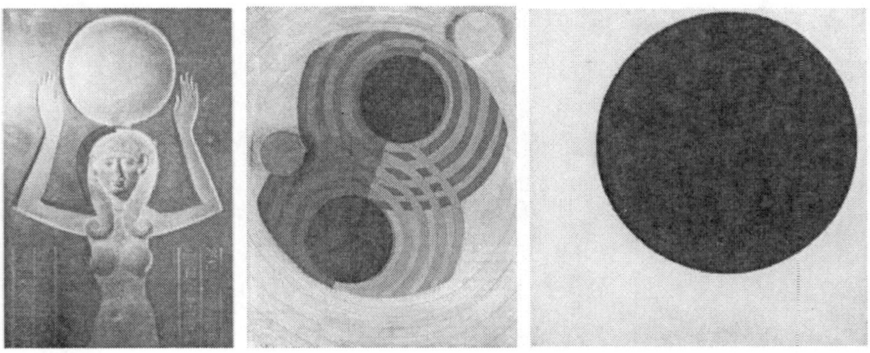

Figure 19. Left. *Goddess Nut with solar disc*, Egyptian art; Middle. Robert Delaunay, *Relief; Rhythms*, 1932; Right. Kazimir Malevich, *Black circle*, 1913.

The circle was present in all periods of art. It is connected with harmony, perfection and whole. It means rhythm and visual "music" (Figure 19). Rhythmic repetition was a special tool for painter's compositions in all times. In landscapes, periodic repetition of trees' foliages has its psychological role from the mystic contents as in the Bosh painting, Figure 20 (left), via metaphysical "stimmung" in De Chirico's "piazzas", Figure 20 (middle), to cosmic harmony and universal rhythm (Fig. 20, right).

Figure 20. Left. H. Bosh, *Adoration of the Magi* (Detail 1500); Middle. G. de Chirico, *Tower*, 1913; Right. Mario Sironi, *Plasticity and Rhythm of Things*, 1914.

Complexity. When periodicity becomes too complicated, and periods start multiplying, one faces the problem of complex dynamics. The Universe is crowded with such dynamics in all ranges of complexity that is known as "controlled chaos". The visual art records such phenomena using its specific language of forms and colors. Some examples are given in Figure 21. Leonardo da Vinci [20], may be inspired by

Figure 21. Left. *Arrival of the Sun*, Eskimo art; Middle. Albrecht Dürer, *The Great Piece of Turf*, 1503; Right. Vincent van Gogh, *Cypresses*, 1889

Pietro del Cosimo, emphasized the power of "messy forms" like stains on old walls, clouds or muddy water in "favoring mind on various discoveries". Thereby he introduced the "blotting method" later extensively used by many artists, including the modern ones [10]. This method helped the artists to invent forms that are more complex then the usual ones.

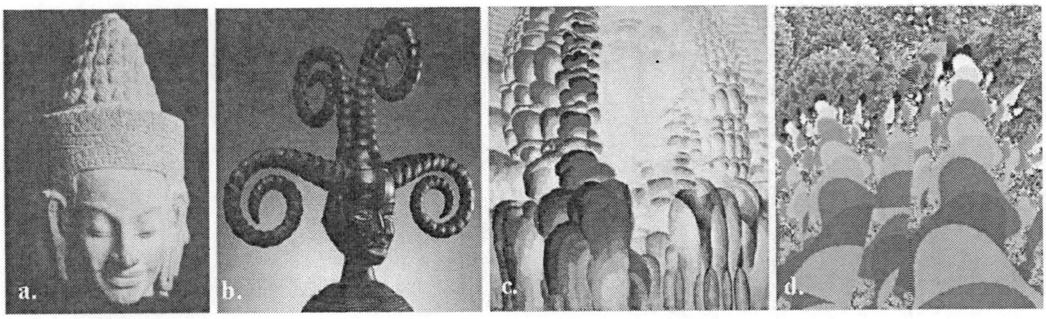

Figure 22. a. *Buddha's head*, Cambodia, XII cent.; b. *Ekoi-Ejagham heads*, Nigeria-Cameroon; c. Frank Kupka, *Hindu Motif*, 1919-1923; d. Computer generated fractal.

Chaos. It is a long way down to chaos. Complex patterns may become increasingly complicated, and degrees of complexity may continue to upgrade. In this way, one comes to fractal forms. The intuition of some artists or some civilizations may be so brilliant to be capable to imagine such forms without any mathematical knowledge, just following its own intuition. Figure 22 offers some illustrations. The hat on Buddha's head (Fig. 22a) bears relief that possesses self-similar geometry of broccoli-like fractal object. Very similar are the braids in the African hairdressing [7] (Fig. 22b). Note that the compositions of Frank Kupka (Fig. 22, c.) or Max Weber (Fig. 23, left) resemble to the low-resolution computer generated fractals (Fig. 22, d. and Fig. 23, right). For some other similarities between fractals and art see [10].

Figure 23. Left. Max Weber, *Interior of the Fourth Dimension*, 1913; Right. Computer generated fractal "Barnsley-m1".

The fantastic stones in the pictures of David Caspar, the snowstorm and the waves in the Hokusai Kakemono pictures are examples of chaotic textures. One can find them in de Kooning's expressionistic figures or abstract forms of Mark Tobey and Jackson Pollock.

According the authors' knowledge, probably the first "fractal" in visual arts was the painting of Salvador Dali (Figure 24). It shows a hallucinating vision of skulls nested inside skulls using "Russian doll" geometry. A slight analysis reveals that the fractal set that corresponds to Dali's work is so called Cantor dust. It is generated by three contractions with approximate contractive factor about 0.21. Using well-known formula [1] the box dimension (in our case it is identical to Hausdorff - Besicovitch dimension) of Visage of War is about 0.705.

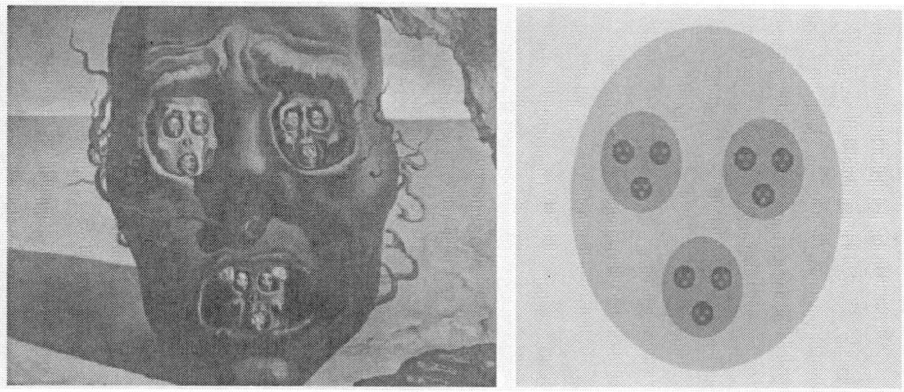

Figure 24. Left. Dali, *Visage of War*, 1940, oil n canvas; right: Cantor dust fractal set with Hausdorff-Besicovitch dimension approximately 0.705.

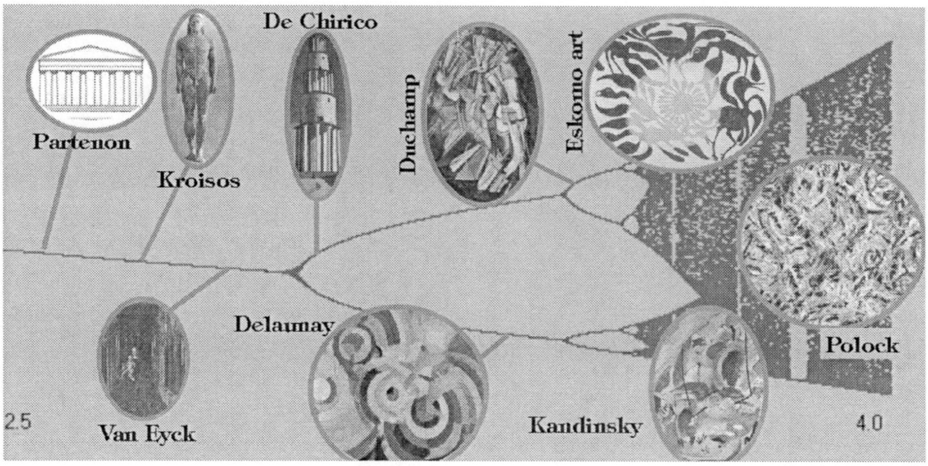

Figure 25. Time-line of art-history as bifurcation diagram.

An attempt of summarizing what is said above is a simplified bifurcation diagram, as Figure 25 shows.

The simpler and more geometric, monumental art belong roughly to dawn of artistic culture. As time goes by, art become increasingly complicated both in formal and iconographic plane while at twentieth century it burst out to a very complex, almost chaotic organization. This route to chaos is known as period doubling or Figenbaum scenario.

6. Conclusion

Art is the human response to the enigma of the Universe. The huge complexity of the Universe is reflected in the human minds. Some very "compressed" and fractal-like psychical contents conserve selected information (mostly in pictorial form) and influence artists' creations. In this way, the products of art, during the mankind history, also resembles on a highly complex corpus that is characterized by the features of any other complex almost-fractal object, like presence of bifurcations, self-similarity or hierarchy of complexity. The aim of this paper is to point that the period-doubling bifurcation dynamics that take place in art-history model its flow, embodying Figenbaum route to chaos. It is much work ahead of us. Can we more exactly describe topology of pre-attractors or even attractors of art-history dynamics? What is with other types of local bifurcations of co-dimension one such as tangential (intermittency), trans-critical or Hopf quasi-periodic bifurcations? What is peculiarity in other arts? Does the real chaos possible? These are only a few among many unsolved questions for future investigations.

Acknowledgment

The authors would like to express their gratitude to professor Nebojša Vilić (Art History) from Skopje for his valuable suggestions.

Appendix

Artworks in Figure 12

Cubism: Juan Gris, *Portrait of Picasso*, 1912, oil, Collection of Mrs. and Mrs. Leigh; Georges Braque, *Maisons a l'Estaque*, 1908;

Purism:Le Corbusier, *Nature morte a la pile d'asiettes*; Fernan Leger, *Woman Holding a Vase*, 1927;

Constructivism: Kasimir Malevich, The Scissors Whetting,1920; Naum Gabo, *Female head*, 1917-20; Vladimir Tatlin, *Female Model*, c. 1910.

Artworks in Figure 13

Solitary House, c. 1898-1900, Haags Gemeentemuseum, The Hague; ***Little Girl***, 1900-01, Haags Gemeentemuseum, The Hague; ***Bos Oele***, 1905-7, The Cleveland Museum of Art; ***Chrysanthemum***, 1906, watercolor and pencil; ***Still Life with Sunflower***, 1907, Detroit Institute of Art; ***Trees along the Gein***, 1907; ***River View with Boat***, c.1908, Rijksmuseum, Amsterdam; ***Molen (Mill)***, 1908; ***Avond (Evening)***; ***Red Tree***,1908, Haags Gemeentemuseum; ***Gray tree***,1911, Haags Gemeentemuseum, The Hague; **Flowering Apple Tree**, 1912; **Trees**, c. 1912, Carnegie Museum of Art, Pennsylvania; ***Tableau No. 2-Composition No. VII***, 1913, Solomon R. Guggenheim Museum; ***Pier and* Ocean**, 1914; ***Ocean 5***, 1915, Charcoal and gouache on paper, Peggy Guggenheim Coll, Venice; ***Composition with planes of color***, 1917; ***Composition in Blue***, 1917; ***Composition with Color Planes and Gray Lines 1***, 1918; ***Composition A***, 1920, Galleria Nazionale d'Arte Moderna e Contemporanea, Rome; ***Composition with Red, Yellow and Blue***, 1921; ***Painting I***, 1926; ***Composition with Yellow***, 1930, Kunstsammlung Nordrhein-Westfalen, Duesseldorf; ***Vertical Composition with Blue and White***, 1936, Dusseldorf; ***Broadway Boogie-Woogie***, 1942-43, Museum of Modern Art, New York;

Artworks in Figure 14

(1) Pablo Picasso: *Reservoir at Horta*, 1909;
(2) Georges Braque: *Candlestick and Playing Cards on a Table*, 1910;
(3) Pablo Picasso: Jeune Fille Devant un Miroir, 1932;
(4) Piet Mondriaan: *Composition with red, yellow and blue II*, 1927;
(5) Barnett Newman: *The word II*, 1954;

References

[1] Barnsley, M. F., *Fractals Everywhere,* Academic Press, 1988.

[2] Barnsley, M. F., *Lecture Notes on Iterated Function Systems*, in: Proc. Symposia in Applied Math., Vol. 39 (R. L. Devaney and L. Keen, Eds.), AMS 1989, pp. 127- 144.

[3] Collingwood, R., George, *Speculum Mentis (The map of knowledge)*, Clarendon Press, Oxford 1924.

[4] Colombier, Pierre du, Muller, Joseph-Emile, in: *Histoire de la peinture*, Fernand Hazan Editeur, Paris 1970.

[5] Cox, Neil:, *Cubism*, Phaidon Press, Ltd, London 2000.

[6] Chaseé, Charles, *Le mouvement symboliste dans l'art du XIX siècle*, H. Floury, Paris 1947.

[7] Eglash, Ron, *African Fractals*, electronic: http://www.rpi.edu/~eglash/eglash.dir/afractal.htm

[8] Janson, H. W., *History of Art*, Abrams, New York 1969.

[9] Kandinsky, Wassily, *Ueber das Geistige in der Kunst*, R. Piper&Co., Verlag München 1912.

[10] Kocić, Lj., *Art elements in fractal constructions,* Visual Math. 4 (2002), no.1, electronic: http://turing.mi.sanu.ac.yu/vismath/ljkocic/index.html or http://members.tripod.com/vismath9/ljkocic/index.html.

[11] Lalo, Charles, *Notions D'Esthetique*, Presses Univ. de France, Paris 1952.

[12] Mandelbrot B., *The Fractal Geometry of Nature*, W. H. Freeman, San Francisco 1982.

[13] Meyer, Y., *Wavelets, algorithms and applications,* SIAM, Philadelphia 1993.

[14] Moon, F., *Chaotic and Fractal Dynamics,* Willey & Sons, Inc., 1992.

[15] Petrović, Đorđe: *Komposition of Architectonic Forms*, Naučna Knjiga, Beograd 1972.

[16] Protić Miodrag B., *Katalog Galerije Milene Pavlović Barilli*, Požarevac (Serbia), 1962.

[17] Read, Herbert, *A Concise History of Modern Painting*, The World of Art Series, Thames and Hudson 1985.

[18] Shearer, R., R., *From Flatland to Fractaland: New Geometries in Relationship to Artistic and Scientific Revolutions*, Fractals, **3**(1995), 617- 625.

[19] Schroeder, Manfred, *Fractals, Chaos, Power Laws*, W. H. Freeman and Co., New York, 1991.

[20] da Vinci, Leonardo, *Treatise on Painting*, 1651.

In: Chaos and Complexity Research Compendium
Editors: F. Orsucci and N. Sala, pp. 195-198
ISBN: 978-1-60456-787-8

Chapter 14

THE MYTH OF THE TOWER OF BABYLON AS A SYMBOL OF CREATIVE CHAOS

Jacques Vicari
University of Geneva (Switzerland)

Drawing according to Cornelisz Antonisz

"We can consider something as a symbol when its linguistic expression lends itself to a task of interpretation because of its double or its multiple meanings"..

Paul Ricoeur, Essay on Freud, 1965, p.19,

1) For forty centuries at least, the «chaos» resulting from the multiplicity of languages has been considered as a necessary brake to the excessiveness of men, the just punishment of their arrogance or a disastrous divine vengeance on mankind.

And what if, with the passing of time, this interpretation should be reversed ?

2) Why did God make the city of men fail by confounding their words, driving them to disperse themselves over the whole Earth ? The reason of this intervention is not explicit. In a Sumerian version, much earlier than the narration of the Bible, the Enki god, who had provoked the Flood, had already intervened to modify the destiny of humanity. Was this for the benefit of man ? The strength of the narration resides precisely in this question.

3) The poem engraved on a square clay tablet (23 x 23 centimeters), well preserved in a museum of Istanbul could contain the beginning of a first answer. On this limited space, the scribe printed a poem of six hundred verses in Sumerian that S. N. Kramer [1943] calls *Enmerkar and the Lord of Aratta.* Here is, in substance, what lines 145 to 155 say :

4) " The whole world, everywhere where it is inhabited speaks to Enlil in a unique language.

That day Enki is at the same time lord, noble and king,

Enki the lord who gives abundance, (whose) words are worthy of confidence, the wise lord who controls the country is the chief of gods, strong in his wisdom, the lord

Changes the speech of men's mouths and (implants) discordin their tongue, which had been one ".

5) We learn here that the confusion of languages is the work of Enki, the wise chief of the gods. During the Flood, another god, Enlil, had tried - with success - to reduce the consequences of the act by allowing one couple to survive. This time Enlil does not intervene. We see that the disappearance of the unique language will be an opportunity for mankind to develop not yet expressed potentialities. Confusion is going to generate history. The chaos will be creative.

6) Further in the narration - in lines 501- 504 - we read that king Enmerkar addresses a message to the Lord of Aratta to tell him that he will be his suzerain. From what is written in line 525, we know that the Lord of Aratta understands the messenger. We must notice that the author of the clay tablet does not speak of any difficulty of comprehension. If the languages had multiplied, how could he be understood ? The answer is simple : Enmerkar invented writing. Writing allowed him to avoid the obstacle created by Enki, to render communication possible and to leave traces.

7) Everyone can understand that a message expressed in a certain language can be transcribed in pictographic characters which can in turn be read in another language, preserving the original meaning. This is still the case today in China : a message in Pekinese, written in ideograms (some of them thousands of years old), can be read in Cantonese without any difficulty, even if the Pekinese and the Cantonese do not speak the same language ! A modern example of the same principle could be the international pictographic signs used in train stations, airports and other places visited by tourists.

8) Today, the narration of Enmerkar and the Lord of Aratta is not only the first known expression of language confusion but again, the first mention of a written communication. We can infer that this innovation had been wished by the gods, presented as wise and benevolent. Confusion will be more than beneficial : it will generate complexity !

9) At the present time, linguists count more than five thousand living languages and just as many dead ones (if not twice as many). Why such a proliferation ? Writing appears to be the first positive consequence of the *confusio babylonica*. But the price to pay will be incalculable : who can estimate the social costs of incomprehension, the economic costs which burden exchanges, the material costs of translations ? And in an evolutionary perspective, who can justify the advantages of this form of biodiversity? It is indeed a «raw challenge», as the philologist and linguist George Steiner [1992, p.13-14] writes. A challenge that he puts forth as follows :

10) « 'After Babel ' argues that it is the constructive powers of language to conceptualize the world which have been crucial to man's survival in the face of ineluctable biologic constraints, this is to say in the face of death. It is the miraculous - I do not retract the word - capacity of grammars to generate counter - factual 'if' propositions and above all, future tenses which have empowered our species to hope, to reach far beyond the extinction of the individual. We endure, we endure creatively due to our imperative ability to say 'No' to reality, to build fictions of alteration, of an dreamt or willed or advanced 'otherness' for our consciousness to inhabit. It is in this precise sense that the utopian and the messianic are figures of syntax».

11) Since Steiner wrote these lines, humanity has been facing another challenge: the never ending increase of knowledge. How can we have access to the multiple scientific works published in so many languages around the world ? José-Luis Borgès [1957] illustrated it in the *Library of Babel* : all books that were, are or will be written, in all idioms, have a place in it. But, at the same time, no one can have a coherent vision of this library, in spite of its logical structure, based on repetition. Because it expands endlessly. For Borgès, the Library of Babel represents the impossible quest of meaning in an expanding universe.

12) A half - century later, we can note that the challenge described by Borgès has been taken up. All books, pictures and sounds can be written in numeric language. Billions of people exchange - instantaneously - millions of items of information written in binary language. They inaugurate a new era of sign circulation which is accompanied by a new form of the scattering of mankind over the Earth : globalization. Let us especially remark that it was necessary that the human population - by multiplication, as well as by increased longevity - be sufficiently dense to expand itself over the whole surface of a well-defined world. As long as this degree of saturation had not been reached, interconnection could not take place.

13) But we already note that the interconnection of all computers facilitates the emergence of new world powers, just as - centuries ago - Persians and Romans were able to project their power from afar thanks to their road networks. When men will prevail over their scattering and the diversity of their languages by adopting a unique code, they will then

perceive that this globalizing unification enslaves them more that it benefits them. Chaos was the final situation of the Babylonian narrative. When men will have returned to the initial situation of a unique language, they will realize that the *confusio babylonica* is a necessary condition for the survival of mankind, as Steiner demonstrated.

14) A few thousand years after Sumer and a few billion more of individuals, we can note that the myth of the Tower of Babylon needs to be analyzed with a new perspective. Today, its meaning can be understood in a different way. Long ago, the incompleteness of the Tower was considered as a constraint, an obstacle to human realizations. Today, this incompleteness can be seen as an ordinary and permanent and necessary condition for a creative process. The myth of the Tower of Babylon shows us its new function, which is to visualize a synthesis and a unity that are pushed to the end of History. It also becomes a tool to understand and anticipate the future of mankind.

Translated from French by François Vicari and Atalia Johnson

Steiner G., "Preface to the 2nd Edition p.xiii-xiv" (July 1991), *After Babel – Aspects of language & translation*, Oxford University Press, 1992, pp.13-14

In: Chaos and Complexity Research Compendium
Editors: F. Orsucci and N. Sala, pp. 199-206
ISBN: 978-1-60456-787-8

Chapter 15

CHAOS AND COMPLEXITY IN ARTS AND ARCHITECTURE

Nicoletta Sala
Accademia di Architettura, Università della Svizzera Italiana
Mendrisio, Switzerland

> "Where chaos begins, classical science stops. For as long as the world has had physicists inquiring into the laws of nature, it has suffered a special ignorance about disorder in the atmosphere, in the fluctuations of the wildlife populations, in the oscillations of the heart and the brain. The irregular side of nature, the discontinuous and erratic side - these have been puzzles to science, or worse, monstrosities."
>
> **James Gleick in *Chaos: Making A New Science***

The dictionary defines the word chaos as "*A condition or place of great disorder or confusion*." It derives by the Greek word *Chàos* that represented the formless and disordered state of matter before the creation of the cosmos. This aspect has been analysed by many Greek philosophers (Orsucci, 2002).

Anaxagoras of Clazomenae (500-428 B.C.) postulated a plurality of independent elements which he called "seeds" (*spermata*) or miniatures of corn and flesh and gold in the primitive mixture; but these parts, of like nature with their wholes (the *omoiomere* of Aristotle (384-322 B.C.) had to be eliminated from the complex mass before they could receive a definite name and character).

This chaotic mixture was controlled by a "mind" or "reason" (*Nous*). The *Nous* set up a vortex in this mixture. They were not, however, the "four roots" conceived by Empedocles of Syracuse (492-430 B.C.), fire, air, earth, and water; on the contrary, these were compounds.

Chaos is then the antithesis of order and it is formally defined as the study of complex non-linear dynamic systems. Complex: a multitude of variables and equations within equations. Non-linear: the equation cannot be solved like your program code. Dynamic: ever-changing, depending upon perspective.

Complexity can occur in natural and man-made systems, as well as in social structures and human beings. Complex dynamical systems may be very large or very small, and in some complex systems, large and small components live co-operatively. A complex system is neither completely deterministic nor completely random and it exhibits both characteristics.

The meteorologist Edward Lorenz discovered the sensitive dependence on initial condition in accidental way, also known as "Butterfly effect", during an atmospheric simulation with a computer in the early 1960 (Lorenz, 1963; Lorenz, 1993).

The complexity is the most difficult area of chaos, and it describes the complex motion and the dynamics of sensitive systems. The chaos reveals a hidden fractal order underlying all seemingly chaotic events. The complexity can occur in natural and man-made systems, as well as in meteorological systems, human beings and social structures.

Chaos theory is closely connected to the fractal geometry, in fact it describes the shapes generated by the complex phenomena. Figure 1 shows a vision from satellite of the Dasht-e Kavir desert (Iran), it is easy to confuse it with a modern painting (from "Ma questa è arte?", 2004).

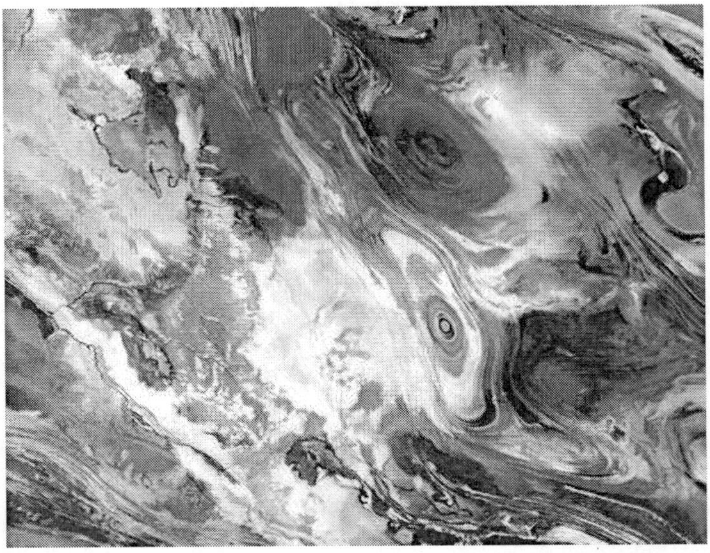

Figure 1. Is it an image from satellite or a modern painting?

Complex dynamical systems may be very small or very large, and in some complex systems small and large components exist in co-operative way. The complexity can also be called the "edge of chaos", it is connected to the fractal geometry, and it can also inspire an aesthetic sense. In fact, in the 1930's the mathematician George Birkhoff (1884-1944) proposed a measure of beauty defined as:

$$M = \frac{O}{C} \tag{1}$$

whereby M stands for "aesthetic measure" (or beauty), O for order and C for complexity. This measure suggests the idea that beauty has something to do with order and complexity.

Modern theory of the complexity is involving different disciplines, for example: Arts and Architecture. To understand this phenomenon we can remember that the Swiss architect Mario Botta affirms: "*The nature should be a part of architecture and the architecture should be a part of the nature, the two terms are complementary. The architecture describes the human's project, the space organisation of life and therefore it is an act of reason, of thought, of work. For this reason it is always "dialog" and comparison with the nature*" [1].

To emphasize the Botta's point of view we can see the figure 2a that shows the genesis of the Corinthian capital[2] described by the Roman architect Marcus Vitruvius Pollio (c. 90-20 B. C.). This is an example of the influence of the nature in the art and in the architecture (Portoghesi, 2000; Sala and Cappellato 2004; Sala, 2004b). The figure 2b illustrates a marble Roman Corinthian capital, embellished with acanthus leaves, (1st century A.D., Rome). Vitruvius is the author of *De Architectura* (probably written between 23 and 27 B.C.) known today as *The Ten Books of Architecture*, a treatise in Latin on architecture, and perhaps the first work about this discipline. His work is divided into 10 books dealing with city planning and architecture in general; building materials; temple construction; public buildings; and private buildings; clocks, hydraulics; and civil and military engines. Vitruvius' book has influenced the Renaissance architecture.

a)

b)

Figure 2. The genesis of the Corinthian capital a) and a Roman Corinthian capital b).

1 *"La natura deve essere parte dell'architettura così come l'architettura deve essere parte della natura; i due termini sono reciprocamente complementari. L'architettura descrive il progetto dell'uomo, l'organizzazione dello spazio di vita e quindi è un atto di ragione, di pensiero, di lavoro. Proprio per questo è sempre "dialogo" e confronto con la natura"* (Sala and Cappellato, 2003, p. 12).

2 This kind of capital has been used originally by the Greeks in a system of supports called the Corinthian order. The Corinthian capital was developed further in Roman times and used often in the medieval period, again, without strict adherence to the rest of the system. The Corinthian capital is more ornate than the Ionic. It is decorated with 3 superimposed rows of carved foliage (acanthus leaves) around the capital. At the comers of the capital there are small volutes. The Corinthian capital is essentially the same from all sides. Adaptations of the Corinthian capital are common in the Middle Ages.

a)

b)

Figure 3. Portoghesi's *Hotel Savoia* (Rimini, Italy) a) that emphasizes the analogy with the chaotic movement of the waves b).

Modern architecture has found inspiration observing the nature and its chaotic and fractal shapes. Figure 3a shows a Portoghesi's building and its analogy with the chaotic movement of the waves (figure 3b) (Portoghesi, 2000). An approach to building design which attempts to view architecture in bits and pieces is the Deconstructivism[3], or Deconstruction. Deconstructivism ideas are borrowed from the French philosopher Jacques Derrida. The basic elements of architecture are dismantled. Deconstructivist buildings may seem to have no visual logic and they can represent the fluxes. They may appear to be made up of unrelated, no-Euclidean shapes abstract forms. Decostructivist architect Zaha Hadid affirms: "*The most*

[3] Deconstruction is certainly not simply a reversal of the process of construction, be it in architectural (physical) or linguistic (conceptual) terms. Derrida himself sustained that Deconstructive architectural thought is impossible, maintaining that "*Deconstruction is not an architectural metaphor*", (Derrida, 1989, p.69).

important thing is motion, the flux of things, a non-Euclidean geometry in which nothing repeats itself: a new order of space". Figure 4 illustrates a project for the *Musée de la Confluences* (2001-2007, Lyon, France) realised by the Austrian architectural group named Coop Himmelb(l)au (Wolf Dieter Prix and Helmut Swiczinsky) that represents a building with complex shapes that passes the limits of the representation imposed by the Euclidean geometry (Zugmann, 2002).

Figure 4. *Musée de la Confluences* (2001-2007, Lyon France), Coop Himmelb(l)au.

The cities are most complex structures created by human societies, and they are currently undergoing profound and rapid changes which influence the quality of life for millions of people. We have to control these changes to preserve or enhance the quality of life, and to ensure environmental sustainability. The morphology and the evolution of the cities can be studied applying fractal geometry and complexity theory now. In the past, three classic theories of urban morphology have been used: the concentric zone theory - urban pattern as concentric rings of different land use types with a central business district in the middle (Burgess, 1925), the sector theory - concentric zone pattern modified by transportation networks (Hoyt, 1939), and the multiple nuclei theory - patchy urban pattern formed by multiple centers of specialized land use activities. These theories have point out their limits. In recent years, some researchers have studied urban form and land use development from a different perspective (Batty and Xie 1994; Batty and Longley, 1994; Lau, 2002; Portugali, 2000; Schweitzer, 1997; Semboloni, 1997; White and Engelen 1993; Wu 1996). These approaches have considered the city as fractal objects or self-regulating, self-organising, and self-evolving living entity composed of numerous tiny cells.

Some of these researches involve the cellular automata (CA), the essence of CA is that local actions lead to complex global behaviours (Batty 2000). Cellular Automata is a spatial modelling technique used to simulate spatial dynamics and the dynamic urban systems, because the urban development is a process of local interactions rather than a global activity (Batty and Xie 1994, White and Engelen, 1993; Wu, 1996; Yeh and Li, 2002). CA works on the principle of self-organisation and continual adaptation.

The art can be interpreted as a way for finding the basics of beauty and harmony that are found in the laws of Nature (Briggs, 1992; Briggs, 1993). Thus the chaos and fractal geometry may help to explain and prove the "rules" of beauty (Sala, 2004a). Next figures 5a and 5b represent respectively a satellite vision of the Kalahari Desert (Namibia) and a Pollock's painting. The analogy is amazing (from "Ma questa è arte?", 2004).

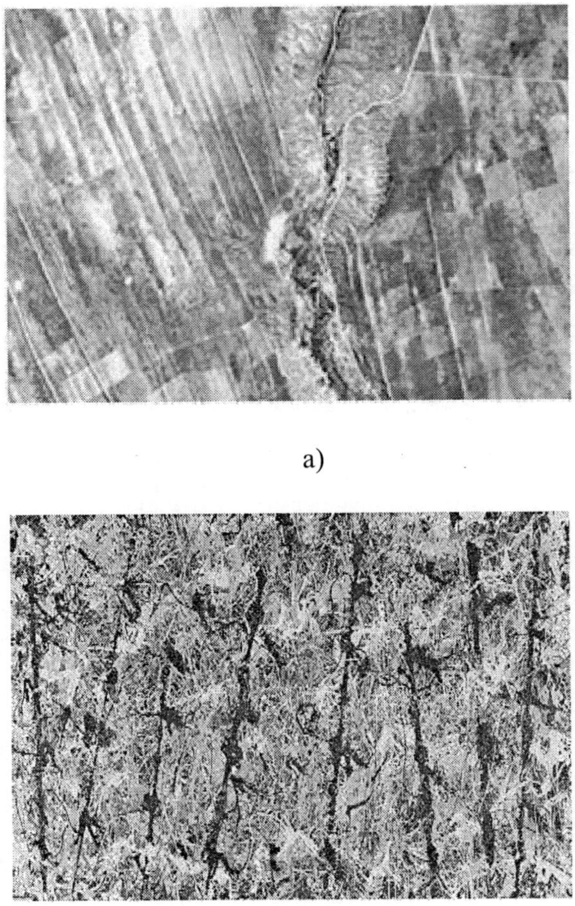

a)

b)

Figure 5. Satellite vision of the Kalahari Desert (Namibia) a) and Pollock's painting b).

The aim of this special issue is to present some connections between chaos, complexity, arts and architecture. Different experts in different fields describe their point of views that involve: Mathematics, Architecture, Arts, Information Technology and Urbanism.

Jay Kappraff presents some interesting aspects connected to the "Complexity and Chaos Theory in Art".

Richard Taylor introduces, in the paper entitled: "Pollock, Mondrian and Nature: Recent Scientific Investigations", his fractal analyses concerning the Pollock's and Mondrian's artistic productions.

Igor Yevin describes the "Visual and Semantic Ambiguity In Art".

Attilio Taverna, italian painter and author of the cover of this issue, shows his artistic point of view on the complexity in the work entitled: "Does The Complexity Of Space Lie In The Cosmos Or In Chaos?"

Manuel Antonio Báez introduces the morphology of the amorphous in his work: "Crystal & Flame: Form & Process, The Morphology of the Amorphous"

Gerardo Burkle-Elizondo, Ricardo Valdez-Cepeda and Nicoletta Sala present their research in the "Complexity In The Mesoamerican Artistic And Architectural Works".

Nikos A. Salingaros shows his point of view on "New Paradigm Architecture".

Ferdinando Semboloni introduces the "Self-organized criticality in urban spatial development".

Xavier Marsault presents a fractal approach on pseudo-urban model based on Iterated Function Systems in his work entitled: "Generation of textures and geometric pseudo-urban models with the aid of IFS".

Renato Saleri Lunazzi describes a "Pseudo-urban automatic pattern generation".

Vladimir E. Bondarenko and Igor Yevin introduce the "Tonal Structure of Music and Controlling Chaos in the Brain".

In the section "Metaphors" Deborah L. MacPherson presents the "Collecting Patterns That Work For Everything".

All papers presented in this issue emphasize that the chaotic shapes, the fractal geometry and the techniques based on soft-computing (for example, Cellular Automata) can help to create a new paradigm in arts and architecture that passes the limits of the Euclidean geometry and it reflects the nature's organisation.

Nicoletta Sala
Co-Editor
International Journal
Dynamical Systems
Chaos and Complexity Letters
Guest Editor of this Special Issue
Chaos and Complexity in Arts and Architecture
Accademia di Architettura
Università della Svizzera Italiana
Mendrisio, Switzerland
E-Mail: nsala@arch.unisi.ch

References

Batty, M. and Longley, P.A. (1994). *Fractal Cities: A Geometry of Form and Function*, London: Academic Press.

Batty, M. and Xie, Y. (1994). From cells to cities. *Environment and Planning B*, **21**(Celebration Issue), pp. 531-548.

Batty, M. (2000). GeoComputation using cellular automata, in S. Openshaw and R.J. Abrahart (eds), *GeoComputation*, London: Taylor & Francis, pp. 95-126.

Briggs, J. (1992). *Fractals The Patterns of Chaos*. London: Thames & Hudson.

Brigg, J. (1993). *Estetica del caos*. Como: Red Edizioni.

Burgess, E.W. (1925). The growth of the city: an introduction to a research project. In: Park R.E., Burgess E.W. and McKenzie R. (eds), *The City*, Chicago: University of Chicago Press, pp. 47–62.

Derrida, J. (1989). Fifty-Two Aphorisms for a Foreword. Papadakis A. (ed) (1989). *Deconstruction;* Omnibus Volume, London: Academy Editions, p. 69.

Gleick, J. (1988). *Chaos: Making A New Science*, New York: Penguin USA

Hoyt, H. (1939). *The Structure and Growth of Residential Neighborhoods in American Cities.* Federal Housing Administration, Washington, DC, USA.

Lau, K.H. (2002). A cellular automation model for urban land use simulation. Available: *http://www.btre.gov.au/docs/atrf_02/papers/54Lau%20Cellular.pdf*

Lorenz, E. (1963). Deterministic nonperiodic flow. *Journal of Atmospheric Science,* **20**, 130-141.

Lorenz, E. (1993). *The Essence of Chaos*. Seattle: WA: University Press.

Ma questa è arte? (2004). *Focus*, n. **136**, pp.148-153.

Orsucci, F. (2002). *Changing Mind.* New Jersey: World Scientific.

Portoghesi, P. (2000). Nature and Architecture. Milano: Skira.

Portugali, J. (2000). *Self-Organization and the City*. Berlin: Springer.

Sala, N., and Cappellato, G. (2003). *Viaggio matematico nell'arte e nell'architettura*. Milano: Franco Angeli (in Italian).

Sala, N., and Cappellato, G. (2004). *Architetture della complessità*. Milano: Franco Angeli (in Italian).

Sala, N. (2004a) Fractal Geometry in the Arts: An Overview Across The Different Cultures. Novak M.M. (Ed.) *THINKING IN PATTERNS Fractals and Related Phenomena in Nature*, Singapore: World Scientific, pp. 177-188.

Sala, N. (2004b). Complexity in Architecture: A Small Scale Analysis. *Design and Nature II: Comparing Design in Nature with Science and Engineering*. WIT Press, pp. 35-44.

Schweitzer, F. (ed.) (1997). *Self-Organization of Complex Structures*, Amsterdam: Gordon and Breach.

Semboloni, F. (1997). An urban and regional model based on cellular automata. *Environment and Planning B*, **24**(2), pp. 589-612.

White, R.W. and Engelen, G. (1993). Cellular automata and fractal urban form: a cellular modelling approach to the evolution of land use patterns. *Environment and Planning A*, **25**(8), pp. 1175-1199.

Wu, F. (1996). A linguistic cellular automata simulation approach for sustainable land development in a fast growing region. *Computers, Environment and Urban Systems*, **20**(6), pp. 367-387.

Yeh, A.G.O. and Li, X. (2002) A cellular automata model to simulate development density for urban planning. *Environment and Planning B*, **29**(3), pp.431-450.

Zugmann, G. (2002). *BLUE UNIVERSE Architectural Manifestos by COOP HIMMELB(L)AU*. Ostfildern-Ruit: Hatje Cantz.

In: Chaos and Complexity Research Compendium
Editors: F. Orsucci and N. Sala, pp. 207-228
ISBN: 978-1-60456-787-8

Chapter 16

COMPLEXITY AND CHAOS THEORY IN ART

Jay Kappraff *

New Jersey Institute of Technology, Newark, NJ 07102

Kauffman and Varela propose the following experiment: Sprinkle sand or place a thin layer of glycerine over the surface of a metal plate; draw a violin bow carefully along the plate boundary. The sand particles or glycerine will toss about in a rapid dance, swarming and forming a characteristic pattern on the plate surface. This pattern is at once both form and process: individual grains of sand or swirls of glycerine play continually in and out, while the general shape is maintained dynamically in response to the bowing vibration.

Hans Jenny in his book *Cymatics* [1] has noted from this experiment:

> "Since the various aspects of these phenomena are due to vibration, we are confronted with a spectrum which reveals patterned figurate formations at one pole and kinetic-dynamic processes at the other, the whole being generated and sustained by its essential periodicity. These aspects, however, are not separate entities but are derived from the vibrational phenomenon in which they appear in their unitariness."

These are poetic ideas, metaphoric notions, and yet they have reflections in all fields from the wave/particle duality of quantum physics, to oscillations within the nervous system to the oscillations and distinctions that we make at every moment of our lives. Complexity and self-organization emerge from disorder the result of a simple process. This process also gives rise to exquisite patterns shown in Figure 1.

* E-mail address: kappraff@aol.com

Figure 1. a) Pattern formed by the vibration of sand on a metal plate.

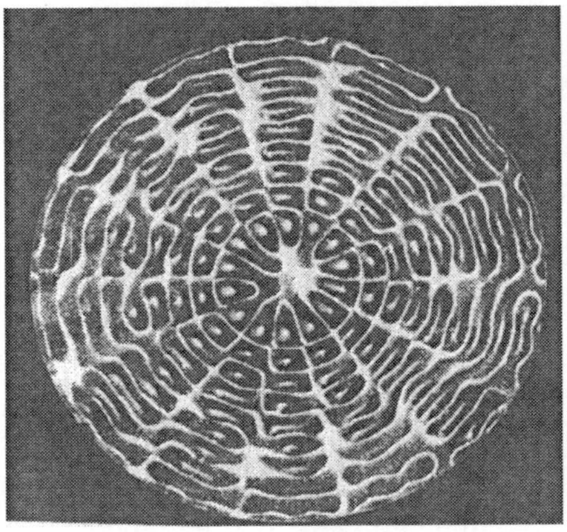

Figure 1. b) Vibration of a thin film of glycerine. From *Cymatics* by Hans Jenny.

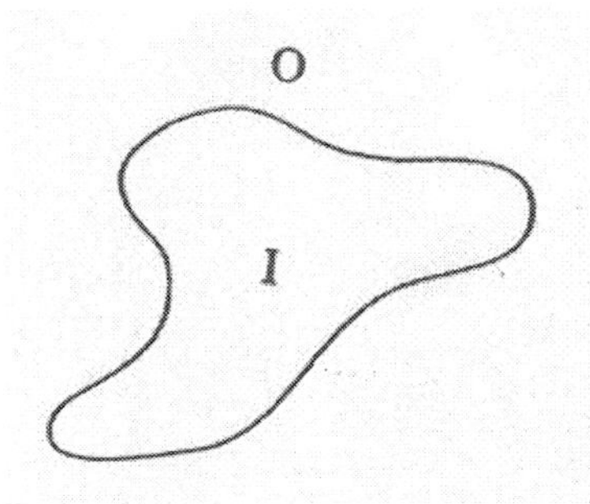

Figure 2. A mark of distinction separating inside from outside.

G. Spencer Brown in his book *Laws of Form* [2] has created a symbolic language that expresses these ideas and is sensitive to them. Kauffman [3] has extended Spencer-Brown's language to exhibit how a rich world of periodicities, waveforms and interference phenomena is inherent in the simple act of distinction, the making of a mark on a sheet of paper so as to distinguish between self and non-self or in and out (see Figure 2). There is nothing new about this idea since our number system with all of its complexity is in fact derived from the empty set. We conceptualize the empty set by framing nothing and then throwing away the frame. The frame is the mark of distinction.

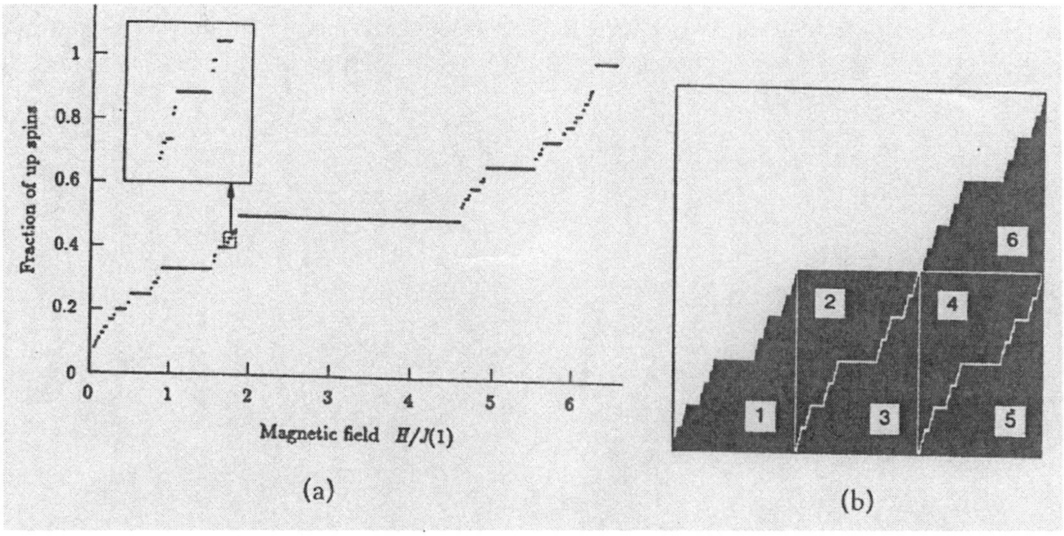

Figure 3. a) The devil's staircase exhibited in the Ising model from Physics; b) The devil's staircase subdivided into six self-similar parts.

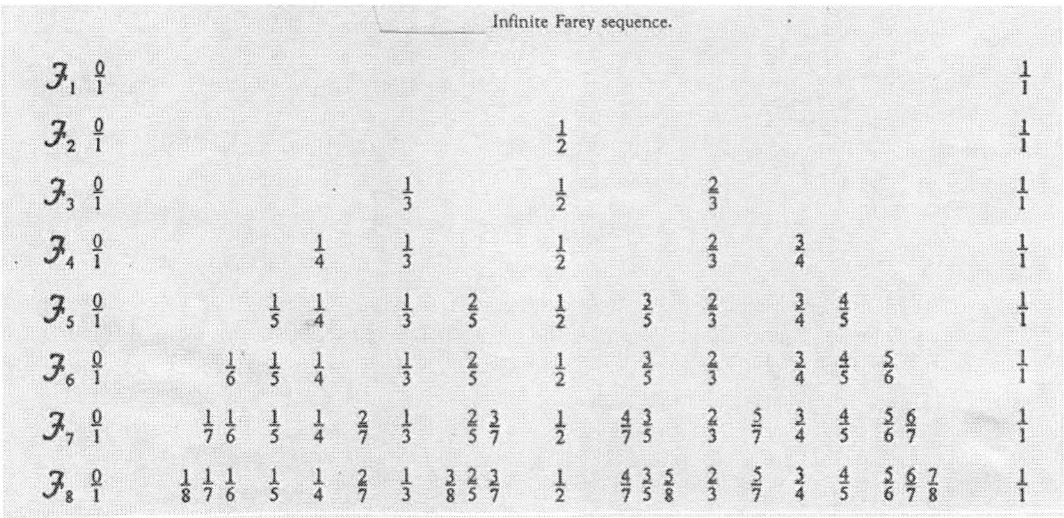
Infinite Farey sequence.

$\mathcal{F}_1$	0/1 1/1
$\mathcal{F}_2$	0/1 1/2 1/1
$\mathcal{F}_3$	0/1 1/3 1/2 2/3 1/1
$\mathcal{F}_4$	0/1 1/4 1/3 1/2 2/3 3/4 1/1
$\mathcal{F}_5$	0/1 1/5 1/4 1/3 2/5 1/2 3/5 2/3 3/4 4/5 1/1
$\mathcal{F}_6$	0/1 1/6 1/5 1/4 1/3 2/5 1/2 3/5 2/3 3/4 4/5 5/6 1/1
$\mathcal{F}_7$	0/1 1/7 1/6 1/5 1/4 2/7 1/3 2/5 3/7 1/2 4/7 3/5 2/3 5/7 3/4 4/5 5/6 6/7 1/1
$\mathcal{F}_8$	0/1 1/8 1/7 1/6 1/5 1/4 2/7 1/3 3/8 2/5 3/7 1/2 4/7 3/5 5/8 2/3 5/7 3/4 4/5 5/6 6/7 7/8 1/1

Figure 4. The first eight rows of the Farey sequence.

I have found that number when viewed properly reveals self-organization in the natural world from subatomic to cosmic scales. The so-called "devil's staircase" shown in Figure 3 places number in the proper framework and reveals a hierarchy of rational numbers in which rationals with smaller denominators have wider plateaus and lead to more stable resonances. The devil's staircase is a representation of the limiting row of the Farey sequence the first eight rows of which is shown in Figure 4. The n-th row is simply a list of all rational fractions with denominators n or less. Notice that row 8 on the interval from 0 to ½ contains all of the critical points on the Mandelbrot set, important for describing chaos theory, where the rationals are fractions of a circle when the Mandelbrot set is mapped from a circle (see Figure 5). On the other hand the interval from ½ to 1 contains many of the tones of the Just musical scale shown on the tone circle in Figure 6, including the tritone (5/7) and the diminished musical seventh (4/7) used in the music of Brahms. Only missing are the dissonant intervals of the semitone and the wholetone [4].

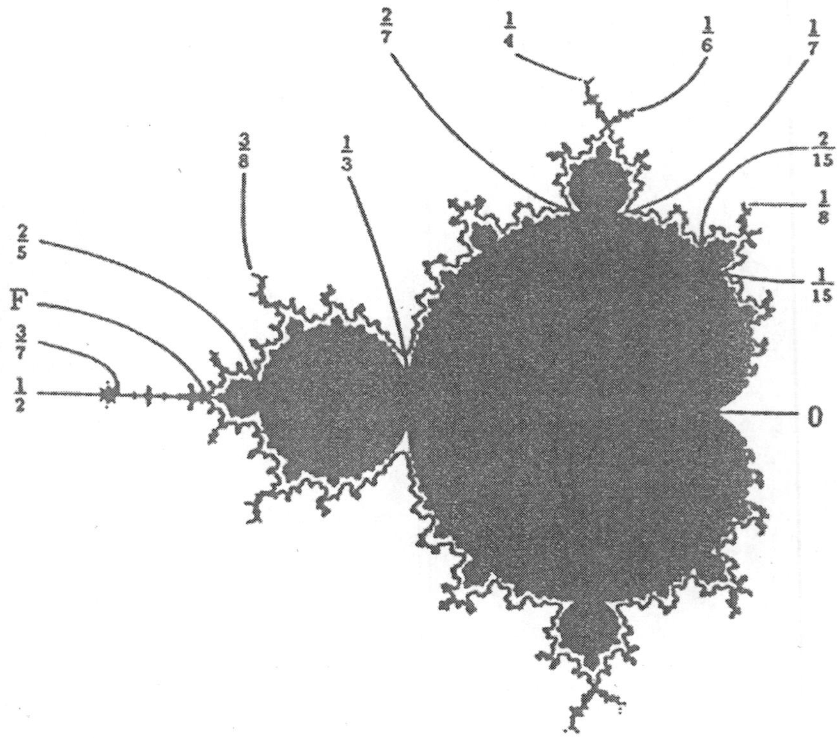

Figure 5. The Mandelbrot set showing critical values of the external angles at fractions from row eight of the Farey Sequence. The fractions determine the period lengths of the iterates z_n for a given choice of the parameter c. The point "F" (Feigenbaum limit marks the accumulation point of the period-doubling cascade. A. Douday: Julia sets and the Mandelbrot set

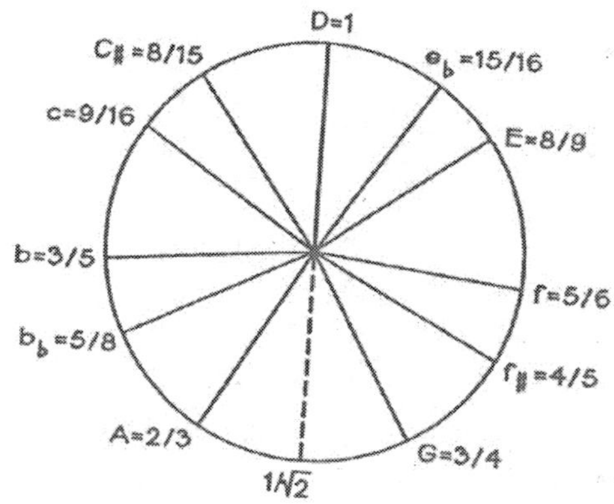

Figure 6. The Just scale shown on a tone circle. Note the symmetry of rising (clockwise) and falling (counterclockwise) scales.

In Figure 7 the number of asteroids in the asteroid belt is plotted against distance from the sun in units of Jupiter's orbital period Notice that sequence of gaps in the belt are at the rational numbers: 1/3, 2/5, 3/7, ½, 3/5, 2/3, ¾ and that these are consecutive entries to rows 6 and 7 in the Farey sequence. I have found (not shown here) that this same Farey sequence also expresses the hierarchy of phyllotaxis numbers that dictate the growth of plants from pinecones to sunflowers [4].

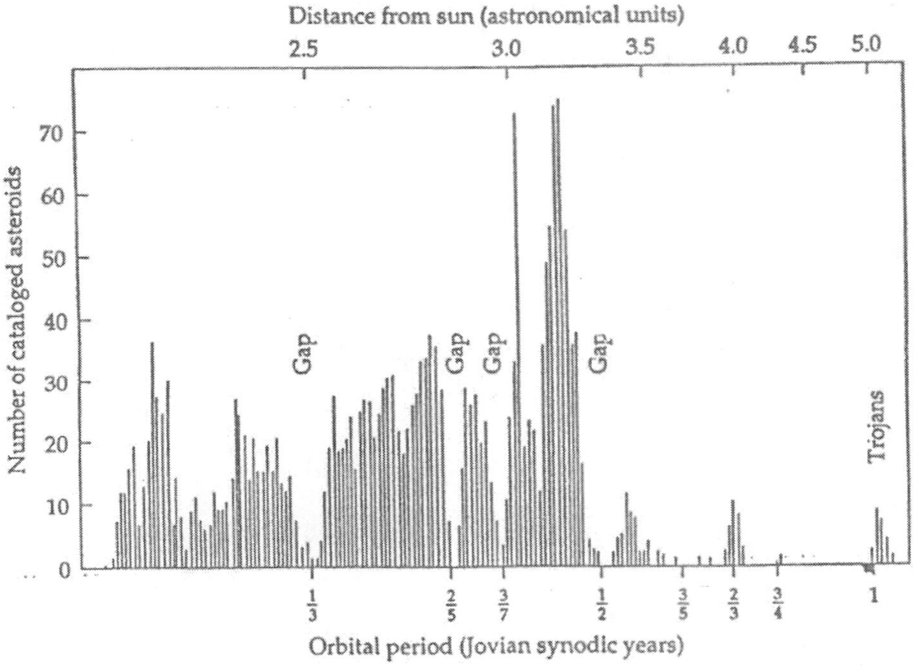

Figure 7. Number of asteroids plotted against distance from the sun (in units of Jupiter's orbital period). Gaps occur at successive points in the Farey sequence. From *Newton's Clock* by I. Peterson Copyright © 1992 by I. Peterson.

We see here that without a telescope or without a living bud or the sound of a musical instrument, our very number system already contains the objects of our observations of the natural world and is capable of reproducing phenomena in all of its complexity. How did this come to pass. Are we observing an objective reality or are we projecting our own organs of perception onto the world? These are deep questions for philosophical study.

From the earliest times humans have tried to make sense of their observations of the natural world even though they often experienced the world as chaotic. Their very existence depended on reliable predictions of such events as the arrival of spring to plant, fall to harvest, the coming and going of the tides, etc. The movement of the heavenly bodies provided the first experiences of regularity in the universe and the application of number to describe these motions may have constituted the earliest development of mathematics. In ancient times astronomy and music were tied together. The earliest cultures were aural by nature and music played an important role as confirmed by the many musical instruments found in burial sites of ancient Sumerians from the third and fourth millennia B.C. There is evidence that the Sumerians were aware of the twelve tone musical scale in which tones were represented by the ratio of integers or rational numbers placed on a tone circle with 12 sectors similar to the positions of the planets in the zodiac [5]. In the East the pentatonic scale of five tones chosen from the twelve was prevalent corresponding to the five observed planets. In the West seven tones was the norm since the sun and moon were added to the planets.

Expressing the musical scale in terms of rational numbers has certain problems associated with it. It was well understood that a bowed length of string has a higher pitch when it is shortened. For example, if a string representing the fundamental tone is divided in half it gives an identically sounding pitch referred to as an octave. Also the inverse of the string length gives the relative frequency, so that the octave has a frequency twice the fundamental. The key interval of the musical scale is the musical fifth gotten by taking a length of string whose tone represents the fundamental tone say D and reducing it to 2/3 or its length. A succession of twelve musical fifths placed into a single octave gives rise to the twelve tone chromatic scale known as "spiral fifths" as shown in Figure 8. Its serpent like appearance leads the ethnomusicologist, Ernest McClain to suggest that this scale lies at the basis of the many serpent myths in all cultures.

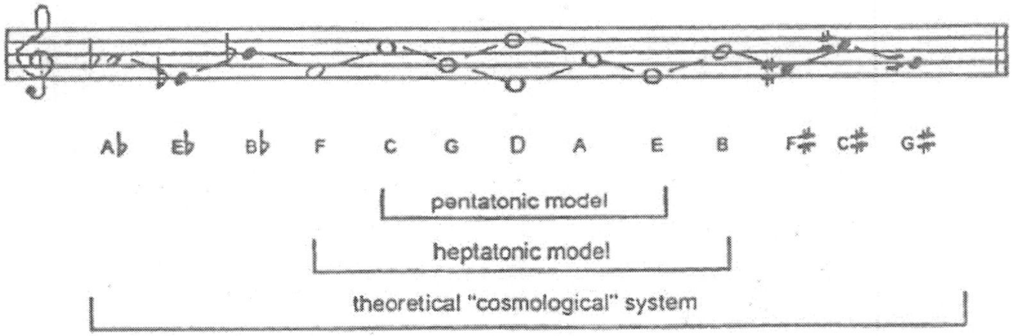

Figure 8. Serpent power: the spiral tuning of fifths. Courtesy of Ernest McClain.

On a piano which is tuned so that each of the intervals of the 12 tone scale are equal in a logarithmic sense (the equal-tempered scale), beginning on any tone and playing twelve successive musical fifths, results in the same tone seven octaves higher. Referring to Figure 8, the first and thirteenth tones in spiral fifths, A_{flat} and G_{sharp}, the tritone or three wholetones located at 6 o'clock on the tone circle, are the same tone in different octaves. However, in terms of rational fifths they differ by about a quarter of a semitone, the so-called Pythagorean comma. This is true because in order for $(2/3)^{12}$ to equal $(1/2)^7$ it would follow that $3^{12} = 2^{19}$ which is certainly false. Unless a limit is placed on the frequency of the tones, the use of rational numbers to represent tone would require an infinite number of tones. This presented ancient civilizations with a kind of 3rd millennium B.C. chaos theory.

Similar problems faced early astronomers as they sought to reconcile the incommensurability of the cycles of the sun and the moon. The solar cycle of 365 ¼ days does not mesh with the lunar cycle of 354 days. A canonical year of 360 days was chosen as a compromise between the two. It turns out that the ratios 365 ¼: 360 and 360:354 are both approximately equal to the Pythagorean comma so that the musical scale had some roots in astronomy. Also if an octave is limited by relative frequencies of 360 to 720 eleven of the tones of the Just scale can be placed as integers within this limit missing only the tritone which you can verify by comparing the intervals of the following sequence with Figure 6 and 9 (the rational numbers represent relative string lengths):

D	Eflat	E	F	Fsharp	G	A	Bflat	B	C	Csharp	D'	
360	384	400	432	450	480	540	576	600	648	675	720	(1)
1	15/16	8/9	5/6	4/5	3/4	2/3	5/8	3/5	9/16	8/15	2	

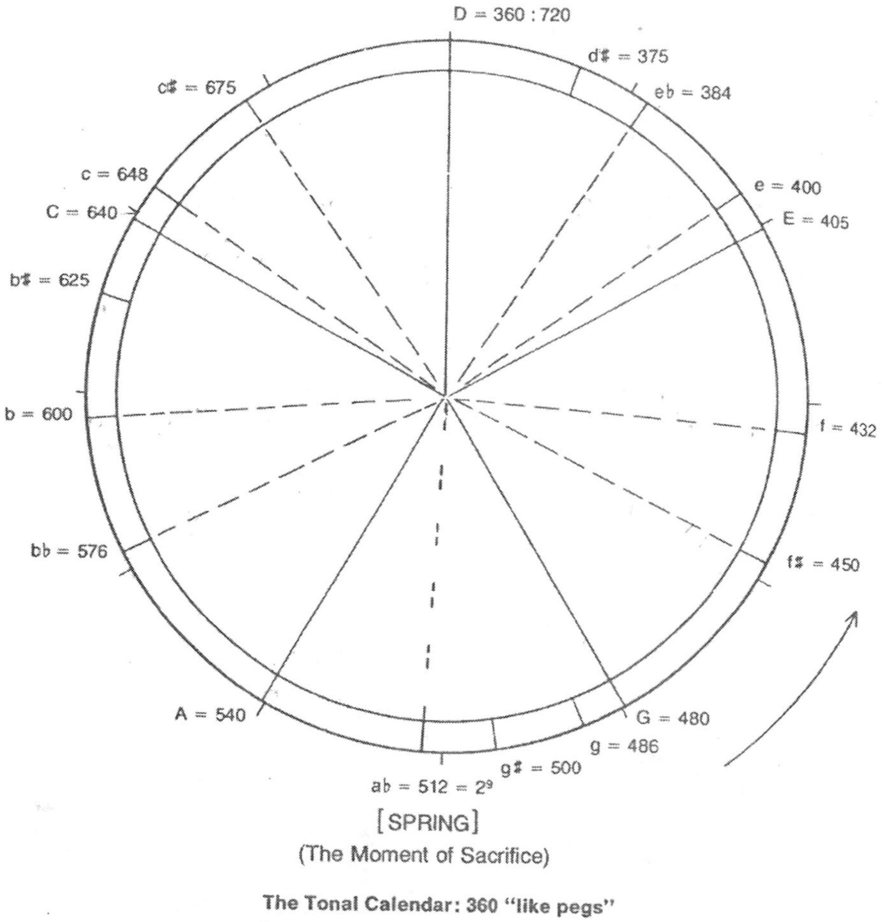

Figure 9. The Just scale shown as integers on a tone circle. Note the symmetry.

All ancient scales were expressed in terms of integers with the integers of the Just scale divisible by primes 2,3, and 5 while the scale of "spiral fifths" were expressed by integers divisible by primes 2, and 3. Notice in Figures 6 and 9 that the tones of the Just scale are placed symmetrically around the tone circle. This is the result of symmetrically placed rational fractions in Sequence 1 being inverses of each other when factors of 2 are cancelled, e.g., 5/6 ≡ 5/3 as compared with 3/5. But factors of 2 result in the same tone in a different octave. Compare the limit of 360/720 with the limit of 286,624/573,268 required for spiral fifths. So the Just scale embodies the two great lessons of the ancient world, the importance of balance and limit in all things. Ernest McClain has traced the use of music as metaphor in the Rig Veda, the works of Plato and the Bible in his books and articles [6],[7],[8].

To ancient mathematicians and philosophers, the concept of rational number was thought to lie at the basis of cosmology, music, and human affairs. On the other hand, while the concept of an irrational number was not clear in the minds of ancient mathematicians, it was understood that rational numbers could be made to approximate certain ideal elements at dividing points of the tone circle into 12 equal sectors, what is now known as the equal tempered scale with $\sqrt{2}, \sqrt[3]{2}, \sqrt[4]{2}$ at 6, 4, and 3 o'clock respectively. The battle between rational and irrational numbers was dramatized by the imagery of the *Rig Veda*. Ernest McClain says [6]:

> The part of the continuum which lies beyond rational number belongs to non-being (Asat) and the Dragon (Vtra). Without the concept of an irrational number, the model for Existence (Sat) is Indra. The continuum of the circle (Vtra) embraces all possible differentiations (Indra). The conflict between Indra and Vtra can never end; it is the conflict between the field of rational numbers and the continuum of real numbers..

This battle between rational and irrational numbers continues into the present where it lies at the basis of chaos theory and the study of dynamical systems. In chaos theory no rational approximation to an irrational number is good enough in terms of yielding closely identical results as I shall demonstrate.

Three decades ago scientists began to realize that many of the phenomenon that they thought to be deterministic or predictable from a set of equations were in fact unpredictable. Changing the initial conditions by as small an amount conceivable led to entirely different results. For example, a rational approximation to an irrational initial condition, no matter how good the approximation, would lead eventually to totally different results. The system of equations predicting weather was one such set of equations. In fact as soon as the equations were more complicated than linear, built into them was chaotic behavior. In other words the fluttering of a butterfly's wings in Brazil could, in principle, over time affect the weather patterns in New York.

The growth of plants is another natural system that appears to exist in a state of incipient chaos [4]. Notice that when the cells of a plant are placed around the stem successively at angles, known as divergence angles, related to the golden mean of $2\pi/\phi$ radians the spiral forms reminiscent of sunflowers appear. Change the divergence angle to a close rational approximation of the golden mean and the spiral is lost and replaced by a spider web appearance (see Figure 10).

Consider the simple map governing the Mandlebrot set [9],

$$z \dashrightarrow z^2 + c \text{ for z and c complex numbers.}$$

Beginning with an initial point z_0 and replacing this in the map leads to the trajectory z_0, z_1, z_2, z_3, ... The Mandelbrot set constitutes all values of c that lead to bounded trajectories. This sensitive dependence on initial conditions holds for values of c outside of the Mandelbrot set. If the value of c is taken internally and away from the boundary of the Mandelbrot set the behavior of the trajectory is simple, leading either to a fixed point or a periodic orbit. The Julia set is the boundary of the set of points of the trajectory that do not escape to infinity. For example, when c = 0, the Julia set is a unit circle. Points outside the Mandelbrot set lead to chaotic behavior of the kind just mentioned. Points near the boundary of the set have the most interesting behavior. One such Julia set for a point near the boundary of the Mandelbrot set is shown in Figure 11. This is somewhat like the state of affairs that exists at the shoreline between land and ocean. The frozen character of the land as opposed to the chaotic nature of the ocean is mediated by the tide pools at the interface between the two. This is where life has its greatest diversity. Stuart Kauffman referred to this region of great differentiation as the "edge of chaos" [10].

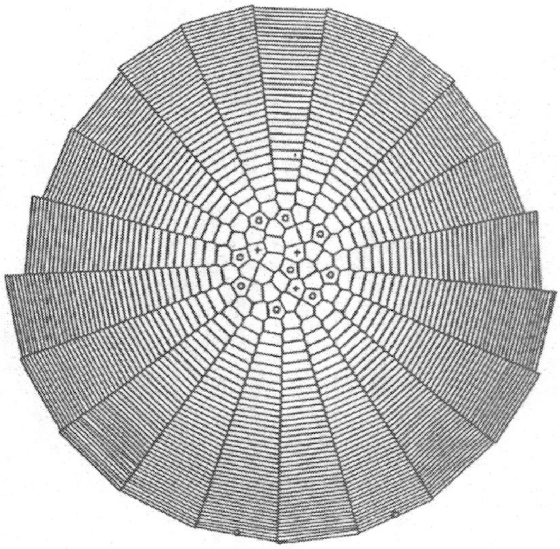

a)

b)

Figure 10. a) A computer generated model of plant phyllotaxis with rational divergence angle $2\pi x 13/21$. Note the spider web appearance; b) irrational divergence angle $2\pi/\phi^2$. Note the daisy-like appearance.

Figure 11. A "dragon" shaped Julia set for a value of c at the boundary of the Mandelbrot set.

There is a strong relationship between chaos and fractals. In fact Julia sets generally have a fractal nature. The study of fractals had its beginning with the research of Benoit Mandelbrot into the nature of stock market fluctuations. However, such structures were noticed earlier by Lewis Richardson in his study of the length of coastlines. Richardson noticed that there was a power law relating the apparent length of coastlines when viewed at different scales. When viewed at a large scale such as the scale of a map, the coastline appears finite. But if the scale is reduced so that all of the idiosycracies of the coastline are evident, the ins and outs of the coastline have no apparent limit and its length is effectively infinite. Furthermore, a small stretch of coastline is similar to the whole when viewed in a statistical sense.

Robert Cogan and Pozzi Escot have shown that music also has a fractal nature [11]. For example they show that musical structures appear and reappear throughout the musical score at different scales. This is the consequence of the music also satisfying a power law referred to as 1/f noise found in the structure of the music of Bach and Mozart [12]. 1/f noise has a spectrum of sound between the spectrum of Brownian motion in which the next note is completely determined from the previous notes resulting in a frozen quality in the music, and white noise in which the tones are randomly chosen leading to a chaotic sound. So we see that good music is again the result of finding the "edge of chaos."

Good art also strives to incorporate the elements of self-similarity although this is generally done subtly. In a great work of art each image must related to the others in terms of its geometry and metaphorical themes. Artists and sculptors have always been inspired by the complex forms of nature. For example the vortices in Van Gogh's famous painting, "Starry Night" in figure 12a appears to be taken directly from the meandering stream winding through separate vortices in Figure 12b. Trains of vortices also appear in the knarled cypress trees found in many of Van Gogh's late paintings such as "St Paul's Hospital, (1889)" of Figure 13a and perfectly embody the bark and knots of the cypress tree in Figure 13b. On the other hand, the design on a palm leaf from New Guinea represent yet another set of vortices shown in Figure 14a and b. Figures 12b, 13b, and 14b were taken from the beautiful photos of complexity in nature found in Theodor Schwenk's book, *Sensitive Chaos* [14].

a)

b)

Figure 12. a) Van Gogh's painting, "Starry Night". About this painting Van Gogh wrote, "First of all the twinkling stars vibrated, but remained motionless in space. Then all celestial globes united into one series of movements…Firmaments and planets both disappeared, but the mighty breath which gives life to al things and in which all is bound up remain [13]."; b) a meandering stream winding through separate vortices. From *Sensitive Chaos* by Schwenk [14].

a)

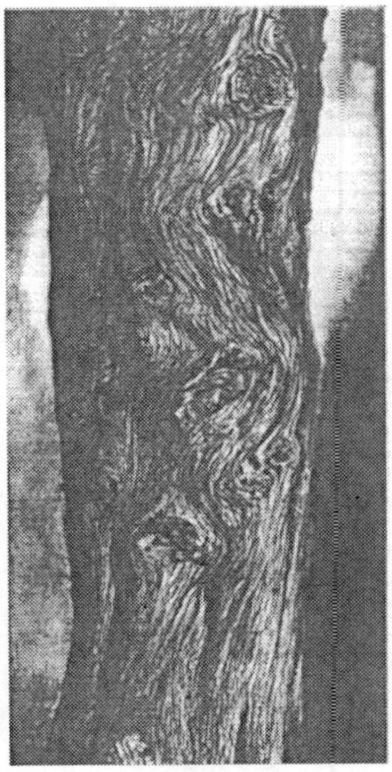

b)

Figure 13. Van Gogh's painting, "St. Paul's Hospital, (1889)". Van Gogh wrote, " The cypress are always occupying my thoughts---it astonishes me that they have not been done as I see them."; b) The bark and knots of a cypress tree from Schwenk [12].

a)

b)

Figure 14. a) Design on a palm leaf (May River, New Guinea) Volkerkundliches Museum, Basel; b) A vortex train from Schwenk [14].

Manuel Baez (see this issue) creates sculptures reminiscent of complex forms from nature out of bamboo sticks and rubber band connectors [15] resulting in structures whose whole is greater than the sum of its parts. Baez describes his system as follows: "These dynamic processes are inherently composed of interweaving elemental relationships that evolve into integrative systems with startling form and structure generating capabilities". Beginning with a simple shape such as a square or pentagon, a module is created which is replicated over and over. Since the sticks are flexible, the model inter-transforms into amazing shapes illustrating the order which exists within apparent chaos. Three structures from his "Phenomenological Garden" all made with 12" and 6" bamboo dowels and rubber bands are shown in Figure 15. They were all generated from a simple square pattern.

Figure 15. The Phenomenological Garden of Manuel Baez.

Bathsheba Grossman invites scientists and mathematicians to send her complex images from their work such as proteins or globular clusters from astronomy or complex geometrical forms and recreates them as three dimensional sculptures in a variety of medias. Her "Cosmological Simulation" (see Figure 16a) was created from simulated scientific data and illustrates the fractal nature of the universe. "Ferritin Protein" (see Figure 16b) is a three-dimensional model in laser etched crystal made from a protein data bank file. Her bronze sculpture "Metatron" is shown in Figure 17. It is made by a lost wax process and created from an operation upon a cube and an octahedron. It appears to be as a singular vortex fixed in time and is evocative to me of frozen music.

Barnsley [16] has shown that fractal images can be created by subjecting an initial seed figure to the following transformations: contractions, translations, rotations, and affine transformations (transformations that transform rectangles to arbitrary parallelograms). For example, Barnsley's fern is created by repeatedly transforming an initial rectangle to three rectangles of different sizes, proportions, and orientations and one line segment as shown in Figure 18. This approach to generating fractals is leading to revolutionary ways of understanding how complex structures arise from simple ones, and it is being applied to many applications from image processing to generation of fractal scenes for movie sets such as that shown in Figure 19 generated by Kenneth Musgrave.

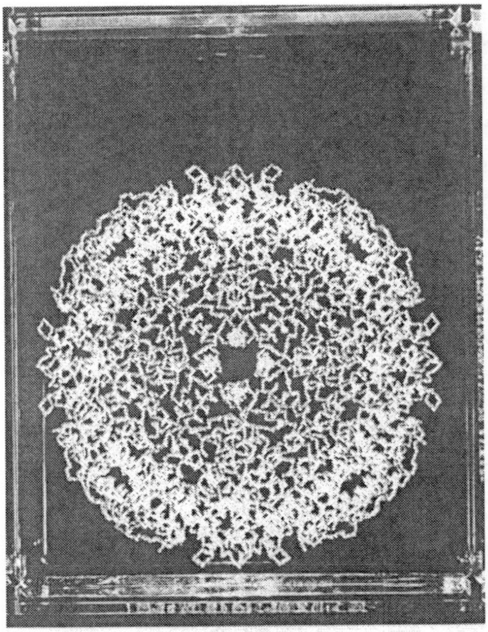

a)

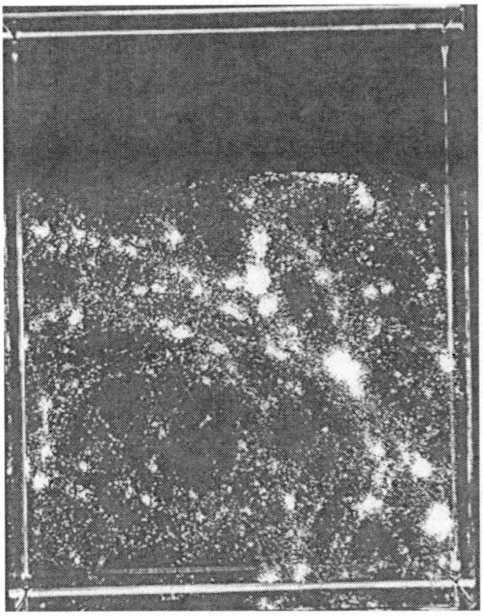

b)

Figure 16. a) Large scale model of a cosmological simulation; b) Ferritin, a symmetrical protein. Courtesy of Bathsheba Grossman.

Figure 17. The Metatron. Courtesy of Bathsheba Grossman.

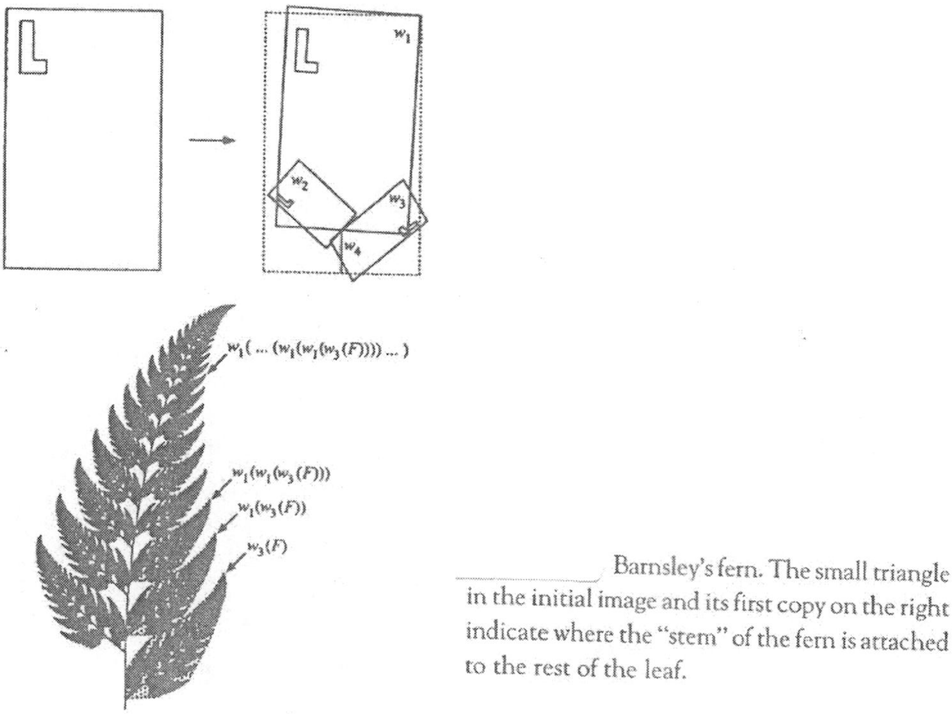

Figure 18. Barnsley's fern. Created by repeated transformation from a rectangular seed pattern.

Figure 19. A fractal scene by Kenneth Musgrave.

Structures and designs with fractal properties appear quite naturally in many cultures. I will present two examples from Ron Eglash's book *African Fractals* [17]. In the western part of the Cameroons lies the fertile grasslands region of the Bamileke. Eglash describes their fractal settlement architecture (see Figure 20).

> "These houses and the attached enclosures are built from bamboo—Patterns of agricultural production underlie the scaling. Since the same bamboo mesh construction is used for houses, house enclosures, and enclosures of enclosures, the result is a self-similar architecture—The farming activities require alot of movement between enclosures, so at all scales we see good-sized openings."

Many of the processional crosses of Ethiopia indicate a threefold fractal iteration (see Figure 21). Eglash suggests that the reason that the iteration stops at three may be for practical reasons. Two iterations is too few to get the concept of iteration across, while more than three presents fabrication difficulties to the artisans.

The twentieth century was a revolutionary time in the history of mathematics and science. First the deterministic nature of physics was replaced by the strange world of quantum mechanics where the outcomes of an experiment depended on probability counter to the intuition of Albert Einstein that "God does not play dice." Then the foundations of mathematics were shaken by Kurt Godel who showed that a mathematical system could not be both consistent and complete while Alan Turing discovered that there was no way of determining whether a computer program would halt once given some initial data.

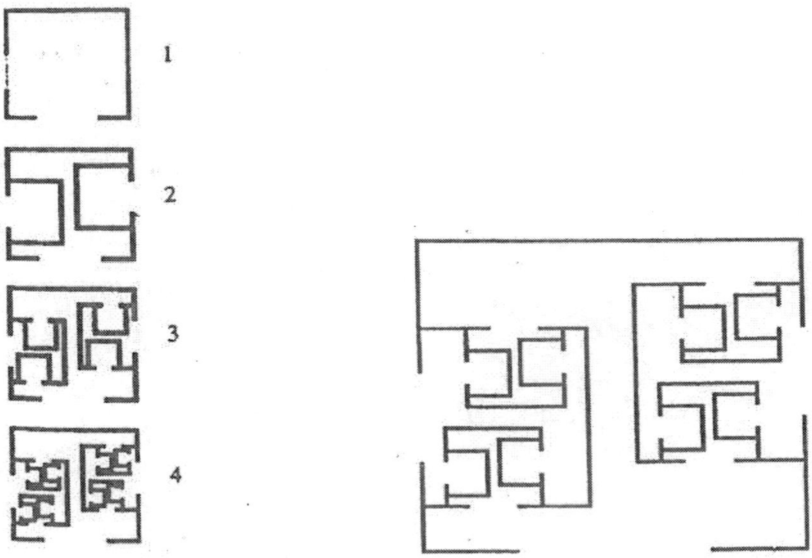

a) b)

Figure 20. a) Fractal simulation of Bamileke architecture. In the first iteration ("seed shape") the two active lines are shown in gray. b) Enlarged view of the fourth iteration. From *African Fractals* by Ron Eglash [15].

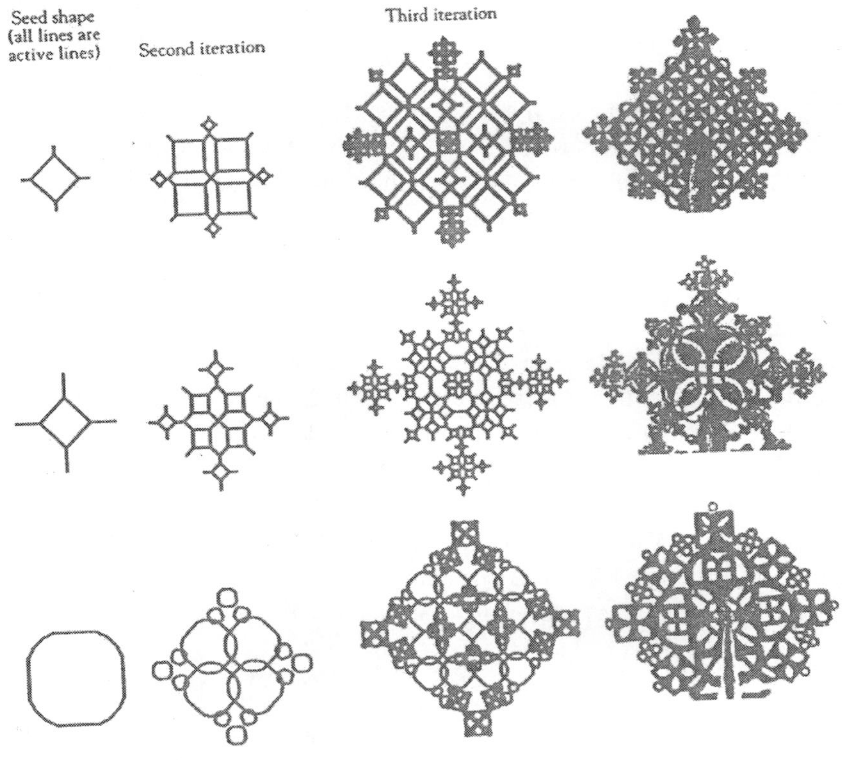

Figure 21. Fractal simulation for Ethiopian processional crosses through three iterations. From *African Fractals* by Ron Eglash [15].

Mathematical and scientific theories are created by observing symmetries of all sorts. This enables the information inherent in the physical system to be compressed into a theory or set of equations. For example, all of the possible motions of celestial or earthbound bodies are governed by Newtons laws which is elegantly stated as F = ma. Knowing only a few facts about the initial motion, in other words only a few bits of information, the theory can predict the ensuing motion. What if the system exhibited no such symmetry? Then each specific instance would have to be observed in its entirety. In other words, no information would have been compressed for us to unlock by a theory. All we could do would be to observe each orbit and record what we saw. Systems generated by rules in which the next state is determined by the flipping of a coin is an example of a system devoid of symmetry. There is no way to determine the final state of the system except by following the coin flips to their conclusion. Similarly in mathematics, a mathematical system is generally compressed by stating several axioms representing a finite number of bits of information from which an unlimited number of theorems follow. Without axioms mathematics would not be concerned with judging truth or falsity but rather with generating patterns.

G.J. Chaitin [18] has recently shown that rather than being an irrelevant curiosity, this state of affairs, reflected in Godel's and Turing's discoveries, is central to the representation of nature by mathematics and science. He created a number from number theory with the property that the determination of its digits was equivalent to flipping coins. We can now say that, it may be that only narrow islands of observation may be derivable from our standard equations and theories. As a result mathematicians have begun to realize that other approaches would be needed to characterize natural phenomena and to coax information from nature. One such program is being explored by Stephen Wolfram in his book *A New Kind of Science* [19].

Wolfram studied the behavior of a large class of systems governed by rules in which the next state of the system was determined by the previous state, so-called cellular automata. In response to simple rules and starting with simple initial conditions, complex forms would emerge such as the one in Figure 22a. Compare this with one of the network of veins of sand created by the interplay of sand and water shown in Figure 22b by Schwenk. Wolfram discovered that all such automata could be classified as being one four types and that naturally occurring systems of growth from plants and animals to blood vessels to crystals, some of which are shown in Figure 23, were themselves cellular automata exhibiting the same properties as the artificial ones he created. Furthermore he discovered an astounding principal which he refers to as the Principal of Computational Equivalence which states that all processes, whether they are produced by human effort or occur spontaneously in nature, can be viewed as computations. Furthermore, in many kinds of systems particular rules can be found that achieve universality, in other words, the ability to function as a computer in all of its generality, e.g., a universal Turing machine. The dramatic discovery of his book was to show that rather than being a rare event, such universality could be created out of simple rules.

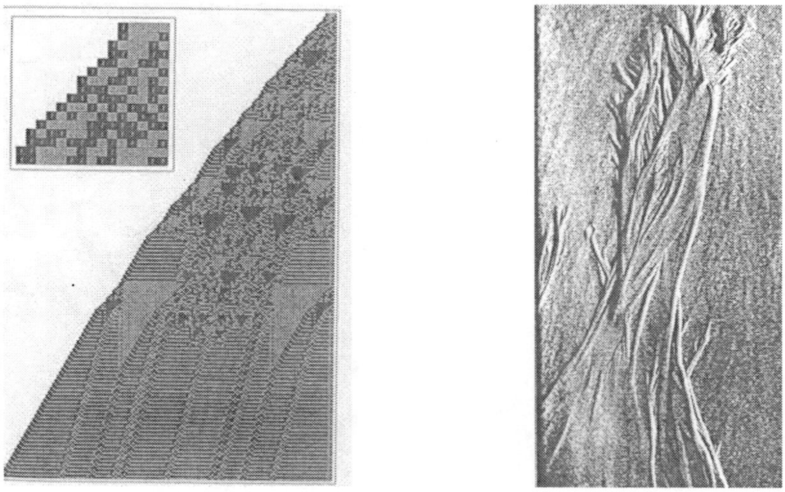

a) b)

Figure 22. a) An example of a system defined by the following rule: at each step, take the number obtained at that step and write its base 2 digits in reverse order, then add the resulting number to the original one. Dark squares represent 1 while light squares 0. For many possible starting numbers, the behavior obtained is very simple. This picture shows what happens when one starts with the number 16. After 180 steps, it turns out that all that survives are a few objects that one can view as localized structures. From *A New Science* by S. Wolfram [19]; b) A network of veins of sand created by the interplay of sand and water. From Schwenk [14].

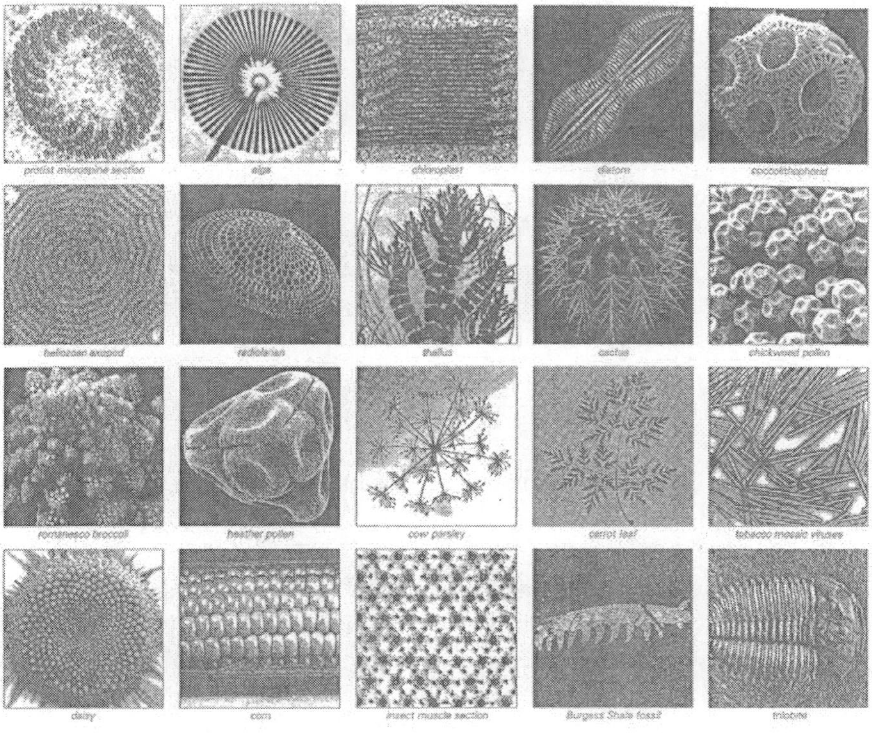

Figure 23. A collection of patterns from nature suggesting natural cellular automata. From *A New Science* by S. Wolfram.

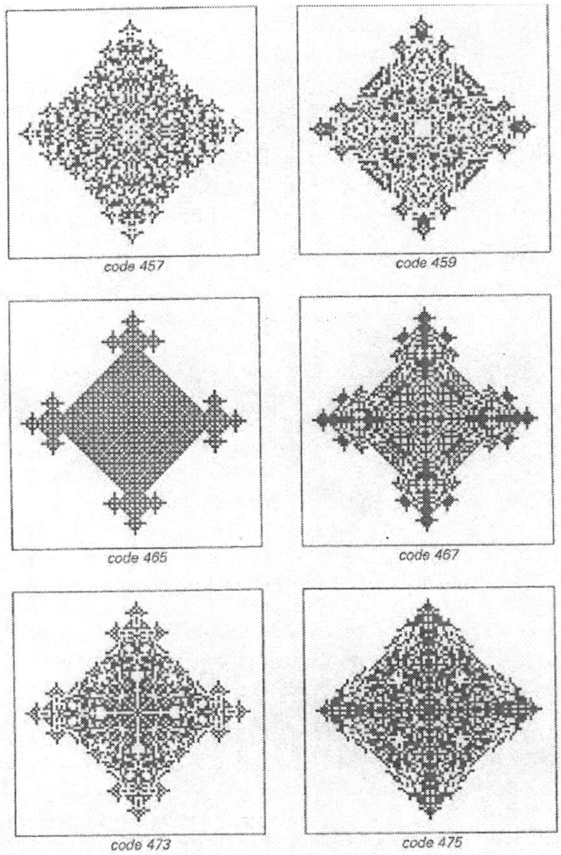

Figure 24. Cellular automata generated by simple rules with the appearance of Ethiopian crosses. From *A New Science* by S. Wolfram [19].

This new approach to science is an invitation for artists and scientists to draw closer to one another. After all, the examples of ornamental art have patterns similar to ones generated by cellular automata. For example, Figure 24 illustrates several eamples generated by cellular automoata reminiscent of the Ethiopian designs of Figure 20. Hans Jenny's and Theodor Schwenk's vibratory patterns offer another link between art, science and nature. Figure 25a from Jenny [1] shows particles of sand in a state of flow being excited by crystal oscillations on a steel plate. Compare this with Figure 25b from Schwenk [14] showing the ripple marks in sand at a beach.

We are heading into an exciting new era of scientific and mathematical explorations in which artists, musicians and scientists will be joining hands to help each other and the rest of us to understand our universe in all of its complexity. More and more the question will be asked: Is it art or is it science? Mathematics will serve as the common language, scientists and engineers will create the technology, and artists and musicians will provide the spirit. These new approaches will suit our age and society much as ancient systems of thought met the needs of those cultures. Just as ancient systems of numerology were incorporated into the myths, religious symbolism and philosophy of those ages, the new science of complexity and chaos theory is certain to spawn its myths and metaphors for our age.

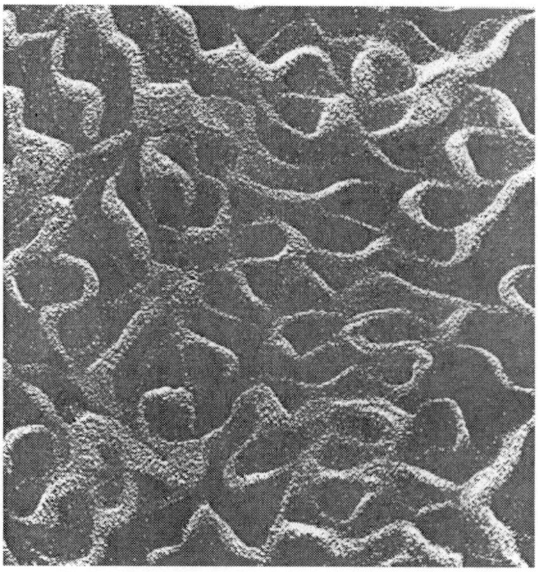

a)

b)

Figure 25. a) Particles of sand in a state of flow excited by crystal oscillations. From Jenny [1]; b) Ripple marks of sand on a beach. From Schwenk [14].

References

[1] Jenny, H., *Cymatics*, Basel: Basilius Press (1967).

[2] Spencer-Brown, G. I, *Laws of Form*, London:George Allen and Unwin, Ltd. (1969).

[3] Kauffman, L.H. and Varela, F.J., "Form Dynamics," *J. Soc. And Bio. Struct.* **3** pp161-206 (1980).

[4] Kappraff, J. *Beyond Measure: A Guided Tour through Nature, Myth, and Number*, Singapore: World Scientific (2003).

[5] McClain, E.G., "Musical theory and Cosmology", *The World and I* (Feb. 1994).

[6] McClain, E.G., *Myth of Invariance*, York Beach, Me.:Nicolas-Hays (1976,1984)

[7] McClain, E.G., *The Pythagorean Plato*, York Beach, Me.:Nicolas-Hays (1978,1984).

[8] McClain, E.G. "*A priestly View of Bible arithmetic in philosophy of science, Van Gogh's Eyes, and God: Hermeneutic essays in honor of Patrick A. Heelan*", ed. B.E. Babich, Boston: Kluwer Academic Publ. (2001).

[9] Peitgens, H-O., Jurgens, H., and Saupe, D., *Chaos and Fractals*, New York: Springer (1992).

[10] Kauffman, S.A., *The Origins of Order: Self Organization and Selection and Complexity*, New York: Oxford Press (1995).

[11] Cogan, R. and Escot, P., *Sonic Design: The Nature of Sound and Music*, Englewood Cliffs, NJ: Prentice Hall (1976).

[12] Gardner, M., "White and brown music, fractal curves and one-over-f fluctuations," Sci. Am., v238, No.4 (1978).

[13] Purce, J., *The Mystic Spiral*, New York: Thames and Hudson (1974).

[14] Schwenk, T., *Sensitive Chaos*, New York: Schocken Books (1976).

[15] Baez, M.A., The Phenomenological Garden, *In On Growth and Form*: *The Engineering of Nature*, ACSA east Central Regional Conference, University of Waterloo, Oct. 2001.

[16] Barnsley, M., *Fractals Everywhere*, San Diego: Academic Press (1988).

[17] Eglash, R., *African Fractals*, New Brunswick: Rutgers Univ. Press (1999).

[18] Chaitin, G.J. "A century of controversy over the foundations of Mathematics," *Complexity*, vol. 5, No. 5, pp.12-21, (May/June 2000).

[19] Wolfram, S. *A New Kind of Science*, Champaign, IL: Wolfram Media, Inc. (2002).

In: Chaos and Complexity Research Compendium
Editors: F. Orsucci and N. Sala, pp. 229-241
ISBN: 978-1-60456-787-8

Chapter 17

POLLOCK, MONDRIAN AND NATURE: RECENT SCIENTIFIC INVESTIGATIONS

Richard Taylor*
University of Oregon, Oregon

Abstract

The abstract paintings of Piet Mondrian and Jackson Pollock are traditionally regarded as representing opposite ends of the diverse visual spectrum of Modern Art. In this article, I present an overview of recent scientific research that investigates the enduring visual appeal of these paintings.

Introduction

Walking through the Smithsonian (USA), it is clear that the stories of Piet Mondrian (1872-1944) and Jackson Pollock (1912-56) present startling contrasts. First, I come across an abstract painting by Mondrian called "Composition With Blue and Yellow" (1935). It consists of just two colors, a few black lines and an otherwise uneventful background of plain white (see Fig. 1). It's remarkable, though, how this simplicity catches the eye of so many passers-by. According to art theory, Mondrian's genius lay in his unique arrangement of the pattern elements, one that causes a profound aesthetic order to emerge triumphantly from stark simplicity. Carrying on, I come across Pollock's "Number 3, 1949: Tiger" (See Fig. 2). Whereas Mondrian's painting is built from straight, clean and simple lines, Pollock's are tangled, messy and complex. This battlefield of color and structure also attracts a crowd, mesmerised by an aesthetic quality that somehow unites the rich and intricate splatters of paint.

* E-mail address: rpt@darkwing.uoregon.edu

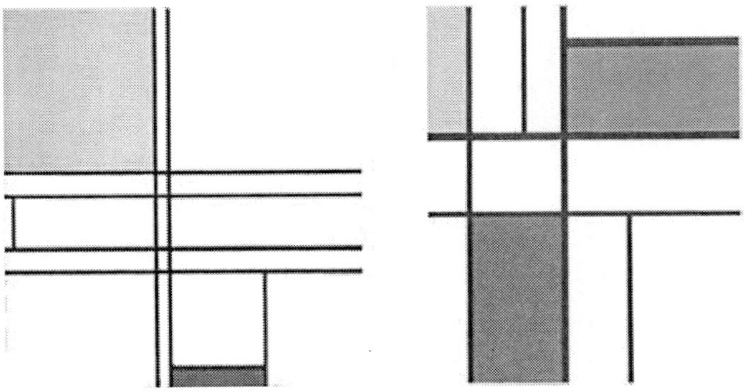

Figure 1. A comparison of Piet Mondrian's "Composition with Blue and Yellow" (1935) with a painting by Alan Lee in which the lines are positioned randomly. Can you tell which is the real Mondrian painting?

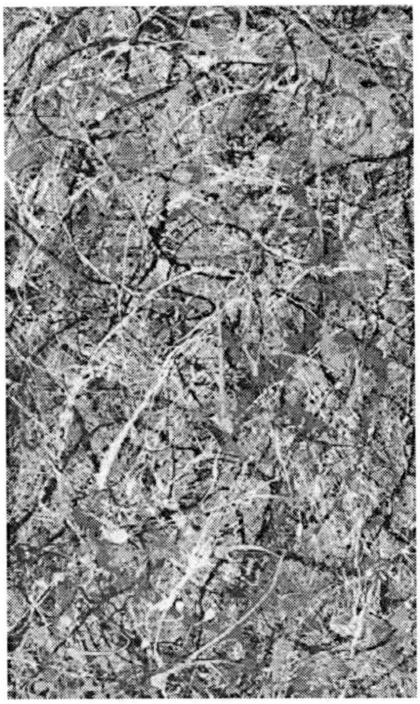

Figure 2. Jackson Pollock's "Number 3, 1949: Tiger."

Both men reached their artistic peak in New York during the 1940s. Although Mondrian strongly supported Pollock, their approaches represented opposite ends of the spectrum of abstract art. Whereas Mondrian spent weeks deliberating the precise arrangement of his patterns [Deicher, 1995], Pollock dashed around his horizontal canvases dripping paint in a fast and spontaneous fashion [Varnedoe et al, 1998]. Despite their differences in the creative process and the patterns produced, both men maintained that their goal was to venture beyond life's surface appearance by expressing the aesthetics of nature in a direct and profound manner. At their peak, the public viewed both men's abstract patterns with considerable scepticism, failing to see any connection with the natural world encountered during their daily lives. Of the two artists, Mondrian was given more credence. Mondrian was a sophisticated intellectual and wrote detailed essays about his carefully composed works. Pollock, on the other hand, was frequently drunk and rarely justified his seemingly erratic motions around the canvas.

Fifty years on, both forms of abstract art are regarded as masterpieces of the Modern era. What is the secret to their enduring popularity? Did either of these artists succeed in their search for an underlying aesthetic quality of life? In light of the visual contrast offered by the two paintings at the Smithsonian, it's remarkable how the passers-by use similar language to discuss their aesthetic experiences. Both paintings are described in terms of 'balance,' 'harmony' and 'equilibrium.' The source of this subtle order seems to be enigmatic, however. None of the gallery audience can define the exact quality that appeals to them. It's tempting to come away from this scene believing that, half a century after their deaths, we might never comprehend the mysterious beauty of their compositions.

Recently, however, their work has become the focus of unprecedented scrutiny from an unexpected source - science. In 1999, I published a pattern analysis of Pollock's work, showing that the visual complexity of his paintings is built from fractal patterns –patterns that are found in a diverse range of natural objects [Taylor et al, 1999]. Furthermore, in an on-going collaboration with psychologists, visual perception experiments reveal that fractals possess a fundamental aesthetic appeal [Taylor, 2001]. How, then, should we now view Mondrian's simple lines?

1. Pollock's Dripped Complexity

First impressions of Pollock's painting technique are striking, both in terms of its radical departure from centuries-old artistic conventions and also in its apparent lack of sophistication! Purchasing yachting canvas from his local hardware store, Pollock simply rolled the large canvases (up to five meters long) across his studio floor. Even the traditional painting tool - the brush - was not used in its expected capacity: abandoning physical contact with the canvas, he dipped the brush in and out of a can and dripped the fluid paint from the brush onto the canvas below. The uniquely continuous paint trajectories served as 'fingerprints' of his motions through the air. During Pollock's era, these deceptively simple acts fuelled unprecedented controversy and polarized public opinion of his work: Was he simply mocking artistic traditions or was his painting 'style' driven by raw genius?

Over the last fifty years, the precise meaning behind his infamous swirls of paint has been the source of fierce debate in the art world [Varnedoe et al, 1998]. Although Pollock was often reticent to discuss his work, he noted that, "My concerns are with the rhythms of nature" [Varnedoe et al, 1998]. Indeed, Pollock's friends recalled the many hours that he spent staring out at the countryside, as if assimilating the natural shapes surrounding him [Potter, 1985]. But if Pollock's patterns celebrate nature's 'organic' shapes, what shapes would these be? Since the 1970s many of nature's patterns have been shown to be fractal [Mandelbrot, 1977]. In contrast to the smoothness of artificial lines, fractals consist of patterns that recur on finer and finer scales, building up shapes of immense complexity. Even the most common fractal objects, such as the tree shown in Fig. 3(a), contrast sharply with the simplicity of artificial shapes.

The unique visual complexity of fractal patterns necessitates the use of descriptive approaches that are radically different from those of traditional Euclidian geometry. The fractal dimension, D, is a central parameter in this regard, quantifying the fractal scaling relationship between the patterns observed at different magnifications [Mandelbrot, 1977, Gouyet, 1996]. For Euclidean shapes, dimension is a familiar concept described by integer values of 0, 1, 2 and 3 for points, lines, planes, and solids respectively. Thus, a smooth line (containing no fractal structure) has a D value of 1, whereas a completely filled area (again containing no fractal structure) has a value of 2. For the repeating patterns of a fractal line, D lies between 1 and 2. For fractals described by a D value close to 1, the patterns observed at different magnifications repeat in a way that builds a very smooth, sparse shape. However, for fractals described by a D value closer to 2, the repeating patterns build a shape full of intricate, detailed structure. Figure 4 demonstrates how a fractal pattern's D value has a profound effect on its visual appearance. The two natural scenes shown in the left column have D values of 1.3 (top) and 1.9 (bottom). Table 1 shows D values for various classes of natural form.

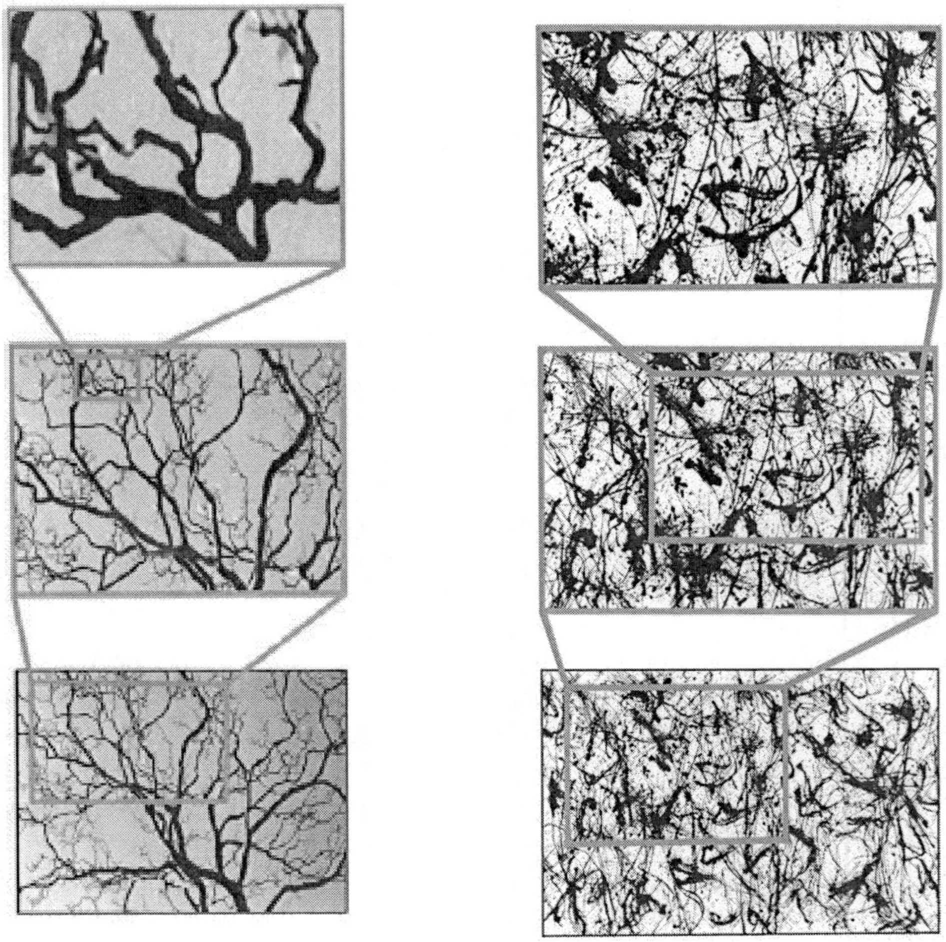

(a) (b)

Figure 3. (a) Trees are an example of a natural fractal object. Although the patterns observed at different magnifications don't repeat exactly, analysis shows them to have the same statistical qualities (photographs by R.P. Taylor). (b) Pollock's paintings (in this case "Number 32, 1950") display the same fractal behavior.

The patterns of a typical Pollock drip painting are shown at different magnifications in Fig. 3(b). In 1999, my research team published an analysis of 20 of Pollock's dripped paintings showing them to be fractal [Taylor et al, 1999]. We used the well-established 'box-counting' method, in which digitized images of Pollock paintings were covered with a computer-generated mesh of identical squares. The number of squares, N(L), that contained part of the painted pattern were then counted and this was repeated as the size, L, of the squares in the mesh was reduced. The largest size of square was chosen to match the canvas size (L~2.5m) and the smallest was chosen to match the finest paint work (L~1mm). For fractal behavior, N(L) scales according to $N(L) \sim L^{-D}$, where $1 < D < 2$ [Gouyet, 1996]. The D values were extracted from the gradient of a graph of log N(L) plotted against log L (details of the procedure are presented elsewhere [Taylor et al, 1999]).

Table 1. D values for various natural fractal patterns

Natural pattern	Fractal dimension	Source
Coastlines: South Africa, Australia, Britain Norway	 1.05-1.25 1.52	 Mandelbrot Feder
Galaxies (modeled)	1.23	Mandelbrot
Cracks in ductile materials	1.25	Louis et al.
Geothermal rock patterns	1.25-1.55	Campbel
Woody plants and trees	1.28-1.90	Morse et al.
Waves	1.3	Werner
Clouds	1.30-1.33	Lovejoy
Sea Anemone	1.6	Burrough
Cracks in non-ductile materials	1.68	Skejltorp
Snowflakes (modeled)	1.7	Nittman et al.
Retinal blood vessels	1.7	Family et al.
Bacteria growth pattern	1.7	Matsushita et al.
Electrical discharges	1.75	Niemyer et al.
Mineral patterns	1.78	Chopard et al.

Recently, I described Pollock's style as ‘Fractal Expressionism’ [Taylor et al, Physics World, 1999] to distinguish it from computer-generated fractal art. Fractal Expressionism indicates an ability to generate and manipulate fractal patterns *directly*. How did Pollock paint such intricate patterns, so precisely and do so 25 years ahead of the scientific discovery of fractals in natural scenery? Our analysis of film footage taken in 1950 reveals a remarkably systematic process [Taylor et al, Leonardo, 2002]. He started by painting localized islands of trajectories distributed across the canvas, followed by longer, extended trajectories that joined the islands, gradually submerging them in a dense fractal web of paint. This process was very swift, with D rising sharply from 1.52 at 20 seconds to 1.89 at 47 seconds. We label this initial pattern as the ‘anchor layer’ because it guided his subsequent painting actions. He would revisit the painting over a period of several days or even months, depositing extra layers on top of this anchor layer. In this final stage, he appeared to be fine-tuning D, with its value rising by less than 0.05. Pollock's multi-stage painting technique was clearly aimed at generating high D fractal paintings [Taylor et al, Leonardo, 2002].

Figure 4. Examples of natural scenery (left column) and drip paintings (right column). Top: Clouds and Pollock's painting *Untitled* (1945) are fractal patterns with D=1.3. Bottom: A forest and Pollock's painting *Untitled* (1950) are fractal patterns with D=1.9. (Photographs by R.P. Taylor).

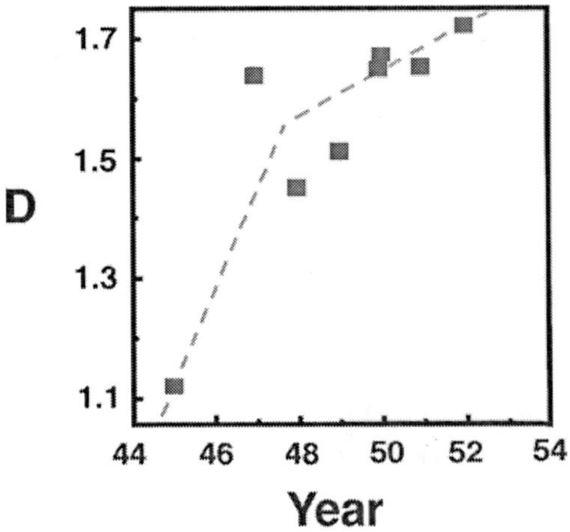

Figure 5. The fractal dimension D of Pollock paintings plotted against the year that they were painted (1944 to 1954). See text for details.

He perfected this technique over a ten-year period, as shown in Fig. 5. Art theorists categorize the evolution of Pollock's drip technique into three phases [Varnedoe, 1998]. In the 'preliminary' phase of 1943-45, his initial efforts were characterized by low D values. An example is the fractal pattern of the painting *Untitled* from 1945, which has a D value of 1.3 (see Fig. 4). During his 'transitional phase' from 1945-1947, he started to experiment with the drip technique and his D values rose sharply (as indicated by the first dashed gradient in Fig. 5). In his 'classic' period of 1948-52, he perfected his technique and D rose more gradually (second dashed gradient in Fig. 5) to the value of D = 1.7-1.9. An example is *Untitled* from 1950 (see Fig. 4), which has a D value of 1.9. Whereas this distinct evolution has been proposed as a way of authenticating and dating Pollock's work [Taylor, Scientific American, 2002] it also raises a crucial question for visual scientists - do high D value fractal patterns possess a special aesthetic quality?

2. Fractal Aesthetics

Fractal images have been widely acknowledged for their instant and considerable aesthetic appeal [see, for example, Peitgen et al, 1986, Mandelbrot, 1989, Briggs, 1992, Kemp, 1998]. However, despite the dramatic label "the new aesthetic" [Richards, 2001], and the abundance of computer-generated fractal images that have appeared since the early 1980s, relatively few quantitative studies of fractal aesthetics have been conducted. In 1994, I used a chaotic (kicked-rotor) pendulum to generate fractal and non-fractal drip-paintings and, in the perception studies that followed, participants were shown one fractal and one non-fractal pattern (randomly selected from 40 images) and asked to state a preference [Taylor 1998, Taylor, Art and Complexity, 2003]. Out of the 120 participants, 113 preferred examples of fractal patterns over non-fractal patterns, confirming their powerful aesthetic appeal.

Given the profound effect that D has on the visual appearance of fractals (see Fig. 4), do observers base aesthetic preference on the fractal pattern's D value? Using computer-generated fractals, investigations by Deborah Aks and Julien Sprott found that people expressed a preference for fractal patterns with mid-range values centered around D = 1.3 [Sprott, 1993, Aks and Sprott, 1996]. The authors noted that this preferred value corresponds to prevalent patterns in natural environments (for example, clouds and coastlines) and suggested that perhaps people's preference is actually 'set' at 1.3 through a continuous visual exposure to patterns characterized by this D value. However, in 1995, Cliff Pickover also used a computer but with a different mathematical method for generating the fractals and found that people expressed a preference for fractal patterns with a high value of 1.8 [Pickover, 1995], similar to Pollock's paintings. The discrepancy between the two investigations suggested that there isn't a 'universally' preferred D value but that aesthetic qualities instead depend specifically on how the fractals are generated.

The intriguing issue of fractal aesthetics was reinvigorated by our discovery that Pollock's paintings are fractal: In addition to fractals generated by natural and mathematical processes, a third form of fractals could be investigated – those generated by humans. To determine if there are any 'universal' aesthetic qualities of fractals, we performed experiments incorporating all three categories of fractal pattern: fractals formed by nature's processes (photographs of natural objects), by mathematics (computer simulations) and by humans (cropped images of Pollock paintings) [Taylor, 2001]. Figure 4 shows some of the images used (for the full set of images, see Spehar et al, 2003). Within each category, we investigated visual appeal as a function of D using a 'forced choice' visual preference technique: Participants were shown a pair of images with different D values on a monitor and asked to choose the most "visually appealing." Introduced by Cohn in 1894, the forced choice technique is well-established for securing value judgments [Cohn, 1894]. In our experiments, all the images were paired in all possible combinations and preference was quantified in terms of the proportion of times each image was chosen. The experiment, involving 220 participants, revealed a distinct preference for mid-range fractals (D=1.3 –1.5), irrespective of their origin [Spehar et al, 2003].

The 'universal' character of fractal aesthetics was further emphasized by a recent investigation showing that gender and cultural background of participants did not significantly influence preference [Abrahams et al, 2003]. Furthermore, based on experiments performed at NASA-Ames laboratory, our recent preliminary investigations indicate that preference for mid-range D fractals extends beyond visual perception: skin conductance measurements showed that exposure to fractal art with mid-range D values also significantly reduced the observer's physiological responses to stressful cognitive work [Taylor et al, 2003, Wise et al, 2003].

Skin conductance measurements might appear to be a highly unusual tool for judging art. However, our preliminary experiments provide a fascinating insight into the impact that art can have on the observer's physiological condition. It would be intriguing to apply this technique to a range of fractal patterns appearing in art, architecture and archeology: Examples include the Nasca lines in Peru (pre-7th century) [Castrejon-Pita et al, 2003], the Ryoanji Rock Garden in Japan (15th century) [Van Tonder et al, 2002], Leonardo da Vinci's sketch *The Deluge* (1500) [Mandelbrot, 1977], Katsushika Hokusai's wood-cut print *The Great Wave* (1846) [Mandelbrot, 1977], Gustave Eiffel's tower in Paris (1889) [Schroeder, 1991], Frank Lloyd Wright's Palmer House in Michigan (1950) [Eaton, 1998], and Frank Gehry's proposed architecture for the Guggenheim Museum in New York (2001) [Taylor, 2001, Taylor, New Architect, 2003].

As for Pollock, is he an artistic enigma? According to our results, the low D patterns painted in his earlier years should be more relaxing than his later *classic* drip paintings. What was motivating Pollock to paint high D fractals? Perhaps Pollock regarded the visually restful experience of a low D pattern as too bland for an artwork and wanted to keep the viewer alert by engaging their eyes in a constant search through the dense structure of a high D pattern. We are currently investigating this intriguing possibility by performing eye-tracking experiments on Pollock's paintings, which are assessing the way people visually assimilate fractal patterns with different D values.

3. Mondrian's Simplicity

Whereas the above research is progressing rapidly toward an appealing explanation for the enduring popularity of Pollock's paintings, the underlying aesthetic appeal is based on complexity. Clearly, Mondrian's simple visual 'language' of straight lines and primary colors plays by another set of rules entirely. In fact, Mondrian developed a remarkably rigorous set of rules for assembling his patterns and he believed that they had to be followed meticulously for his paintings to display the desired visual quality. The crucial rules concerned the basic grid of black lines, which he used as an artistic 'scaffold' to build the appearance of the painting. Mondrian used only horizontal and vertical lines, which he believed "exist everywhere and dominate everything." In one of the more notorious exchanges in Modern Art history, he argued fiercely when colleague Theo Van Doesburg proposed that they should also use diagonal lines. Mondrian passionately believed that the diagonal represented a disruptive element that would diminish the painting's balance. So strong was his belief that he threatened to dissolve the 'De Styl' art movement that had formed around his painting style. Mondrian wrote to him declaring, "Following the high-handed manner in which you have used the diagonal, all further collaboration between us has become impossible."

Although Mondrian's theory of line orientation has legendary status within the art world, only recently have his aesthetic beliefs been put to the test. Whereas Pollock's paintings are being used as novel test beds for examining peoples' responses to visual complexity, scientists are becoming increasingly interested in Mondrian's paintings because of their visual simplicity. In terms of neurobiology, it is well-known that different brain cells are used to process the visual information of a painting containing diagonal lines than for one composed of horizontal and vertical lines [Zeki, 1999]. However, as neurologist Semir Zeki points out, whether these changes in brain function are responsible for the observer's aesthetic experience is "a question that neurology is not ready to answer" [Zeki, 1999]. In 2001, one of my collaborators, Branka Spehar, performed visual perception experiments aimed at directly addressing the link between line orientation and aesthetics. She used images generated by tilting 3 Mondrian paintings at different orientations [Spehar, 2001, Taylor, Nature, 2002]. The 4 orientations included the original one intended by Mondrian, and also 2 oblique angles for which the lines followed diagonal directions. Spehar showed each picture through a circular window that hid the painting's frame. This removed any issues relating to frame orientation, allowing the observer to concentrate purely on line orientation. Using the 'forced choice' technique, she then paired the 4 orientations of each painting in all possible combinations and asked 20 people to express a preference within each pair. The results revealed that people show no aesthetic preference between the orientations featuring diagonal lines and those featuring horizontal and vertical lines. Spehar's results clearly question the importance of Mondrian's vertical-horizontal line rule.

Mondrian's obsession with the orientation of his lines extended to their position on the canvas. He spent long periods of time shifting a single line back and forth within a couple of millimetres, believing that a precise positioning was essential for capturing an aesthetic order that was "free of tension" [Deicher, 1995]. Australian artist Alan Lee recently used visual perception experiments to test Mondrian's ideals [Lee, 2001, Taylor, Nature, 2002]. Lee created 8 of his own paintings based on Mondrian's style. However, he composed the patterns by positioning the lines randomly. He then presented 10 art experts and over 100 non-experts with 12 paintings and asked them to identify the 4 of Mondrian's carefully composed patterns and the 8 of his random patterns (see Fig. 1). Lee's philosophy was simple – if Mondrian's carefully located lines delivered an aesthetic impact beyond that of randomly positioned lines, then it should be an easy task to select Mondrian's paintings. In reality, both the experts and non-experts were unable to distinguish the two types of pattern. Line positioning doesn't influence the visual appeal of the paintings!

Could this surprising result mean that, despite Mondrian's time-consuming efforts, his lines were nevertheless random just like Lee's? To test this theory, I performed a pattern analysis of 22 Mondrian paintings and this showed that his lines are *not* random. For random distributions, each line has an equal probability of being located at any position on the canvas. In contrast, my analysis of 170 lines featured in the 22 paintings show that Mondrian was twice as likely to position a line close to the canvas edge as he was to position it near the canvas center. In addition to dismissing the 'random line theory,' this result invites comparisons with traditional composition techniques. In figurative paintings, artists rarely position the center of focus close to the canvas edge because it leads the eye's attention off the canvas. If Mondrian's motivations were to apply this traditional rule to his line distributions, he would have avoided bunching his lines close to the edges. Another compositional concept applied to traditional artworks is the Golden Ratio (sometimes referred to by artists as the "Divine Proportion"). According to this rule, the aesthetic quality of a painting increases if the length and height of the rectangular canvas have the ratio of 1.61 (a number derived from the Fibonacci sequence). Whereas the shapes of Mondrian's canvases don't match this ratio, a common speculation is that he positioned his intersecting lines such that the resulting rectangles satisfy the Golden Ratio. However, this claim has recently been dismissed in a book that investigates the use of the Golden Ratio in art [Livio, 2002].

4. Discussion

These recent scientific investigations of Mondrian's patterns highlight several crucial misconceptions about Mondrian's compositional strategies. According to the emerging picture of Mondrian's work, the lines that form the visual scaffold of his paintings are not random. However, their positioning doesn't follow the traditional rules of aesthetics, nor does it deliver any appeal beyond that achieved using random lines. The aesthetic order of Mondrian's paintings appears to be a consequence of the presence of a scaffold and it's associated colored rectangles, rather than any subtle arrangement of the scaffold itself. In other words, the appeal of Mondrian's visual language isn't affected by the way the individual 'words' are assembled! What, then, were his reasons for developing such strict 'grammatical' rules for his visual language?

Mondrian wrote extended essays devoted to his motivations, and these focussed on his search for an underlying structure of nature [Mondrian, 1957]. This is surprising because, initially, his patterns seem as far removed from nature as they possibly could be. They consist of primary colors and straight lines - elements that never occur in a pure form in the natural world. His patterns are remarkably simple when compared to nature's complexity. However, his essays reveal that he viewed nature's complexity with distaste, believing that people ultimately feel ill at ease in such an environment. He also believed that complexity was just one aspect of nature, its least pure aspect, and one that provides a highly distorted view of a higher natural reality. This reality, he argued, "appears under a veil" - an order never directly glimpsed, that lies hidden by nature's more obvious erratic side. He believed that any glimpse through this "veil" would reveal the ultimate harmony of the universe. Mondrian wanted to capture this elusive quality of nature in his paintings.

Despite the differences in their chosen visual languages, both Pollock and Mondrian aimed to capture the underlying structure of the natural world on canvas. Declaring "I am nature," Pollock focused on expressing nature's complexity. Remarkably, he painted fractal patterns 25 years before scientists discovered that nature's complexity is built from fractals. Furthermore, based on the fractal aesthetic qualities revealed in the perception experiments, current research is aimed at reducing people's physiological stress by incorporating fractal art into the interior and exteriors of buildings [Taylor et al, 2003, Wise et al, 2003]. These scientific investigations enhance Pollock's artistic standing in the history of Modern Art, with his work interpreted as a direct expression of nature's complexity. Now that science has caught up with Pollock, how should we view Mondrian's alternative view of nature?

The recent investigations of Mondrian's patterns indicate that peoples' aesthetic judgments of his visual language are insensitive to the ways that his language is applied. It's tempting to conclude that Pollock succeeded in the quest for natural aesthetics and that Mondrian failed. However, this interpretation doesn't account for the enduring popularity of Mondrian's patterns. Perhaps he succeeded in glimpsing through nature's "veil" with an unmatched clarity and was able to move his lines around with a subtlety well beyond our current scientific understanding of nature? Just as art can benefit from scientific investigation, so too can science learn from the great artists.

Acknowledgments

I thank my collaborators B. Spehar, C. Clifford, B. Newell, A. Micolich, D. Jonas, J. Wise and T. Martin.

References

Abrahams, F.D., Sprott, J.C., Mitina, O., Osorio, M., Dequito, E.A. and Pinili, J.M., (2003), Judgments of time, aesthetics, and complexity as a function of the fractal dimension of images formed by chaotic attractors, in preparation (private communication).

Aks, D, and Sprott, J, (1996), Quantifying aesthetic preference for chaotic patterns, *Empirical Studies of the Arts*, 14, 1.

Burrough, P.A., (1981), Fractal dimensions of landscapes and other environmental data, *Nature,* 295 240-242.

Cambel, A.B., (1993), *Applied Chaos Theory: A Paradigm for Complexity,* Academic Press, (London).

Castrejon-Pita, J.R., et al, (2003), Nasca lines: a mystery wrapped in an enigma, to be published in *Chaos*.

Chopard, B., Hermann, H.J., and Vicsek, T., (1991), Structure and growth mechanism of mineral dendrites, Nature, 309 409.

Cohn, J., (1894), Experimentelle unterschungen uber die gefuhlsbetonung der farben helligkeiten, und ihrer combinationen. *Philosphische Studien* 10 562

Deicher, S., (1995), *Mondrian*, Taschen (Koln)

Eaton, L.K.,(1998), *Architecture and Mathematics* (Ed. K. Williams).

Family, F., Masters, B.R. and Platt. D.E., (1989), Fractal pattern formation in human retinal vessels. *Physica D*, 38 98.

Feder, J. (1988), *Fractals*, Plenum (New York).

Gouyet, J.P. (1996), *Physics and fractal structures*, Springer (New York).

Kemp M. (1998) Attractive attractors, Nature; 394:627.

Lee, A. (2001), Piet Mondrian: Buridan's Ass and the Aesthetic Ideal, *Proceedings of the International Conference on "The Art of Seeing and Seeing of Art,"* Australian National University, (Canberra).

Livio, M. (2002), *The Golden ratio*, Broadway Books, (New York).

Louis, E., Guinea F. and Flores, F., (1986), The fractal nature of fracture, *Fractals In Physics* (eds. L. Pietronero and E. Tossati), *Elsevier Science* 177.

Lovejoy, S., (1982), Area-perimeter pelation for pain and cloud areas, *Science*, 216 185.

Mandelbrot, B.B., (1977), *The Fractal Geometry of Nature*, W.H. Freeman and Company, (New York).

Mandelbrot B. B. (1989) Fractals and an art for the sake of art. *Leonardo*; Suppl: 21-24.

Matsushita, M. and Fukiwara. H., (1993), Fractal growth and morphological change in bacterial colony formation, In *Growth patterns in physical sciences and biology*, (eds. J.M. Garcia-Ruiz, E. louis, P. Meaken and L.M. Sander), Plenum Press (New York).

Mondrian, P., (1957) *Life and Work*, Abrams (New York).

Morse, D.R., Larson, J.H., Dodson, M.M., and Williamson, M.H., (1985), Fractal dimension of anthropod body lengths, *Nature*, 315 731-733.

Niemeyer, L., Pietronero, L., and Wiesmann, H.J., (1984), Fractal dimension of dielectric breakdown, *Physical Review Letters*, 52 1033.

Nittmann, J., and Stanley, H.E., (1987), Non-deterministic approach to anisotropic growth patterns with continuously tunable morphology: the fractal properties of some real snowflakes, *Journal of Physics A* 20, L1185.

Peitgen, P.O., Richter, P.H. (1986), The beauty of fractals: images of complex dynamic systems. Springer-Verlag (New York).

Pickover, C., (1995), *Keys to Infinity*, Wiley (New York) 206.

Potter, J., (1985), To a violent grave: an oral biography of Jackson Pollock, G.P. Putman and Sons (New York).

Richards, R., (2001), A new aesthetic for environmental awareness: Chaos theory, the beauty of nature, and our broader humanistic identity, *Journal of Humanistic Psychology*, 41 59-95.

Schroeder, M., (1991), *Fractals, Chaos and Power Laws*, W.H. Freeman and Company, (New York)

Skjeltorp. P., (1988), Fracture experiments on monolayers of microspheres, *Random Fluctuations and Pattern Growth* (ed. H.E. Stanley and N. Ostrowsky) Kluwer Academic (Dordrecht).

Spehar, B., (2001), An Oblique Effect in Aesthetics Revisited, *Proceedings of the International Conference on "The Art of Seeing and Seeing of Art,"* Australian National University, (Canberra).

Spehar, B., Clifford, C., Newell, B. and Taylor, R.P., (2003), Universal aesthetic of fractals, to be published in *Chaos and Graphics*, 37

Sprott, J.C., (1993), Automatic Generation of Strange Attractors, *Computer and Graphics*, 17, 325

Taylor, R.P. (1998), Splashdown, *New Scientist*, 2144 30.

Taylor, R.P., (2001), Architect reaches for the clouds, *Nature*, 410, 18.

Taylor, R.P., (2002), Spotlight on a visual language, *Nature*, 415, 961.

Taylor, R.P. (2002), Order in Pollock's chaos, *Scientific American*, 116, December edition.

Taylor, R.P., (2003) Fractal expressionism-where art meets science, *Art and Complexity Elsevier Press* (Amsterdam).

Taylor, R.P. (2003), Second nature: the magic of fractals from Pollock to Gehry, *New Architect*, July issue.

Taylor, R.P., Micolich, A.P., and Jonas, D., (1999), Fractal analysis of Pollock's drip paintings *Nature*, 399, 422.

Taylor, R.P., Micolich, A.P., and Jonas, D, (1999), Fractal expressionism, *Physics World*, 12, 25-28.

Taylor, R.P., Micolich, A.P., and Jonas, D., (2002), The construction of Pollock's fractal drip paintings, 35 203 *Leonardo*, MIT press.

Taylor, R.P., Spehar, B., Wise, J.A., Clifford, C.W.G., Newell, B.R. and Martin, T.P., (2003), Perceptual and physiological responses to the visual complexity of Pollock's dripped fractal patterns, to be published in the Journal of *Non-linear Dynamics, Psychology and Life Sciences.*

Van Tonder, G.J., Lyons, M.J., and Ejima, Y., (2002), *Nature* 419, 359.

Varnedoe, K., and Karmel, K., 1998, *Jackson Pollock*, Abrams (New York).

Wise, J.A., and Taylor, R.P., (2003), Fractal design strategies for environments, to be published in the *Proceedings of the International Conference on Environmental Systems*

Werner, B.T., (1999), Complexity in natural landform patterns, *Science*, 102 284.

Zeki, S., (1999) *Inner Vision*, Oxford University Press.

In: Chaos and Complexity Research Compendium
Editors: F. Orsucci and N. Sala, pp. 243-254
ISBN: 978-1-60456-787-8

Chapter 18

VISUAL AND SEMANTIC AMBIGUITY IN ART

Igor Yevin[*]

Mechanical Engineering Institute, Russian Academy of Sciences,
4, Bardina, Moscow, 117324 Russia. 72

Abstract

Non-linear theory proposed different models perception of ambiguous patterns, describing different aspects multi-stable behavior of the brain. This paper aims to review the phenomenon of ambiguity in art and to show that the mathematical models of the perception of ambiguous patterns should regard as one of the basis models of artistic perception. The following type of ambiguity in art will be considered. Visual ambiguity in painting, semantic (meaning) ambiguity in literature (for instance, ambiguity which V.B.Shklovsky called as "the man who is out of his proper place"), ambiguity in puns, jokes, anecdotes, mixed (visual and semantic) ambiguity in acting and sculpture. Synergetics of the brain revealed that the human brain as a complex system is operating close to the point of instability and ambiguity in art must be regarded as important tool for supporting the brain near this critical point that gives human being possibilities for better adaptation.

Non-Linear Models Perception of Ambiguous Patterns

In perception psychology, multi-stable perception of ambiguous figures is often considered as a marginal curiosity. Nevertheless, this phenomenon is one of the most investigated in psychology. The first description of ambiguity was given by Necker in 1832. The most known examples of ambiguous figures are specially designed patterns such Necker' cube, "young girl-old lady" and so on. But visual and semantic ambiguity is very often connected also with that the available visual or semantic information is not sufficient by itself to provide the brain with its unique interpretation. The brain uses past experience, either its own or that of our ancestors to help interpret coming insufficient and therefore ambiguous information. Many patterns in our every day life, in a way, are ambiguous patterns, but using

[*] E-mail address: yevin@online.ru, Phone: (095) 57604

additional information, we usually resolve or avoid ambiguity [1]. Nikos Legothetis recently shown that resolution of ambiguity is an essential part of consciousness job [2].

This paper aims to review and to familiarize with the present state the phenomenon ambiguity in art and to show that the mathematical models of the perception of ambiguous patterns should regard as the basic models of artistic perception.

Ambiguous patterns are examples of two-state, bimodal systems in psychology. When we perceive ambiguous figure, like the fourth picture in the row on Figure 1, the perception switches between two interpretations, namely "man's face" or "kneeling girl" because it is impossible for the brain to recognize both interpretations simultaneously. Just like for any bifurcative state, it is impossible for ambiguous figure to predict what namely interpretation will appear first. G.Caglioti from Milan Politectic Institute firstly paid attention, that ambiguous figures are cognitive analogue of critical states in physics.

Various authors pointed out that perception of ambiguous figures possess non-linear properties, and that multistabile perception could be modeled by catastrophe theory methods [3,4,5]

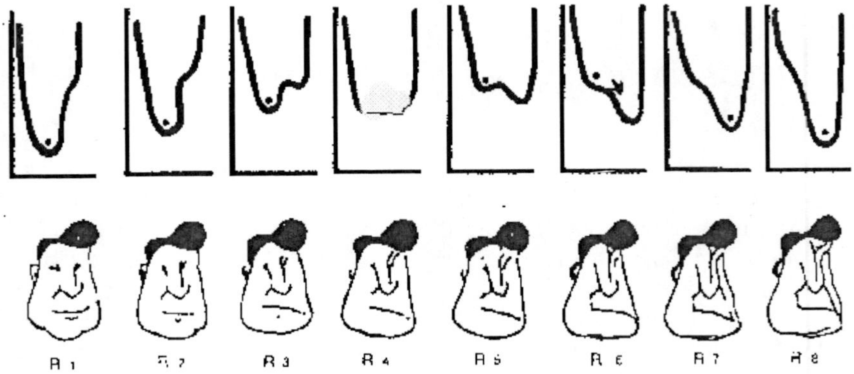

Figure 1. Ambiguous patterns are two-state systems. Their perception one can model by using elementary catastrophe "cusp".

The switch between two interpretation could be described by elementary catastrophe "cusp"

$$x^3 - bx - a = 0$$

where a and b are control parameters and x is the state variable. The first parameter a is called the *normal factor* and quantitatively describes the change in bias in the drawing in a "shape space" from a man's face to a woman's figure.

Because this model may be used for description of perception double meaning situations, it is reasonable to develop the idea of "shape space" on "meaning space" firstly introduced by Ch.Osgood [6].

The second parameter b is called the *splitting factor* or *bifurcation factor* and describes how much the amount of details is presented in the ambiguous figure.

The state variable x is presented as a scale from +10 ("looks a lot like a man's face") to ~ 10 ("looks a lot like kneeling girl"). For this model we could formally represent potential function

$$V = \frac{1}{4}x^4 + b\frac{1}{2}x^2 + ax$$

which depicted on Figure 1, and consider catastrophic jump from one image to another as non-equilibrium phase transition. It is worth to note, that unlike to physical sciences, where potential function usually deduces from fundamental laws or standard theories, in mathematical models in psychology and others "soft sciences" potential function is hypothesized and really is considered as potential energetic function, which should be minimized. In this case it might be also considered as Lyapunov function in Hopfield's model of pattern recognition.

Actually, during the viewing of ambiguous figures, perception lapses into sequence of alternations, switching every few seconds between two or more visual interpretations.

Ditzinger and Haken offered an approach to the description of such oscillation under recognition of ambiguous figures [7]. Each pattern is described in this model as a vector in the space of quantitative parameters. There is a procedure for selecting non-correlated parameters, which enable to reduce an information volume. The most informative parameters are the order parameters (all they peculiarities occur near critical points, as in the case of order parameters near phase transition [7]).

Pattern recognition procedure is the following. First, pattern-prototypes are stored in the computer memory. Then, the pattern that should be recognized is inputted. The recognition dynamics is built in such a way, that its vector evolves in a parameter space to the most similar pattern stored in the computer memory.

The prototype patterns are encoded by $V_i (i = 1,...,M)$. It is assumed that all these vectors are linearly independent. The components of every vector encode the features of the patterns.

A pattern to be recognized is encoded by a vector $Q(0)$ and is inputted in a computer memory at $t = 0$ A dynamic of pattern recognition is constructed so that $V_i (i = 1,...,M)$, that is the initial vector $Q(t)$, is pulled into one of prototype patterns V_k with which it mostly coincides.

Recognized pattern is presented as the linear combination of prototype patterns

$$Q(t) = \sum_{j=1}^{M} d_i(t)V_i + \xi(t)$$

where $d_i(t)$ is the order parameter, characterizing the degree to which a pattern is recognized, and $\xi(t)$ is a residual, uncorrelated with V_i.

The dynamic of pattern recognition is described as a gradient process in networks with only M neurons according to

$$\dot{d}_i(t) = \lambda d_i - (B + C)d_i \sum_{j \neq i}^{M} d_j^2 - Cd_i^3,$$

$$\lambda_i > 0, B > 0, C > 0, d_i(0) = V_i^{'} Q(0)$$

This system has only the attractors of the type (0, 0,..., $d_k \neq 0$,...0). It can be shown that they must be either saddle points or nodes, but not limit circles (oscillations).

Figure 2. Image ambiguity: "young girl" – "old lady".

Ditzinger and Haken offered synergetic model of the perception of ambiguous patterns, describing dynamical features of such perception. It is based on the model of pattern recognition described above, and the model of the saturation of attention. The recognition of ambiguous patterns is reduced to inputting only two patterns-prototypes (e.g., "young girl" and "old lady") into computer memory with the order parameters d_1 and d_2. In this case the dynamics of pattern recognition is described in the following way:

$$\dot{d}_1 = d_1(\lambda_1 - Ad_1^2 - Bd_2^2)$$

$$\dot{d}_2 = d_2(\lambda_2 - Bd_1^2 - Ad_2^2)$$

$$\dot{\lambda}_2 = g(1 - \lambda_2 - d_2^2)$$

$$\dot{\lambda}_1 = g(1 - \lambda_1 - d_1^2)$$

where the overdot means $\frac{d}{dt}$, λ_1 and λ_2 are time dependent attention parameters, and A, B, and g are constants. The last two equations describe the saturation of attention in the perception of prototype patterns. As analysis shows, the oscillation of perception occurs when the appropriate relations between constants are satisfied [7]. The recognition of ambiguous patterns has very profound and various analogies with numerous artistic phenomena. This model perception of visual ambiguous patterns also could be applied on the case of meaning ambiguity, because meaning perception also includes such phenomena as saturation of attention and the concept of the order parameter [8].

Visual Ambiguity in Art

Let us first consider specially designed visual ambiguity in art. Painting by Giuseppe Arcimboldo "The Librarer" is one of the first examples of such type ambiguity in painting. At first sight we recognize face, but a closer look reveals just an arrangement of different books.

Figure 3. Giuseppe Arcimboldo "*The Librarer*"

The most famous example of ambiguity in painting is, of course, *Mona Lisa* by Leonardo. In *The Story of Art* Ernest Gombrich said:

> "Even in photographs of the picture we experience this strange effect, but in front of the original in the Paris Louvre it is almost uncanny. Sometimes she seems to mock at us, and then again we seem to catch something like sadness in her smile."

> "This is Leonardo's famous invention the Italians call "*sfumato*" - the blurred outline and mellowed colors that allow one form to merge with another and always leave something to our imagination. If we now turn to the "Mona Lisa", we may understand something of its mysterious effect. We see that Leonardo has used the means of his "*sfumato*" with the utmost deliberation. Everyone who has ever tried to draw or scribble a face knows that what we call its expression rests mainly in two features: the corners of the mouth, and the corners of the eyes. Now it is precisely these parts which Leonardo has left deliberately indistinct, but letting them merge into a soft shadow. That is why we are never quite certain in which mood Mona Lisa is really looking at us. Her expression always seems just elude us" [9, p.228].

The ambiguity of Mona Lisa's smile one can compare with ambiguous images like "young girl - old lady". The oscillation in the perception of that painting can be described by Ditzinger-Haken's model.

Figure 4. Ambiguity of *Mona Lisa's* smile.

Figure gives an example other kind of visual ambiguity, when the human face and part of his figure is designed from. An example of such ambiguity is *Disappearing Bust of Voltaire* by Salvador Dali.

Figure 5. Ambiguity of Voltaire bust in Salvador Dali's painting *Disappearing Bust of Voltaire*.

Semantic Ambiguity of Visual Scenes

Let us consider the following painting by J. Vermeer [11].

Why depicted scene is semantically ambiguous? Because the available information is not sufficient and this scene offers huge amount of meaning interpretations.

Undoubtedly, there is some relationship between the man and the woman. But is he her husband or a friend? Did he actually enjoy the playing or he think that she can do it better?

Is the woman really playing - she is after all standing - or she is concentrating on something else, pėrhaps something he told her, perhaps announcing a separation or a reconciliation?

All these and many others scenarios have equal validity.

There is a humorous book called "Captions Courageous" by Reisner and Capplow attempting reinterpretation of famous masterpieces in painting – with more or less wit [12]. This possibility to create new interpretations for famous paintings which are perceived as comic is connected with insufficient information.

Figure 6. Jan Vermeer. A lady at the Virginals with a Gentleman

Semantic Ambiguity in Plot Development and in Comic Situations

A significant type of ambiguity in art means the possible existence in artwork (most often in position of main hero) of two different states, one of them may be hidden until a certain time. A commonplace example of this form of instability exists in numerous book and movie plots in which a spy or Secret Service agent is hiding his identity while maneuvering about in hostile camp. At any moment, he may be unmasked, and the agent's task is to extend his secret identity as long as possible.

In well-known American movie “ROBOCOP” the main character is simultaneously a robot, incarnating an idea pitiless and perfect machine of revenge, and a human being, capable on deep and tender feelings. Another, less- banal example, ambiguity of social nature - what V.B.Shklovsky describes as "the man who is out of his proper place" - is also widely presented in art [13]. The main character Hlestakov in the play by N.Gogol “Inspector General” obviously one may describe using this kind of ambiguity.

In Apuleius's "Golden Ass" the main character is, of course, out of his proper place because the ass in reality is a man.. The plots of such tales like "The Ugly Duckling" by H.Andersen and "The Beauty and the Beast" also are of the same type of ambiguity, sustained over the entire period of the plot.

In the majority of the novels by Agatha Kristy we deal with semantic ambiguity, as almost any character of these novels could appear as the murderer. This state of semantic ambiguity is skillfully supported by the author down to an outcome of the plot: “You know, that I never deceive. I simply speak something such, that it is possible to interpret double” - once confessed A.Kristy.

Without ambiguity of natural languages, the existence of poetry is impossible. According to A.N.Kolmogorov, entropy of language H contains two terms: meaning capacity h_1 - capability to transmit some meaning information in a text of appropriate length, and flexibility of language h_2 - a possibility to transmit the same meaning by different means [14]. Namely h_2 is a source of poetic information, and the ambiguity of language is one of the causes of it's flexibility. Languages of science usually have $h_2 = 0$, they exclude ambiguity, and cannot be used as a material for poetry. Rhythm, rhymes, lexical and stylistic norms of poetry will put some restrictions on a text. Measuring that part of the ability to carry information spent on those restrictions (denoted as β), A.N.Kolmogorov formulated the law, according to which poetry is possible if $\beta < h_2$. If the language has $\beta \geq h_2$, than poetry is impossible.

We know that the brain resolves a visual ambiguity by means of oscillation. A semantic ambiguity (the ambiguity of meaning) is a result of ambiguous words or whole sentence. Semantic ambiguity, wide spread in comic situations, also resolves by oscillations.

Ambiguity of humor is often a clash of different meanings. It involves double or multiple meanings, sounds, or gestures, which are taken in the wrong way, or in incongruous ways.

Here is D.D.Minayev's epigram:

> "I am a new Byron" - you proclaim yourself. I can agree with you: The British poet was lame
> The rhymes of yours are also lame."

The method used in this epigram is connected with a comparison based on different distant meanings (Byron was the lame, and a vain poet was also a lame, but in his rhymes).

The situation described in this epigram is common to a lot of semantically ambiguous comic situations, which contain two states. One state we should call a state with high social status. This position is honorable and sometimes brings profit. The second state we should call a state with low social status. Everybody avoids occupying it. In the aforesaid example, the state with the high social status ("a good poet") we connect with words "a new Byron". Another poet is trying to get this state. But the author of the epigram unexpectedly transfers a poet to the second state with a low social status. This state we connect with the words "the rhymes of yours are also lame". Such an unexpected leap is achieved by using the same word ("lame") for totally different states.

So, a feeling of comic is very often connected with sudden transition from a state of high social status to a state of low social status, or the other way round. Is it a single transition? Does it happens only once? Of course not. It is a multistabile perception of meaning. The rhythmical, repeating nature of laughter (ha-ha-ha, etc.) shows that such transitions are repeated. Evidently, a laughing person mentally oscillates every time from the state of high social status to the state of low social status and *vice versa*, by comparing them. As a result, the rhythmical laughter is generated by the nervous system.

Let us consider also the following anecdote about Sherlock Holmes and Dr.Watson.

> Sherlock Holmes and Dr. Watson are going camping. They pitch their tent under the stars and go to sleep. Sometime in the middle of the night Holmes wakes Watson up.
>
> "Watson, look up at the stars, and tell me what you deduce."
>
> Watson says, "I see millions of stars, and if there are million of stars, and if even a few of those have planets, it's quite likely there are some planets like Earth, and if there are a few planets like Earth out there, there might also be life."
>
> Holmes replied: "Watson, you idiot, somebody stole our tent".

We see, that Watson and Holmes offered two different semantic interpretations of the same visual picture of star sky and if Watson gave namely one of possible interpretation of picture of star sky, Holmes paid attention on semantic context of this picture and connected it with their rest position.

The origin of the oscillatory character of laughter should be connected with the fundamental property of the distributed neuron set, i.e. as the oscillation occurring in the perception of ambiguous patterns. According to Ditzinger-Haken's model of recognizing of ambiguous patterns, stable limit cycles can be formed in systems of usual nonlinear differential equations for those variables, which describe the visual perception (e.g. attention). Evidently, this is the common characteristic of distributing neuron sets. That's why it is manifested not only in evolutionary low stages (the ancient visual-morphologic structure of nervous and psychological activity of a human being), but also in its latest stages as well (in the semantic-analytical structures of the left cerebral hemisphere).

Comic situations are very often connected with polysemantic, i.e. semantically ambiguous, situations. Another situation of perception of ambiguous patterns occurs in a parody of a famous person by some actor. On one hand, we can recognize the manners, gestures, style and voice of that famous person. On the other hand, we see quite a different person. The same method is used in literary and poetic parodies. Every time we are dealing with a bimodal, double-meaning situation. As a result, we have the oscillation of perception, and laughter is one of the external manifestations of this oscillation.

One can assume that in ambiguous comic situations oscillations occur between two semantic images. The phenomena of synchronization are typical for a self-organizing process in an active medium (and the nerve substance is an active medium). From that, we can conclude that the period of oscillation between semantic patterns coincides with the period of outward macroscopic oscillations, manifested as laughter with the duration of about 0.1 sec. This value is much smaller than the oscillation period, which occurs when recognizing ambiguous figures (1-5 sec.).

Why does laughter occurs in the perception of double-meaning situations, and not in the visual perception of ambiguous patterns? We can explain this by essentially different periods of the corresponding oscillations. In the visual perception this period is approximately equal to t=10 sec., and in the perception of the ambiguity of meaning this period is about t=0.1 sec. That difference could be explained by the fact that a much smaller mass of nerve substance is involved in creating semantic patterns, compared with constructing visual patterns. This is because visual information is processed in the massive and ancient visual cortex, and semantic patterns are interpreted in compact Broke-Vernike zone in the left brain hemisphere. Anecdotes, jokes and sketches deliberately are created as short as possible (laconic), in order to reduce the time needed for the saturation of attention in the process of recognition.

Mixed Ambiguity

Ambiguity of Sculpture

We have considered visual ambiguity in painting (see also [10]) and semantic ambiguity in jokes, anecdotes and puns. Let us consider mixed (visual and semantic) ambiguity, taking an example from sculpture art. Sculpture involves an ability to depict representatives of living nature (most often man and animals) from materials of inanimate nature (wood, stone, bronze, etc).

In creativity of different sculptures can be observed a prevalence of one of these phase with respect to another. In Michelangelo's works we see triumph of alive and even spiritual under inert matter of stone. Gombrich wrote in book "The Story of Art": "While in "*The Creation of Adam*" Michelangelo had depicted the moment when life entered the beautiful body of a vigorous youth, he, now, in the " *Dying Slave*", chose the moment when life was just fading, and the body was giving way to the laws of dead matter. There is unspeakable beauty in this last moment of final relaxation and release from the struggle of life - this gesture of lassitude and resignation. It is difficult to think of this work as being statue of cold and lifeless stone…".

It is interesting to note, that ambiguity of sculpture art influences on literature, because the plots of some works of arts in literature are based on the idea of animated statue - that is, the transition "inanimate-animated" (such as opera "Don Giovanni" by Mozart, "Bronzer Horseman", "Stone Guest" by A.Pushkin) and of course in ancient legend about sculptor Pygmalion.

Ambiguity of Dolls

In the essay "Dolls in system of culture" Yu.Lotman marks ambiguous (as well sculpture) nature of this cultural phenomenon closely connected to ancient opposition alive and dead, spiritual and mechanical. At the same time, as against a sculpture, the doll demands not contemplation but play. It serves as a certain stimulator provoking creativity[15].

Ambiguity of Acting

Like any human being, an actor has in his everyday life some set of rather stable physiological and psychological personal properties: sex, appearance, timbre of voice, gait, temper, and so on. The acting involves it's ability to create a second phase, a "role" phase, different from the original physiological and psychological nature of the actor. In other words, a bimodal "actor-role" state created may be compared with ambiguous patterns, for instance, the pattern where we see in turn "young girl" or "old lady". One may say that in this case young girl will "play the role" of old lady and *vice versa*.

In acting, one can observe the existence of two polar types of actors:

1) An actor as a bright, brilliant individuality, eccentric person with the original appearance, and so on (Alain Delon, Arnold Schwarzenegger). It is rather easy to make a parody of such actors;
2) An actor with prominent outstanding abilities for transformation and reincarnation (Laurence Olivier, Alec Guiness). In that case, it is very difficult to make a parody.

Yu.Lotman note, that in the cinema more, than at the theatre the spectator sees not only role, but also actor [15, p.658]. Observing play of the famous actor we alternately focus our attention or on guise (image) of actor familiar to us on other movies, or on peculiarities of a role, which the actor plays. Such oscillations of attention is the reason, that with the reference to acting we use a word "play".

In the case of acting the prototypes are, for instance, "Laurence Olivier" (the image of actor) and "Othello" (the image of character). Therefore, according to the common law of perception of ambiguous patterns, the oscillation of our attention takes place, and we see in turn either an actor or his role.

Just as like bimodal nature of sculpture art begets plots about animated statue, bimodality of actor art gives a possibility to use a phase transition called "character invasion" for plot development [16].

The main hero of the film "A Double Life" plays the role of Othello for so long time that it begins to affect to his psychic activity, making him more and more jealous of his beloved, and like the stage character, he strangles her and then kills himself. In the film "Jesus of Montreal" the actor playing the role of Jesus Christ becomes transformed into a Christ-like figure [16].

As a rule, all bimodal metastable states in the end of movies turn into stable, onemodal states as a result of bifurcation.

Conclusion

In ordinary speech, and especially in scientific communication, in general we try to avoid ambiguity. By contrast, in humor, one of the aims is to create ambiguous situations to provoke laughing. And in art as a whole ambiguity is an indispensable, necessary part.

> "...art is supposed to have multiple meanings. It self-defeating to increase one aspect of meaning. The more a single meaning dominates a work, the less it is a work of art. Something that has one and only one meaning – no matter how interesting or important that meaning is - - is no longer a work of art" [17, p.46]

Synergetics and the theory of complexity revealed that the human brain operate near unstable point, because only near criticality the human brain could create new forms of behavior. Ambiguity in art is an important tool maintaining the brain near this unstable, critical point.

References

[1] P. Kruse, M. Stadler, *Ambiguity in Mind and Nature.: Multistable Cognitive Phenomena.* Springer, Berlin, 1995.

[2] N.L.Legothetis. Vision: *A Window on Conciousness*. Scientific American. November, 1999 pp.69-75

[3] T. Poston, I. Stewart, Nonlinear Model of Multistable Perception. *Behavioral Science.,* **23** (5), 1978, 318-334.

[4] I.N. Stewart, P.L. Peregoy, Catastrophe Theory Modeling in Psychology. *Psychological Bulletin*, **94**(21), 1983, 336-362.

[5] L.K. Ta'eed, O. Ta'eed, J.E.Wright, Determinants Involved in the Perception of Necker Cube: an Application of Catastrophe Theory. *Behavioural Science*, **33**, 1988, 97-115

[6] Osgood, Ch., Suci, G., Tannenbaum P., 1958, *The Measurement of Meaning*, University of Illinois Press

[7] H. Haken, Principles of Brain Functioning. Springer, Berlin, 1996.

[8] W. Wildgen, *Ambiguity in Linguistic Meaning in Relation to Perceptual Multistability.* In P.Cruse and M.Stadler [1].

[9] E. Gombrich, *The Story of Art*. Phaidon, New York, 1995.

[10] G. Caglioti, *Dynamics of Ambiguity*. Springer Berlin, 1992.

[11] S.Zeki. *Inner Vision.* Oxford University Press. 1999

[12] Reisner B. and Kapplow H. *Captions Courageous*. Abeland-Schuman, 1954

[13] V.B.Shklovsky. *Tetiva*. Moscow, 1967.(In Russian)

[14] A.N. Kolmogorov, *Theory of Poetry*. Moscow, Nauka, 1968, 145-167 (in Russian)

[15] Yu.Lotman. *About Art.* St Petersburg, 1998 (In Russian)

[16] Neuringer C. and Willis R. The Cognitive Psychodynamics of Acting: Character Invasion and Director Influence. *Empirical Studies of the Arts*. v.13, N1, 1995 p.47

[17] C.Martindale. *The Clockwork Muse*. Basic Books. 1990.

In: Chaos and Complexity Research Compendium
Editors: F. Orsucci and N. Sala, pp. 255-257
ISBN: 978-1-60456-787-8

Chapter 19

DOES THE COMPLEXITY OF SPACE LIE IN THE COSMOS OR IN CHAOS?

Attilio Taverna*
Painter

The art of painting, as we have already known for a long time, is first and foremost an aesthetic inquiry on the nature of space. It's easy to understand why. The state of being of an aesthetic experience such as a painting, always needs an extension, sometimes of a surface, often of a double dimension, always of some kind of phenomenology of space. Here is the ultimate reason why.

In our modern times, even in the case of drawing the structure of a chip, or when we shoot a real event with a video camera, we use an extension as a support. So to say we are using an idea of space already known to us, in the same way in which we use the net. We can use it only because there's an idea of pluri-dimensional space in it that we identified as fundamental: cyber-space, precisely/exactly.

But what is the space? Can we say that we know it for sure?

Even Plato in the Timeo's dialogue, the big Greek cosmogonic tale of 25 centuries ago, said that space has a *bastard nature*. He also admonished that space is the condition of possibility of being of all phenomena but at the same time it cannot become a phenomenon. That means that space is the *conditio sine qua non* for a phenomenon to appear but it cannot appear in the way phenomenon do. That's the reason why it has a bastard nature: it allows appearance but doesn't appear.

So now, it becomes clear how the idea of space is something immersed in the ontological oscillation, which is something irrepressible. As we've already seen, it's space's own nature that allows the decline, in the visible manifestation, of the horizon of beings. This nature of space is the condition of possibility of every phenomenon's apparition.

25 centuries after Plato, the German philosopher Immanuel Kant, attempted to solve the enigma of the nature of space in his "Critique of the Pure Reason" said: "the space is not an empirical concept, drawn by external experiences… it is, instead, a necessary representation a

* E-mail address: attiliotaverna@libero.it

priori, which serves as a fundament to all the other external intuitions". He would conclude by saying that the intuition of space is the original shape of sensibility.

Space and time are the pure forms, a priori, of sensitivity.

And when, at the beginning of the last century, space and time joined together thanks to physics-mathematics in only one quadri-dimensional being called spacetime, the aesthetical experience of painting became, as a result of physics, aesthetical search on the nature of spacetime.

That's all about philosophy. But talking also about science with an observation from Albert Einstein on the genesis of the theory of relativity, we can understand how the true nature of space is inevitably implicated with the formal and ideal systems which we call geometry. Albert Einstein, in fact, would say: suddenly I realized that geometry had a physical meaning….

After this consideration and intuition of the great physicist, who had revolutionized the knowledge of reality, how can we not ask ourselves about the meaning of the ideal forms of geometry, such as the curvature of spacetime, for instance, - which is a geometrical form produced by men- clash/coincide with one of the fundamental forces of nature, the gravitational force? Even better, gravity is the curvature of spacetime. And so?

How can't we wonder also about another question: What's the form in ontology?

Art is not, and cannot be considered, unrelated to this question. And its own history testifies and documents this fact.

Art has conducted this query maybe since the beginning of man's history. And painting realizes a vision of this possible question on the nature of spacetime, its possible form, before being any other form of aesthetic query, as we have already said.

My aesthetical experience fed on this query as well. The nature of space, in my opinion, is a kind of chromatic polyphony of ideal and formal opportunities, not necessary axiomatic, as the systems of Euclidean geometries and not-Euclidean, but conceived as ideal opportunities of never-ending geometries existing in an unfinished space.

We can't forget that while Albert Einstein was conceiving the theory of relativity and made us aware of the physical meaning of geometrical forms, philosophy was analyzing with rigour the formal and primary idealizations of geometry. We have to remember Edmund Husserl's studies. He is another popular German philosopher, who, at the beginning of last century, thought geometry was an ideal and eidetic dimension, defining it as the *visual language of idealities non-in-chains concepts.*

So to say that the whole phenomenology was subjected to the causal principle, while this formal and ideal dimension called geometry was not subjected to it. From now on we could think of space as a dimension hanging on the greatest freedom of thinking and form joined together that man has ever possessed.

The complexity of any possible notion of space becomes dizzing. And modernity took charge of this demonstration. Any other possible example would be superfluous.

At the same time the mathematic notion of chaos contributed to change the idea of space which was crystallizing in the geometric systems consolidated/established axiomatically.

We can add something else: if by chaos we mean the inability to foretell the future evolutions of every kinetic non-linear system, with the not completely known conditions of the initial system, we have to admit that the unpredictability of every future evolution of every system is totally open to a description made by endless ideal and unknown formality, from a formal and geometrical core of unpredictable descriptions.

So what comes to light as geometrical language, the "visual language of idealities non-in-chains concepts ", is not a knowledge of the past, but is something ineluctable, a necessary knowledge of the future.

We have also to underline, as useful indication, that in the theoretical contemporary physic some theories are elaborated – for instance the one of the *superstrings* - and these theories need many dimensions of space to explain their mathematical compatibility and their theoretic correctness. To say that the reality of spacetime is not exactly what happens in front of our senses and that we are used to see and express everyday.

Even if art doesn't want to find the foundation of the world, because this is not in his epistemic status and this result belongs to the purpose of hard sciences, physic for example, nevertheless art carries out the world as a fundament. That's its vocation. That's its destiny.

And if the reality of spacetime gives up as foundation, so to say as the condition of possible apparition of any possible apparition, should art be excluded from this query on the foundation? No, centairly. That would be impossible. Great narrations of aesthetics would never stop to question everything, even better, on the everything, because the specific task of art is aesthetic query to the very limit of possibility. Since ever.

Conclusion and question: if reality, the reality of spacetime, is possible to be described in the physic-mathematic sciences by an idea of a very complex space multi-dimensionality that escape any visibility and any chance of daily visibility,

Who can see these possible concepts of space that are the real space described by science- if not an aesthetical experience that found its foundation on the artistic praxis right on this lyrical query on the nature of spacetime?

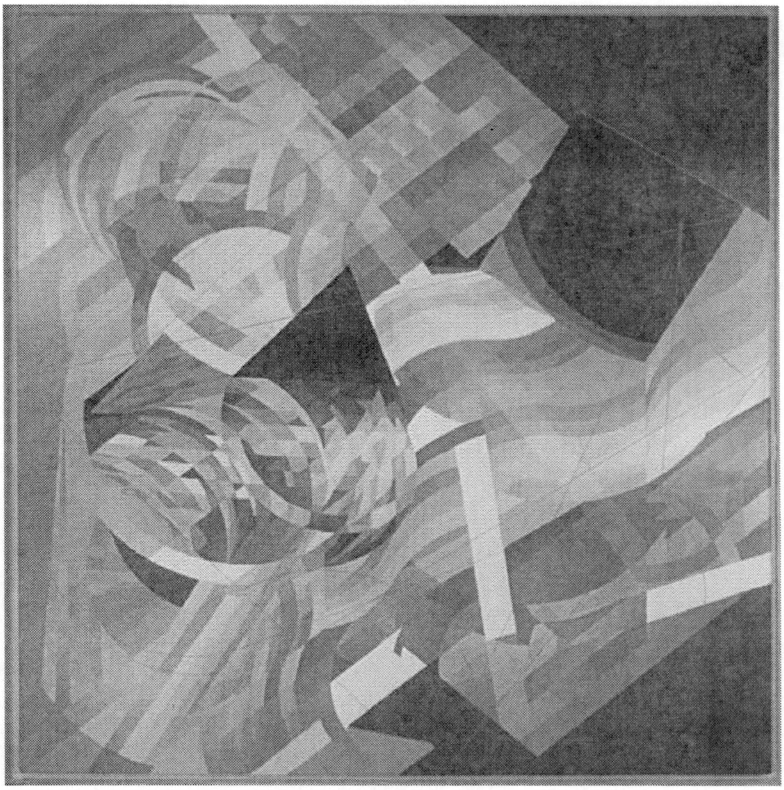

In: Chaos and Complexity Research Compendium
Editors: F. Orsucci and N. Sala, pp. 259-277
ISBN: 978-1-60456-787-8

Chapter 20

CRYSTAL AND FLAME/FORM AND PROCESS THE MORPHOLOGY OF THE AMORPHOUS

Manuel A. Báez*
Form Studies Unit, Coordinator, School of Architecture, Carleton University, Ottawa, Ontario K1S 5B6 Canada

"Philosophy is written in this enormous book which is continually open before our eyes (I mean the universe), but it cannot be understood unless one first understands the language and recognizes the characters with which it is written. It is written in a mathematical language and its characters are triangles, circles, and other geometric figures. Without knowledge of this medium it is impossible to understand a single word of it; without this knowledge it is like wandering hopelessly through a dark labyrinth."

Galileo Galilei, "The Assayer" (1623) [1]

"Why is geometry often described as "cold" and "dry?" One reason lies in its inability to describe the shape of a cloud, a mountain, a coastline, or a tree. Clouds are not spheres, mountains are not cones, coastlines are not circles, and bark is not smooth, nor does lightning travel in a straight line.

More generally, I claim that many patterns in Nature are so irregular and fragmented, that, compared with Euclid – a term used in this work to denote all of standard geometry – Nature exhibits not simply a higher degree but an altogether different level of complexity. The number of distinct scales of length of natural patterns is for all practical purposes infinite.

The existence of these patterns challenges us to study those forms that Euclid leaves aside as being 'formless,' to investigate the morphology of the 'amorphous.' Mathematicians have disdained this challenge, however, and have increasingly chosen to flee from nature by devising theories unrelated to anything we can see or feel."

Benoit B. Mandelbrot [2]

"We are living in a world where transformation of particles is observed all the time. We no longer have a kind of statistical background with permanent entities floating around. We see that irreversible processes exist even at the most basic level which is accessible to us. Therefore it becomes important to develop new mathematical tools, and to see how to make

* E-mail address: mbaez@ccs.carleton.ca

the transition from the simplified models, corresponding to a few degrees of freedom, which we have traditionally studied in classical dynamics or in quantum dynamics, to the new situations involving many interacting degrees of freedom."

Ilya Prigogine [3]

Abstract

This paper presents the work and research produced through an on-going architectural project entitled *The Phenomenological Garden.* The project seeks to investigate the morphological and integrative versatility of fundamental processes that exist throughout the natural environment. Work produced by students in workshops incorporating educational methods and procedures derived from this research will also be presented. This evolving project is a systematic investigation of the versatile and generative potential of the complex processes found throughout systems in Nature, biology, mathematics and music. As part of the Form Studies Unit in the School of Architecture at Carleton University, the work seeks to investigate how complex structures and forms are generated from initially random processes that evolve into morphologically rich integrated relationships.

The morphological diversity revealed by this working and teaching method offers new insights into the complexity lurking within nature's processes and bridges the theoretical gap between Galileo Galilei's conception of nature, as revealed above, and the modern theories of Chaos and Complexity as exemplified by Benoit Mandelbrot and Ilya Prigogine. This working process also offers insights into the conceptual and philosophical aspirations of such key central figures as Antoni Gaudi, Louis Sullivan, Frank Lloyd Wright, and Buckminster Fuller in the early formative period of modern architecture, and more recently, the architect/engineer Santiago Calatrava. The implications of these developments are relevant to the study of morphology as well as to the field of architecture at a time when it is addressing the concepts and themes emerging out of our deeper understanding of dynamic and complex phenomena in the physical world.

Introduction

Through the aid of modern computer visualization and analyzing techniques, we have recently acquired deeper insights into the ways energy is interwoven into dynamic systems and structures of startling beauty and versatility that often recall the patterns and motifs found throughout the natural and man-made environment. The elemental cellular patterns that emerge from these processes inherently contain information and are themselves dynamic *events-in-formation*. An understanding and appreciation of our innate relationship with this phenomenon can be achieved through hands-on systematic "readings" of the complex characteristics of these emergent cellular units and their assemblages.

These fertile, self-organizing and regulatory systems and patterns inherently exist within and generate the rich realm of natural phenomena. Simultaneously, they are also composed of and generate elemental inter-active relationships that gradually evolve into versatile integrative systems. When the versatility and generative potential of these systems and their interrelated cellular patterns are systematically analyzed, they can yield new insights into the emergence of complex morphological structure and form.

The intrinsic nature of the patterns generated by these dynamic processes reveals that they are cellular configurations of highly ordered relationships. Through these apparently static patterns and stable forms flow the highly dynamic undulations of an energetic process. These emergent complex networks are fluently encoded patterns of potentiality offering a multitude of possible or alternative "readings." The cellular units comprising these patterned morphogenetic inter-activities innately contain the intrinsic attributes of the versatile processes that generate them. We are inextricably part of and surrounded by this rich and dynamically complex matrix of natural phenomena. The probing of the inherent nature of this pro-creative matrix can lead to an insightful understanding of the *reciprocal relationship* between matter, developmental processes, growth and form. Rich and exciting educational methodologies are also offered through new procedures and techniques that would inherently allow for intuitive learning through self-discovery.

Background

Galileo Galilei's metaphor of the book of nature reflects the new philosophical direction of his time while, simultaneously, following an ancient tradition regarding the nature of the physical universe. He emphasizes the importance of understanding the nature of the characters through which the language of this book is written. At the time, it was believed that all-encompassing scientific knowledge could be achieved solely through the quantifiable and visual aspects of the material world and its organizing parts. Galileo's vision reflects the influence of the work of Plato, most notably his *Timaeus* where we find an emphasis on the primary importance of the elementary geometric units or ideas behind the material world. This was in sharp contrast to the Aristotelian philosophy dominating the Western world up until the Scientific Revolution in the sixteenth and seventeenth centuries. Prior to this, the world was envisioned as a living organism where spirit, substance and form were inextricably interrelated. This new mechanistic vision culminates with René Descartes' analytic method and eventually Isaac Newton's grand synthesis of Newtonian mechanics. This vision would prevail and dominate Western science until the early part of the twentieth century. The first major influential challenge to this mechanistic vision came in the late eighteenth and nineteenth centuries from the Romantic Movement in literature, art and philosophy.

Primordial Seeds

The Romantic Movement, as exemplified by J. W. von Goethe, had a profound influence on the American architect Louis Sullivan and, subsequently, Frank Lloyd Wright through the strong German cultural presence in late nineteenth century Chicago, the transcendentalism of Ralph Waldo Emerson, and the writings of the philosopher Herbert Spencer. For Sullivan and Wright, the creative process was seen as a transcendental experience similar to natural growth and development. Reminiscent of Goethe's botanical observations, Sullivan made references to "the germ of the typical plant seed with its residual powers."[4] In the primary geometric figures, Sullivan saw primordial seeds with "residual power" to grow and generate organic forms. He illustrated the development of his own ornament through the morphological transformations of these primary units (see Figure 1). To Sullivan these were the primary generative units of a "plastic" and "fluent geometry" containing "radial energy" and "residual power" capable of projecting outwards or inwards through the inherent "energy lines" or axes of the units.

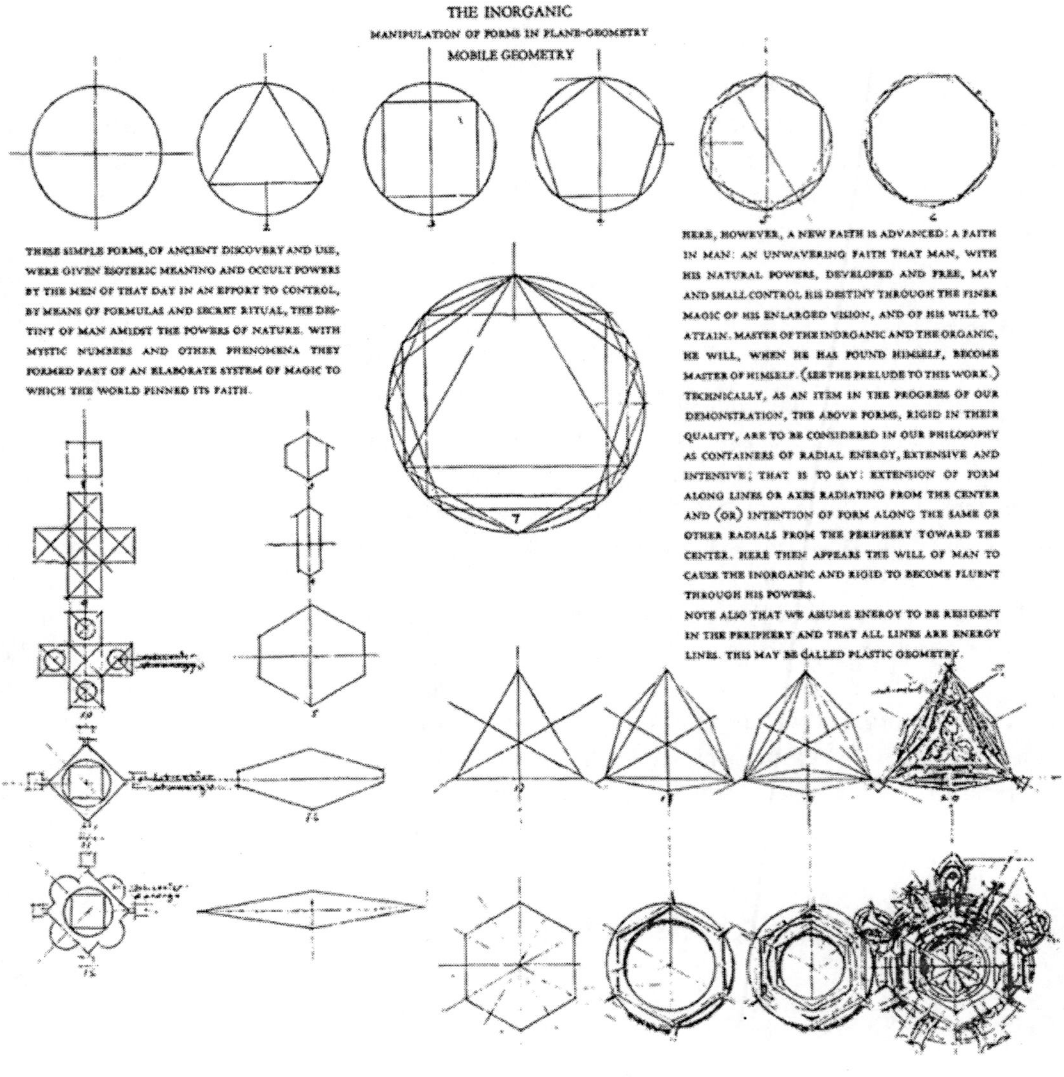

Figure 1. Louis Sullivan [4], Manipulation of forms in plane geometry.

This dynamic, generative and comprehensive vision of nature inspired the work and ideas of both Sullivan and Wright. They both incorporated a basic unit system of working that would undergo systematic morphological permutations, limited only by the designer's imagination. Wright would state:

> "All the buildings I have ever built, large and small, are fabricated upon a unit system—as the pile of a rug is stitched into the warp. Thus each structure is an ordered fabric. Rhythm, consistent scale of parts, and economy of construction are greatly facilitated by this simple expedient—a mechanical one absorbed in a final result to which it has given more consistent texture, a more tenuous quality as a whole."[5]

Louis Sullivan and Frank Lloyd Wright both envisioned an organic, versatile, vibrant and integrative design process. Recent developments in modern science and in the early part of the twentieth century reveal a similar conception regarding the complex nature of the physical world.

A. Complex Tissue of Events

During the early part of the twentieth century, a fundamental conceptual shift was underway regarding our comprehension of the physical world and the principles involved in its organizing and structuring processes. Fundamentally, the nature of matter was revealed to consist of an irreconcilable yet intrinsic paradoxical contradiction. At the heart of this dilemma was the nature of form, structure, developmental organization, and emergent patterns. Measurable or numerically quantifiable form and position were inextricably linked and reciprocally related to the highly complex behaviour of the dynamic inter-actions of energy. Subsequently, it was revealed that through these highly complex processes emerge three-dimensional networks or patterns of probable or possible alternatives. According to the German physicist Werner Heisenberg:

> "The world thus appears as a complicated tissue of events, in which connections of different kinds alternate or overlap or combine and thereby determine the texture of the whole." [6]

The intrinsic nature of this dynamic conception consists of the realization and comprehension of patterns as highly complex networks of organizational texture and potentiality. Understanding these inherent characteristics would provide the necessary insights in order to probe deeper into this new paradoxical conceptualization.

The contradictory nature of matter is a recurring theme that's encountered when contemplating the relationships between substance and form, subject and object, as well as unity and multiplicity. In the history of biology, this ancient dilemma is found to be inextricably associated with the understanding of the forms of living organisms and their growth or developmental processes. In physics and biology, at the most elementary level, nature's processes are essentially the inter-relationships between things in a myriad of different orders of magnitude. We are inextricably part of and surrounded by Heisenberg's encoded "*tissue of events*." The probing of the inherent nature of this fluently textured tissue, can lead to an insightful understanding of the nature of patterns and their correlation with matter, developmental processes, growth and form. In the words of Gregory Bateson:

> "We have been trained to think of patterns, with the exemption of those in music, as fixed affairs. It is easier and lazier that way but, of course, all nonsense. In truth, the right way to begin to think about the pattern which connects is to think of it as primarily (whatever that means) a dance of interacting parts and only pegged down by various sorts of physical limits and by those limits which organisms characteristically impose." [7]

This dynamic conception envisions emergent networks as fluently encoded records or events that contain the in-forming and expressive potential of their generative processes. Modern computer visualization and simulation techniques are providing deeper insights into the richness of these networks that are embedded within Heisenberg's "*tissue of events*" and Bateson's "*dance of interacting parts.*" More profound fundamental insights are offered into the earlier developments regarding the nature of the physical world. Again, within the realm of complex phenomena, we encounter "objects" or confined spatial forms that "attract" or resolve the dynamic inter-actions of energy. The emergent spatially confined activity is the mediation or resolution of the conflicting inter-actions. These processes reveal a wealth of detail and self-similarity at almost infinite scales of organization. Revealed in greater depth within this complexity is the fundamental role of the relationships between interacting parts in different orders of magnitude along with their emergent patterns and behaviour.

In biology, a fundamental characteristic of these complex systems is that there is permanence to the overall macro behaviour while, simultaneously, the constituent parts are continuously dying out and being replaced. The human body is one of these complex systems similar to ant colonies or beehives. Hundreds of different cell types make up the overall complexity of the body. Approximately 75 trillion of these cells are actively at work in our body. In a matter of seconds, thousands of these cells have died and billions have been completely replaced within a week. This high turnover rate does not affect our overall conscious awareness of a "permanent" body. Contained within each cell nucleus is the entire genome for an organism with individual cells reading only a small portion of that information. The interactive context within which the individual cell finds itself, will determine the tiny portion of information that it will read. Through this multi-cellular communication process, cells self-organize into more sophisticated structures. Cells can detect the overall state of their surroundings as well as any changes within that state such as gradient fluctuations. Through this process, cells eventually self-organize into complex collectives leading to more complicated and sophisticated interactions. Throughout this decentralized process, local interactions and communication leads to the emergence of coordinated collective behaviour at different levels or scales of interactivity.

We find other complex systems, forms and structures lurking within vastly differing scales of observation. Within the vast expanse of outer space, we encounter dynamically organized operations of light energy that remind us, through its spiral structures, of forms and patterns lurking within our immediate environment. The efficiency and incredible adaptability of this elemental form is further revealed through its use by nature in the highly versatile double-helix structure of DNA. Other dynamic and complex patterns can be generated through vibrations in a liquid or a fine powder and when a dense liquid is evenly heated in a pan. In all of these examples, the dynamic events activated within the medium resolve themselves or eventually mediate into resonant, highly charged and encoded networks of energy. Within these potent patterns of phenomenal inter-activity and their cellular units, we encounter a correlation between "stable" form and dynamic inner structure.

This interrelationship between scales and between matter, process, and form, found both in physics and in biology, is not just encountered within the realm of appearances. D'Arcy Thompson was well aware of this and describes the quest to understand this interrelationship as "*the search for community of principles or the essential similitudes*."[8] Most essential regarding such a quest, the anthropologist Gregory Bateson reminds us, is *"the discarding of magnitudes in favor of shapes, patterns, and relations."*[9] Within the realm of the organizing principles of integrated, highly adaptable and structured relationships, we encounter scaleless order or, perhaps more significantly, a multitude of possible scales or magnitudes.

Works-in-Process

My work and research has been inspired by the broad implications of the developments described above. Multiple-exposure photography was used in the initial phases of the work as a way of generating a series of images entitled *Multiples*. The resulting improvised images would emerge from the purely visual intermingling or blending of a repeated image or module (see Figure 2). Subsequently, a more physical, materially based and dynamic process was required and eventually conceived through the use of the rotary motion generated by a potter's wheel. Intrinsic forms lurking within the spinning wheel's spiral vortex were cast by securing a metal cylinder containing hot water and wax to the wheel. This process generated a series of forms reminiscent of seashells and biological shapes. Figure 3 shows two views of two of these wax forms. The potter's wheel was also used to spin a suspended cotton string into initially stable and sequential wave-formations that become turbulent at higher speeds. This project, entitled *Ariadne's Thread/Rumi's Ocean* [10], was inspired by scientific investigations of dynamic phenomena. It was recorded from different vantage points, generating a wealth of morphological formations and generative working procedures, as well as insights into the correlation between reference frame and perception. Figure 4 shows several of the forms generated with the spinning string. The whirling string shown on the left is spinning at a rate whereby it casts shadows of itself on its generated surface.

Figure 2. Manuel A. Báez, *Multiple #1.*

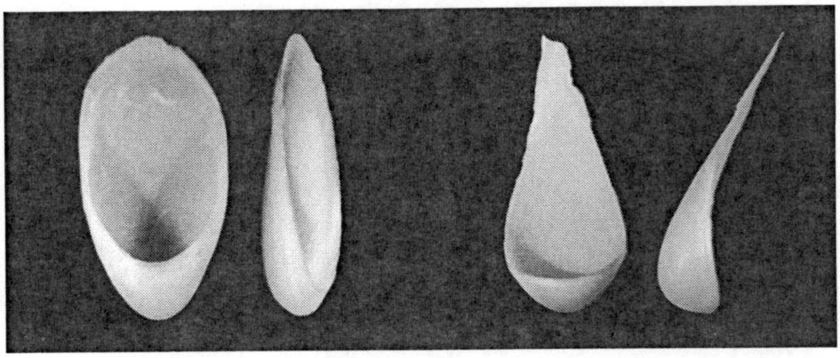

Figure 3. Manuel A. Báez, Wax Forms cast with a potter's wheel, 7" high x 3" wide.

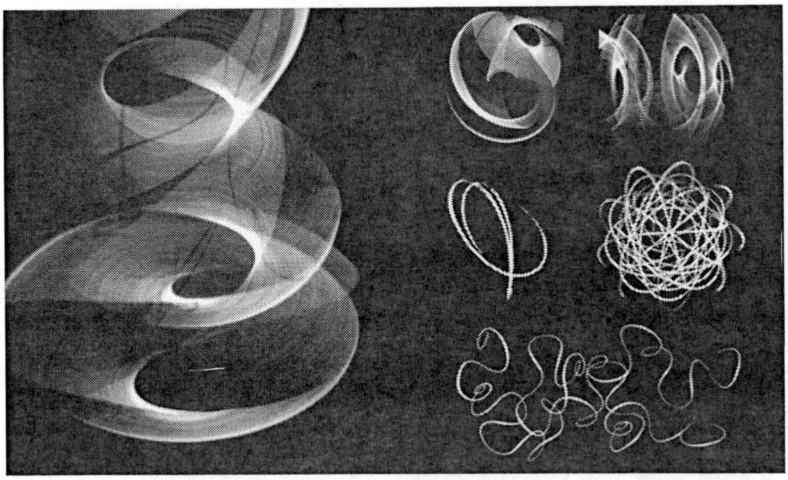

Figure 4. Manuel A. Báez, *Ariadne's Thread/Rumi's Ocean*, String & Potter's Wheel, 1993-present. Left & upper right: String Formations; lower right: Collaged Motion Drawings; middle: Calligraphic String Drawing; middle right: Multiple Exposure String Drawing or "*Ariadne's Ball of Thread.*"

Through extensive research and analysis of the work generated from the projects described above, and the conceptual developments that inspired them, the dynamic versatility of several elemental forms were explored by incorporating a flexible joint as part of an assembling process. These elemental relationships can be found within the inner structure of nature's resolutions to dynamic phenomena. The underlying woven stress patterns found superimposed and interacting within the inner structure of bones, is a biological example of one way nature resolves a dynamically complex structural situation. Elemental shapes, such as a triangle, square, pentagon, etc., were considered as dynamic relationships instead of to static diagrams. The joints consist of two bamboo dowels joined together with rubber bands, thus allowing for a high degree of flexibility. Through a variety of different arrangements of these joints, very versatile cellular units have been conceived and their form generating potential explored through the construction of cellular membranes or fabrics. The flexibility of the joints and their three-dimensional relationships, both within an individual cell and throughout the cellular membrane, generates a wealth of forms and structures through the emergent transformative and organizing properties of the integrated assembly. These properties recall and re-generate the inherent characteristics of the natural phenomena that inspired their conception.

The Garden of Phenomeno-logical Paths

The most extensive exploration incorporated into the *Phenomenological Garden* project has been that of a square geometric relationship. Gradually, it becomes apparent that this is an extremely versatile relationship between joints. The cellular membrane is constructed with 12" and 6" bamboo dowels and rubber bands. The upper left-hand corner of Figure 5 shows the fabric along with several improvised studies. The upper right-hand corner shows an inherently coiling structure that's approximately 30 feet in overall length and 2 feet wide. The forms and structures that can be discovered and developed through the process will be determined by how the initial fabric is probed and segmented into its inherent patterns.

As stated above, the three-dimensional joint relationship, as an integrated assembly, contains and is inter-active *in-formation*. What can be revealed from this information depends on the methods and/or means of inquiry. The encoded information or *potentiality* has a multitude of possible readings or interpretations. Through ones increasing experience and familiarity with the working process, more expressive forms and intricate structures can be conceived. One literally feels the stresses being worked on and with, along with the inherent in-forming potential of the membrane. This is a random exploration of the interactions without any preconceived goals. This type of exploration allows for the discovery of unanticipated patterned arrangements and their resulting interactive emergent behaviour. The resulting pattern detection and subsequent "readings," allow for the development of more sophisticated coordination and regulated structuring. Sensually fluid curves begin to emerge, as well as very organic or biological forms and structures. The experience is that of a process whereby one feels, follows and flows with, while guiding the versatile form generating properties of the dynamic relationship.

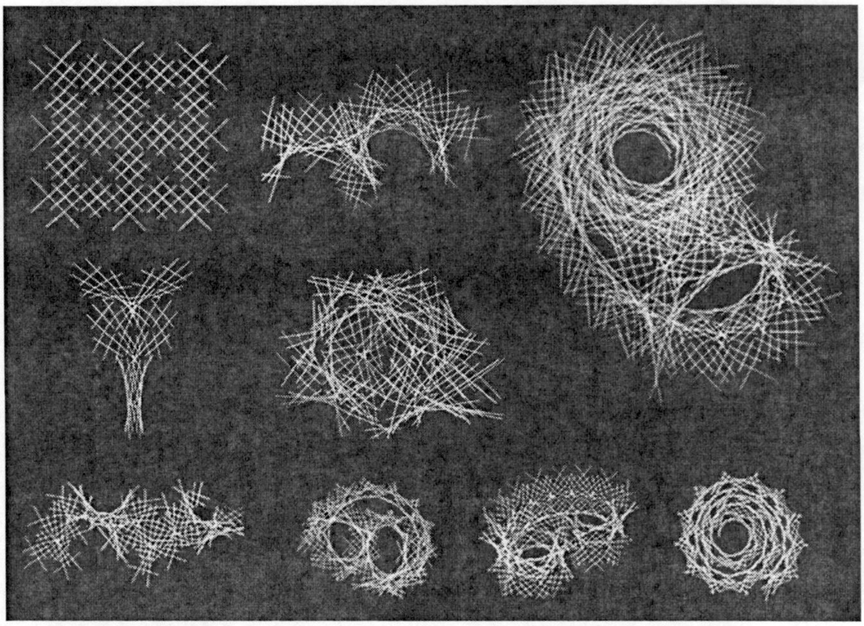

Figure 5. Manuel A. Báez, *Suspended Animation Series*, 1994-present. Form Studies with square cellular units, 12" and 6" bamboo dowels joined together with rubber bands. Upper left-hand corner shows a portion of the membrane used throughout all fabrications shown in figures 5, 6 and 7.

The sculptural forms shown in the upper right-hand corners of Figures 5, and 6 are made from the same coiling structure. The inherent properties and versatility of this structure has been explored by uncoiling and re-arranging it into different configurations (see Figure 5, lower right). Again, the organic looking forms and structures are all generated from the emergent properties of the assemblies. Figure 6 shows an installation done at Cranbrook Academy of Art in Bloomfield Hills, Michigan, USA. By then, the fluent expressiveness of the fabric and working process, along with its possibly limitless capabilities, had become apparent. The installation was part of a symposium that I conceived and was invited to organize at Cranbrook Academy of Art for the Sybaris Gallery in Royal Oak, Michigan. The symposium, entitled *Metaphoric Interweavings,* explored the interrelationships and similarities between weaving, musical composition and architecture through the use of a modular compositional process: artist Lissa Hunter lectured on her work, basketry and weaving; classical pianist Marina Korsakova-Kreyn gave a lecture/performance on the intricate structure of musical compositions by J. S. Bach; and professor of architecture Gulzar Haider lectured on the use of *muqarnas* as modules in spatial transformations in Islamic architecture. *Mugarnnas* is a system of projecting niches used for spatial transition zones and for architectural decoration.

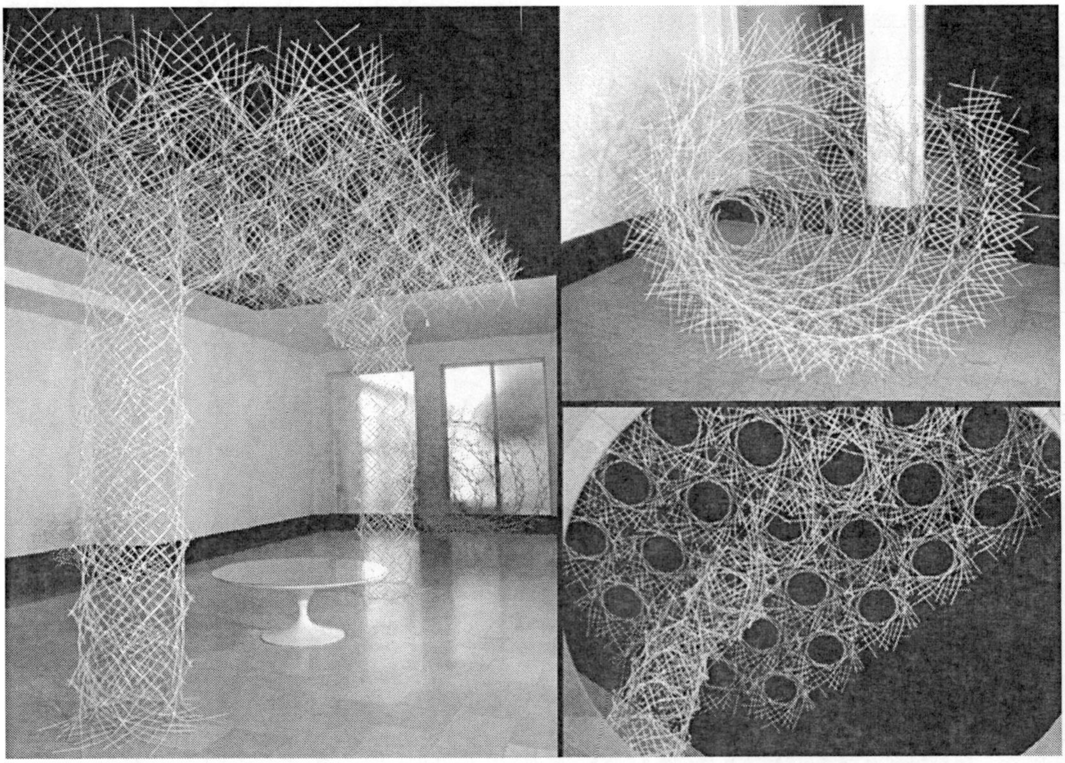

Figure 6. Manuel A. Báez, *Phenomenological Garden*, Installation for the *Metaphoric Interweavings* Symposium at Cranbrook Academy of Art, Bloomfield Hills, Michigan, USA, 1998. Upper right: 4' - 6" high sculptural form, lower right: reflected ceiling view of the installation through the mirrored central table (lower left).

The installation in Figure 6 initiated the *Phenomenological Garden* project. It was entirely constructed using the same square cellular unit and membrane shown in Figure 5. Two supporting columns are gradually transformed into an intricately patterned ceiling structure. The majority of the patterns that emerged were unconsciously assembled and a rich variety of them are revealed as one walks around the installation or looks into the mirrored central table (Figure 6 lower right). A different vantage point will reveal an entirely different pattern, at times familiar, but quite often completely unexpected.

As the project has evolved, the multiplicity of shadows cast by these constructions has become increasingly more relevant to the theme of the work. They have added another layer to the multiple readings and interpretations. Figure 7 shows an installation at the Network Gallery of Cranbrook Academy of Art. The shadows played a major role in this installation along with the three-dimensional sculptural possibilities of the working process. A series of improvised sculptural weavings and freestanding structures cast their shadows on the walls and floor of the Gallery. Again, different vantage points reveal different aspects of the woven structures.

Figure 7. Manuel A. Báez, *Phenomenological Garden*, Installation at the Network Gallery of Cranbrook Academy of Art, Bloomfield Hills, Michigan, USA, 1999. Improvised sculptural weavings and freestanding structures constructed with the membrane shown in Figure 5.

The Crossings Workshop

The *Phenomenological Garden* is a project that has been evolving and will continue to do so as the explorations develop. Other cellular joint relationships have been studied along with their emergent properties. Figure 8 shows some of the work produced by students in my

Crossings Workshop at Carleton University. The Workshop incorporates the educational potential of the research and work as a way of introducing the students to the rich potential of the working process and the developments that have inspired its conception.

Figure 8. Crossings Workshop *Suspended Animation Series,* Cellular Form Studies. Works by Carleton architecture students: Mariam Shaker, Diana Park, Sherin Rizkallah, Daniel Cronin and Sharif Kahn.

The left side of Figure 8 shows a structure constructed using a square cellular unit. By suspending it from the ceiling, the gradual effect of gravity is clearly demonstrated in the subtle, progressive undulations of the structure. To the middle and lower right of this structure are two different arrangements of the same structure constructed with a seven-sided (heptagonal) module. This structure is also shown in Figure 9 and is particularly interesting because, through different configurations of the same structure, the diversity of possible forms is clearly shown. Equally interesting and diverse are the organic looking shadow "drawings" shown in Figure 9. In the upper right-hand corner of Figure 8, are two other structures constructed with a square cellular unit and, again, they clearly demonstrate the different possibilities contained within the same cell. On the lower right-hand corner is a structure constructed using a five-sided (pentagonal) cellular unit. The numerous intrinsic assembling procedures lead to unexpected overall patterns and dynamic arrangements that generate new and diverse developmental directions for the assembling process.

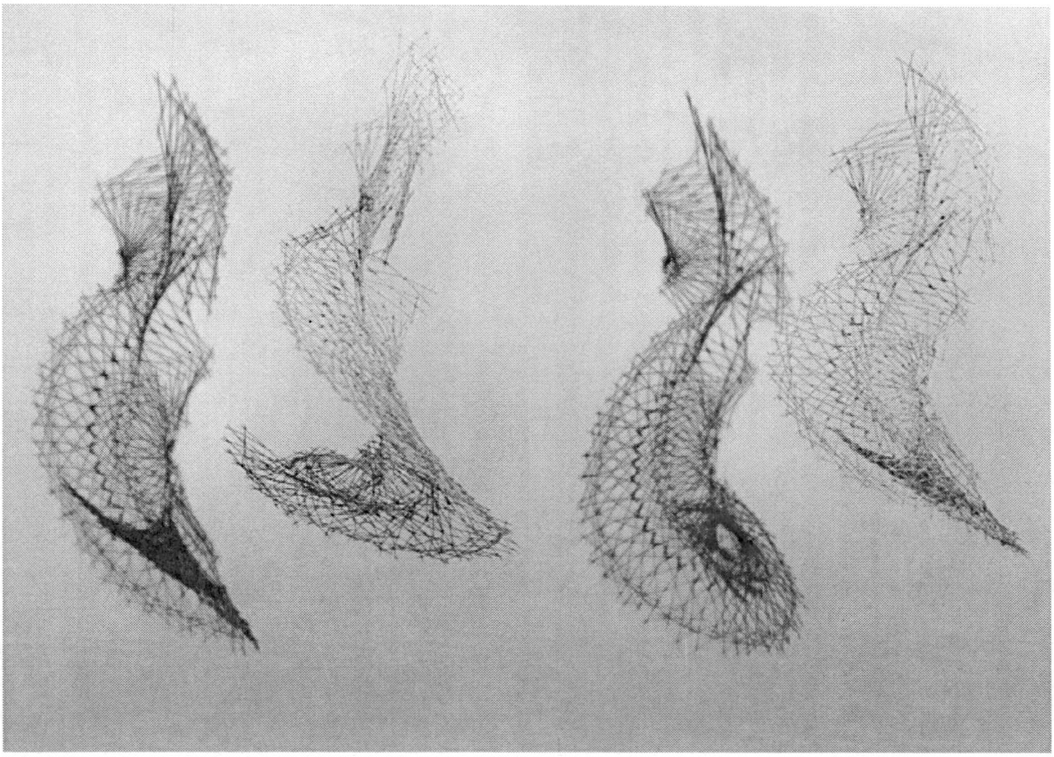

Figure 9. Crossings Workshop *Suspended Animation Series*, Cellular Form Studies and Shadow "drawings" (heptagonal cellular units). Work by Diana Park.

An Intermingling of X, Y, and Z Co-ordination

The cellular unit shown in Figure 10 is constructed with 12" and 5" bamboo dowels that are joined together, again, with rubber bands. The unit is composed of three surfaces (or planes) at right angles to each other with each surface being defined by four 12" dowels assembled into a grid of two pairs at right angle to each other and four 5" dowels, one at each end of the 12" pairs (Figure 10, lower right). The three surfaces have a high degree of transformability due to the flexibility of the joints and each surface defines one of the X, Y and Z coordinate directions in three-dimensional space. Each surface can fully collapse along the two orthogonal diagonals of the assembled grid. Individually, each surface can fully collapse along the two orthogonal diagonals of the assembled grid. Three-dimensionally, this cubic cellular unit (or module) is composed of multiple "*interacting degrees of freedom*" through the combination of 42 flexible joints. From another perspective, this complex intermingling is also the interactions of the three flexible hyperbolic paraboloids within the three-dimensional assembly. Figures 11 and 12 show several configurations that can be developed from this dynamic interplay.

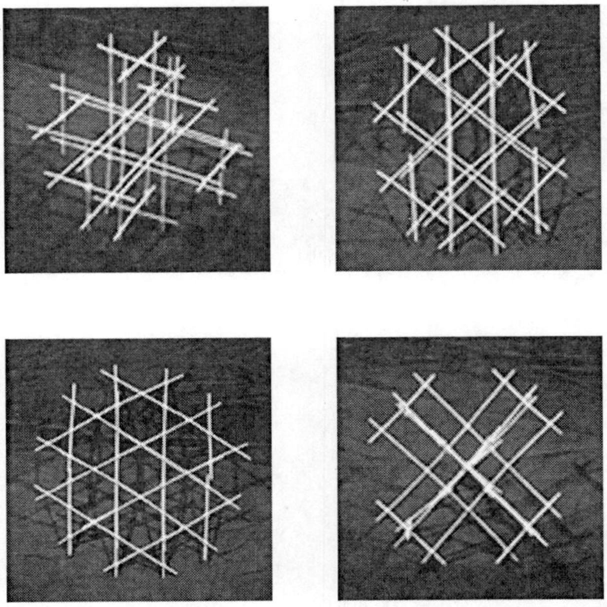

Figure 10. Crossings Workshop *Suspended Animation Series*, views of X, Y & Z Coordinates Cellular Unit: Three intersecting planes at right angles to each other. Lower right: clearly shows one of the planes with the central diagonal edges of the other two. Upper right and lower left: show views through the four diagonals of the cubic assembly.

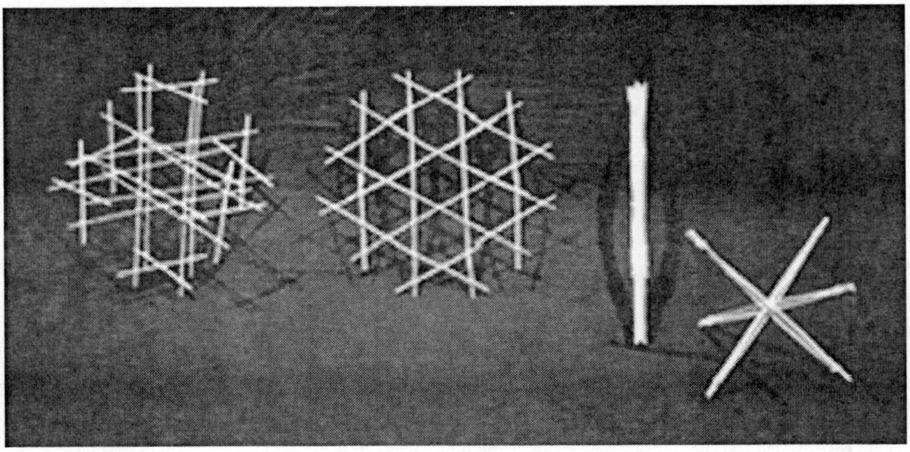

A B C D

Figure 11. Crossings Workshop , X, Y & Z Coordinates Cellular Unit and several of its basic transformations. **A**: The Cellular Unit. **B**: Flattened assembly along one of the four diagonals of the cubic assembly. **C**: Collapsed assembly centered around one of the four diagonals. **D**: Collapsed X, Y and Z axes with 5" dowels removed.

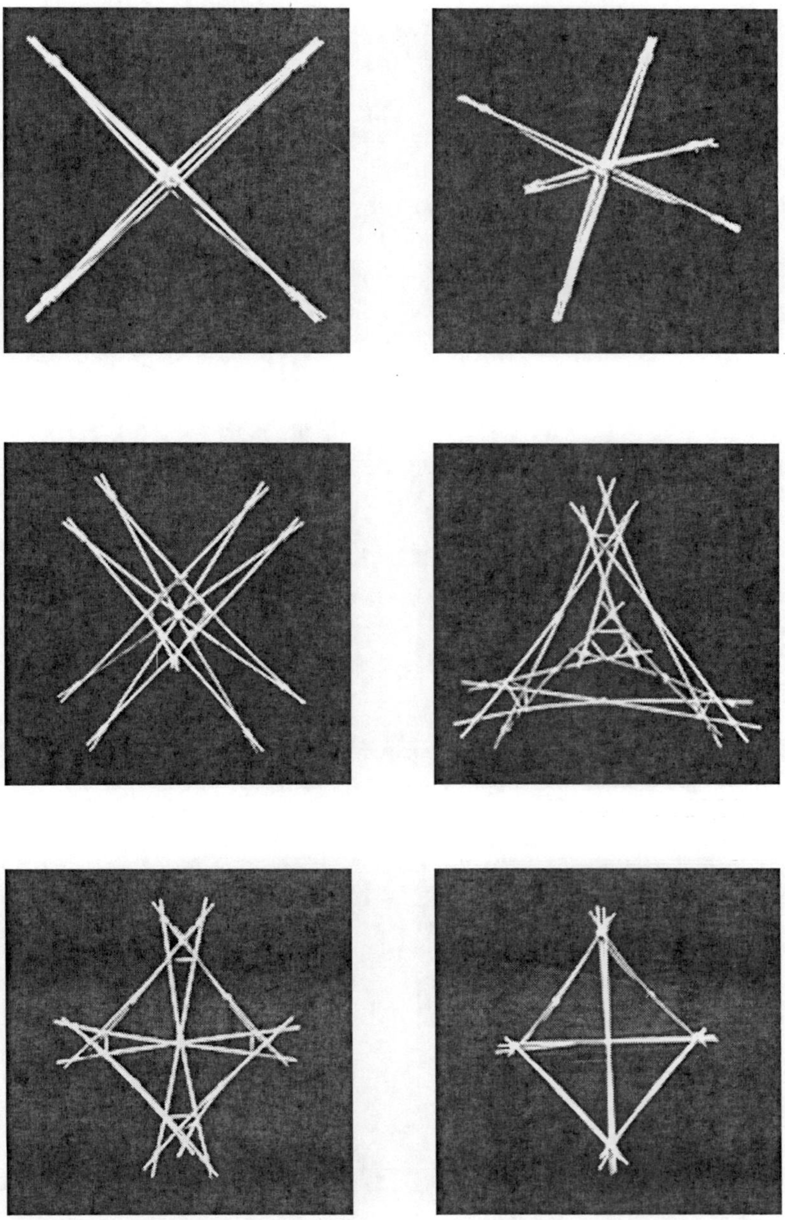

Figure 12. Crossings Workshop, different stages of the cellular unit shown in Figure 11 as it completely collapses into the X, Y and Z axes (upper left and right) and gradually expands into a tetrahedron (from left to right starting from the top).

Figures 13, 14 and 15 show several forms and structures that can emerge as the assembling process gradually evolves into more complex configurations. Figures 13 shows two axial views of the same construction. This particular assembling process generated a dodecahedron that was not preconceived nor initially anticipated. Cellular units (as shown in Figure 11, left side) were assembled together using their inherent interacting properties as the guiding principles. Within the resulting three-dimensionally dynamic pattern of the form one can discern the complex interweaving of the rich geometric properties of the dodecahedron:

cubes, tetrahedrons, octahedrons, icosahedrons and golden rectangles (to name a few) in a reciprocally complex relationship. Several of these shapes can be discerned in the two views provided. The left side of Figure 14 shows another construction generated through the same process as in Figure 13 and also reveals the same level of complex multilayering of forms. The different modifications to the original unit in Figure 10 lead to the emergence of totally different complex patterns and dynamic properties.

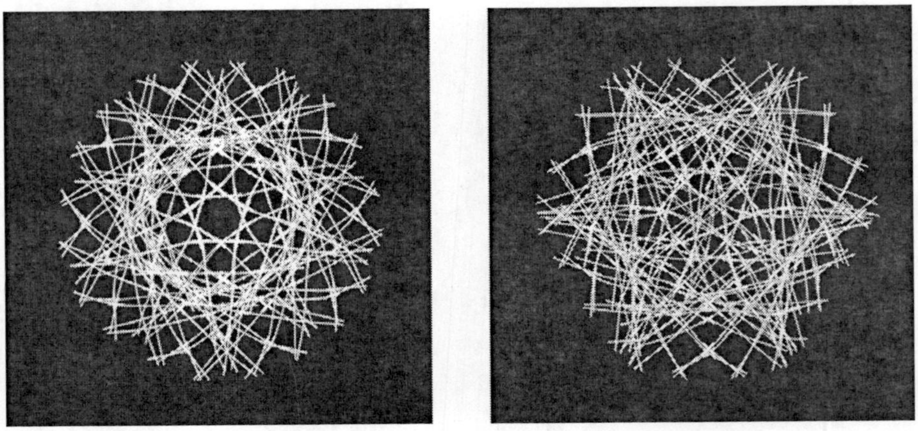

Figure 13. Crossings Workshop *Suspended Animation Series*, two views of the same construction using the cellular unit shown in Figure 10. The construction is a dodecahedron that emerged from the assembling process. Throughout the structure and the generated patterns one can discern the cubes, tetrahedrons, octahedrons and icosahedrons that are intrinsically embedded within the dodecahedron.

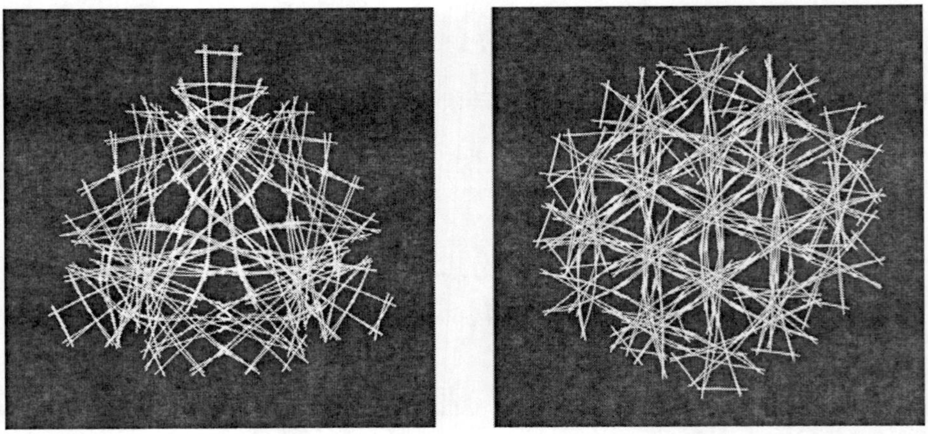

Figure 14. Crossings Workshop *Suspended Animation Series*, Cellular Constructions. Left: Constructed with the same cellular unit as in Figure 13 and exhibits the same properties. Right: Constructed using a variation on the cellular unit used in Figure 13. Different patterns are revealed throughout these constructions from different points of view.

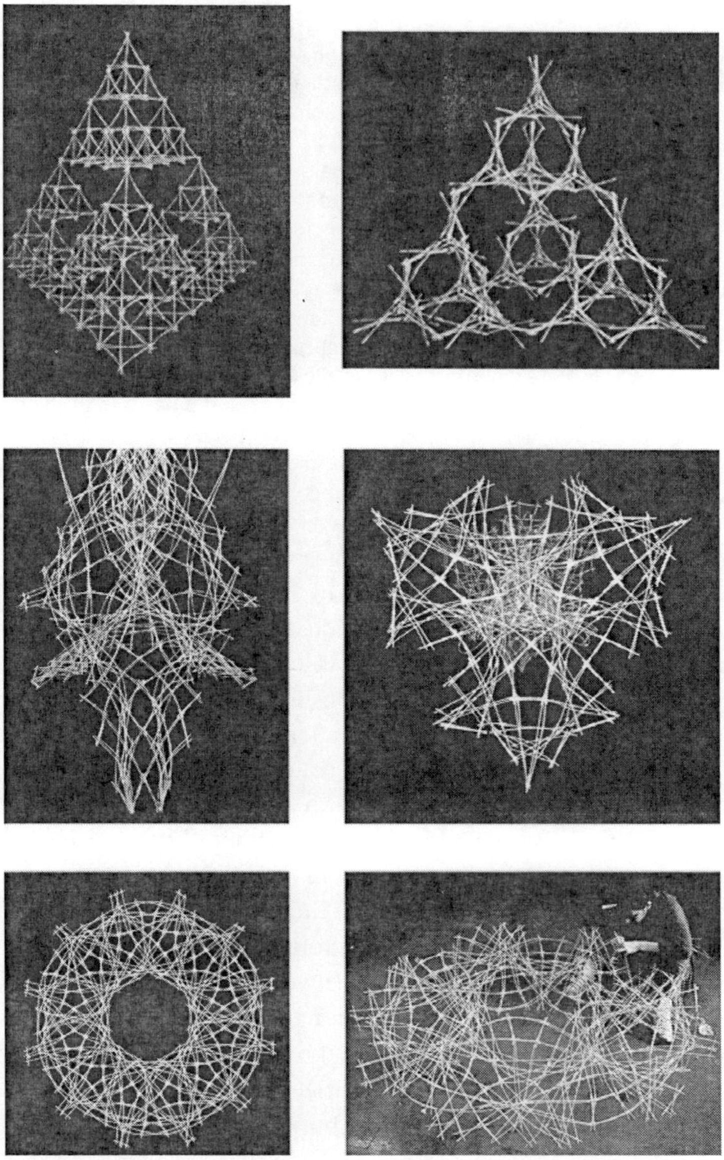

Figure 15. Crossings Workshop *Suspended Animation Series*, Cellular Constructions. Upper left and right, by Dan Levin and Michael Lam, constructed with the cellular unit in Fig. 12; Upper left, with the units fully expanded and upper right, with the units almost fully collapsed. Middle left and right, by Michael Putman, Patrick Bisson and Rheal Labelle, with the cellular unit in Fig. 10. Lower left and right, by Ana Lukas, with the cellular unit in Fig. 10.

Conclusion

In *Six Memos for the Next Millennium*, the Italian writer Italo Calvino offers us the following observations and advise:

> "The crystal, with its precise faceting and its ability to refract light, is the model of perfection that I have always cherished as an emblem, and this predilection has become even more meaningful since we have learned that certain properties of the birth and growth of crystals resemble those of the most rudimentary biological creatures, forming a kind of bridge between the mineral world and living matter.
>
> Among the scientific books into which I poke my nose in search of stimulus for the imagination, I recently happened to read that the models for the process of formation of living beings "are best visualized by the crystal on one side (invariance of specific structures) and the flame on the other (constancy of external forms in spite of relentless internal agitation)."
>
> What interests me here is the juxtaposition of these two symbols, as in one of those sixteenth-century emblems . . . . Crystal and Flame: two forms of perfect beauty that we cannot tear our eyes away from, two modes of growth in time, of expenditure of the matter surrounding them, two moral symbols, two absolutes, two categories for classifying facts and ideas, styles and feelings. . . . I have always considered myself a partisan of the crystal, but the passage just quoted teaches me not to forget the value of the flame as a way of being, as a mode of existence. In the same way, I would like those who think of themselves as disciples of the flame not to lose sight of the tranquil, arduous lesson of the crystal." [11]

The richness of nature's processes challenges our imagination because of its complex simplicity. This paradox has inspired the work of J. W. von Goethe, Louis Sullivan, Frank Lloyd Wright and countless other creative individuals. Italo Calvino was also inspired by this tradition and was well aware of the modern developments in science. These developments, along with the history of science and its relationship with literature and philosophy, were a source of inspiration for his creative imagination. To Calvino, the Crystal and Flame symbolize the paradoxical and contradictory nature of matter as revealed to us in the twentieth century. This correlation between form and process, as well as, simplicity and complexity has been revealed to us periodically throughout history. "This is common to all our laws;" states the physicist Richard Feynman, "they all turn out to be simple things, although complex in their actual actions."[12] Benoit Mandelbrot elaborates on this paradox and the complexity of fractal geometry: "The effort was always to seek simple explanations for complicated realities. But the discrepancy between simplicity and complexity was never anywhere comparable to what we find in this context."[13] The work-in-progress presented here inherently addresses this fundamental paradox through an integrative working process. Such a process can offer new directions to the fields of morphology, architecture and other disciplines at a time when the ideas emerging out of our deeper understanding of complex phenomena are being embraced for conceptual inspiration. The way towards the rich realm of diversity, as nature shows us, is through simple fundamental rules that eventually lead to a paradox of constrained and versatile freedom.

References

[1] As quoted by Italo Calvino in (1999) *Why Read the Classics?,* New York: Pantheon Books.

[2] Mandelbrot B. B. (1983) *The Fractal Geometry of Nature*, New York: W. H. Freeman and Co.

[3] Buckley P. and Peat F.D., editors (1996) *Glimpsing Reality: Ideas in Physics and the Link to Biology*, Toronto: University of Toronto Press.

[4] Sullivan L. (1924) *A System of Architectural Ornament According with a Philosophy of Man's Powers*, New York: Eakins Press.
[5] "The Life-work of American Architect Frank Lloyd Wright," (1965) *Wendigen*, New York: Horizon Press.
[6] Heisenberg W. (1958) *Physics and Philosophy*, New York: Harper Torch Books.
[7] Bateson G. (1980) *Mind and Nature*, New York: Bantam Books.
[8] Thompson D.W. (1992) *On Growth and Form*, Complete Revised Edition, New York: Dover Books.
[9] Bateson G. (1980) *Mind and Nature*, New York: Bantam Books.
[10] Ariadne is the mythological Greek guide to the labyrinth of chaos and the individual life. Jalai al-Din Rumi is the Great Persian mystic poet of the thirteenth century and the creator of the whirling, circular dance of the Mevlevi dervishes.
[11] Calvino I. (1988) *Six Memos for the Next Millennium,* Cambridge, Mass.: Harvard University Press.
[12] Feynman R. (1967) *The Character of Physical Law*, Massachusetts: The M. I. T. Press.
[13] Quoted in: Peitgen H., Jurgens H., Saupe D., Zahlten C. (1990) "*Fractals: An Animated Discussion*," VHS/color/63 minutes, New York: Freeman.

Bibliography

Bateson G. (1980) *Mind and Nature*, New York: Bantam Books.

Buckley P. and Peat F.D., editors (1996) *Glimpsing Reality: Ideas in Physics and the Link to Biology*, Toronto: University of Toronto Press.

Capra F. (1996) The Web of Life: *A New Scientific Understanding of Living systems*, New York: Anchor Books Doubleday.

Calvino I. (1988) *Six Memos for the Next Millennium,* Cambridge, Mass.: Harvard University Press.

Feynman R. (1967) *The Character of Physical Law*, Massachusetts: The M. I. T. Press.

Heisenberg W. (1958) *Physics and Philosophy*, New York: Harper Torch Books.

Johnson S. (2001) *Emergence: The connected Lives of Ants, Brains, Cities, and Software*, New York: Scribner.

Mandelbrot B.B. (1983) *The Fractal Geometry of Nature*, New York: W. H. Freeman and Co.

Peitgen H., Jurgens H., Saupe D., Zahlten C. (1990) "*Fractals: An Animated Discussion*," VHS/color/63 minutes, New York: Freeman.

Prigogine I. (1980) *From Being to Becoming*, San Francisco: Freeman.

Prigogine I., Stengers I. (1984) *Order out of Chaos*, New York: Bantam Books.

Sullivan L. (1924) *A System of Architectural Ornament According with a Philosophy of Man's Powers*, New York: Eakins Press.

Thompson D.W. (1992) *On Growth and Form*, Complete Revised Edition, New York: Dover Books.

In: Chaos and Complexity Research Compendium
Editors: F. Orsucci and N. Sala, pp. 279-287
ISBN: 978-1-60456-787-8

Chapter 21

COMPLEXITY IN THE MESOAMERICAN ARTISTIC AND ARCHITECTURAL WORKS

Gerardo Burkle-Elizondo [a]
Universidad Autónoma de Zacatecas. Unidad de Postgrado II. Doctorado en Historia.
Ave. Preparatoria s/n, Col. Hidráulica. CP 98060, Zacatecas, Zac. México
Ricardo David Valdez-Cepeda [b]
Universidad Autónoma Chapingo. Centro Regional Universitario Centro Norte.
Apdo. Postal 196, CP 98001, Zacatecas, Zac. México
Nicoletta Sala [c]
Accademia di Architettura, Università della Svizzera italiana,
largo Bernasconi 2 CH – 6850 Mendrisio, Switzerland

Abstract

It has been demonstrated that scribers, artists, sculptors and architects used a geometric system in ancient civilizations. There appears such system includes basically golden rectangles distributed in a golden spiral fashion. In addition, it is clear that we do not know the sequence in which the lines or pictures were originally traced or drawn. By this way, the artistic and architectural works can be considered as static objects and so they may be characterized by an inherent dimension. The aim of this paper is to introduce a description of the complexity presents in the Mesoamerican artistic and architectural works (e.g., tablets from Palenque and other sites, Maya stelae, Maya hieroglyphs, pyramids, palaces and temples, calendars and astronomic stones, codex pages, murals, great stone monuments, astronomic stones and ceramic pots). Our findings indicate a characteristic higher fractal dimension value for different groups of Mesoamerican artistic and architectural works. Results could be suggesting that Mesoamerican artists and architects used specific patterns and they preferred works with higher (1.91) box and information fractal dimensions.

Keywords: Archeology, Golden Figures, Mesoamerican Tablets, Stelae and Pyramids, Fractals, Fractal dimension.

---

[a] E-mail address: burklecaos@hotmail.com
[b] E-mail address: vacrida@hotmail.com
[c] E-mail address: nsala@arch.unisi.ch

Introduction

The scientific perception of reality has changed through the centuries. For example, the Baroque style liked a mathematical curve, the ellipse; in that time, the ellipse became popular and was used in physics, astronomy, engineering and art (Hilgemeier, 1996; Stierlin, 2001; Sala and Cappellato, 2003); so in the mind of a cultivated person, the planets traveled along perfect ellipses, Kepler's laws, and people were certain about the stability of the solar system. However, it has been discovered that systems of orbiting bodies have rational proportions of orbital periods that become unstable sooner or later but this phenomena can be modeled for near future prediction taking into account our limited knowledge of the initial conditions. Contrary to this, with art produced by humans, there is no form to know the sequence in which the lines or pictures were originally traced or drawn. This means there are no equations or temporal information useful to characterize ancient artistic and architectural works when treated as complex systems. This means geometric analysis and mathematics used in art composition and design of buildings are not yet clearly elucidated, although at least some serious studies deserve be mentioned. Roman and Greek architects liked circles and golden rectangles Also, Egyptians used the an approximation of the golden rectangle in art, architecture and hieroglyphics (www.geocities.com/CapeCanaveral/Station/8228/arch.htm).

Martínez del Sobral (2000) studied Mesoamerican art, sculptures, codex, and pyramids and urban architectural designs, and she have demonstrated the strong influence of golden measures on them, whilst de la Fuente (1984) pointed out Olmeca monumental heads were made under the basis of golden rectangles as harmonic units. These growing golden rectangles appear to be distributed following a golden spiral. In addition, both authors have demonstrated that in the prehispanic world, a system like this was used by scribers (named 'tlacuilos'), artists, sculptors and architects making of it a standardized technique in composition, and these abilities and knowledge were transferred from one generation to another, like a tradition.

By this way, the Mesoamerican artistic and architectural works can be considered as static objects (Miller, 1999; Stierlin, 2001), and so they may be having an inherent dimension. Therefore, the fractal dimension is an experimentally accessible quantity that might be related to the aesthetic of the pattern(s) of these works. Then it would be interesting to know if the artists and architects preferences were different for groups or types of work in the ancient Mesoamerican culture.

In this paper, we present a fractal analysis of some Mesoamerican artistic and architectural works, and a comparison among them taking into account different groups or types of work.

Material and Methods

To determine to degree of the complexity in the Mesoamerican arts, we collected 90 images (Table 1) of Mesoamerican artistic and architectural works by reviewing literature on archeology. From the 90 figures, 61 correspond to the Maya culture (MC) during late preclassic (300 b. C. to 250 a. C.), and early and late classic (250 to 700 b. C.) periods, developed at Mexican Chiapas and Yucatán states, and Guatemala and Honduras; 26 to the Aztec or Mexican culture (AC) during classic and epiclassic periods (300 to 1100 a. C.), developed at Mexican Central Highplains; two to the ancient Olmec culture (OC) developed from 1350 to 900 b. C., at Mexican (Veracruz, state); and one to the Toltec culture (TC), developed from 700 to 1100 a. C. corresponding to the first step of Nahua civilization, at Mexican Hidalgo State. All these 90 images have been digitized using a Printer-Copier-Scanner (Hewlett Packard®, Model LaserJet 1100A) and saved in bitmap (*.bmp) format on a Personal Computer (Hewlett Packard®, Model Pavilion 6651). Thereafter, these images were analyzed with the program Benoit, version 1.3 [9, 10] in order to calculate Box (D_b), Information (D_i), and Mass dimensions (D_M), and their respective standard errors and intercepts on log-log plots. It was taken under consideration that the information dimension differs from the box dimension in that it weigths more heavily boxes containing more points. Figure 1 shows a partial fractal analysis realized by the program Benoit®.

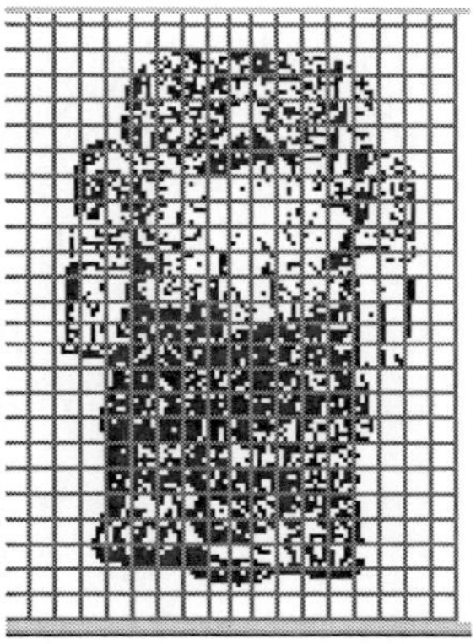

Figure 1. Partial fractal analysis realized by the program Benoit® .

Box Dimension

The box dimension is defined as the exponent D_b in the relationship:

$$N(d) \approx \frac{1}{d^{D_b}} \tag{1}$$

where N(d) is the number of boxes of linear size d (number of pixels in this study), necessary to cover a data set of points distributed in a two-dimensional plane. The basis of this method is that, for objects that are Euclidean, equation (1) defines their dimension. One needs a number of boxes proportional to 1/d to cover a set of points lying on a smooth line, proportional to $1/d^2$ to cover a set of points evenly distributed on a plane, and so on. Applying the logarithms to the equation (1) we obtain: $N(d) \approx -D_b \log(d)$.

Information Dimension

In the definition of box dimension, a box is counted as occupied and enters the calculation of N(d) regardless of whether it contains one point or a relatively large number of points. The information dimension effectively assigns weights to the boxes in such a way that boxes containing a greater number of points count more than boxes with less points.

The information entropy I(d) for a set of N(d) boxes of linear size d is defined as

$$I(d) = -\sum_{i=1}^{N(d)} m_i \log(m_i) \qquad (2)$$

where m_i is:

$$m_i = \frac{M_i}{M} \qquad (3)$$

where M_i is the number of points in the i-th box and M is the total number of points in the set.

Consider a set of points evenly distributed on the two-dimensional plane. In this case, we will have

$$N_d = \frac{1}{d^2} \qquad (4)$$

and if it is considered that $m_i = d^2$. So equation (2) can be written as

$$I(d) \approx -N(d)\left[d^2 \log\left(d^2\right)\right] \approx -\frac{1}{d^2} = \left[2\, d^2 \log(d)\right] = -2\log(d) \qquad (5).$$

For a set of points composing a smooth line, we would find $I(d) \approx -\log(d)$. Therefore, we can define the information dimension D_i as in:

$$I(d) \approx -D_i \log(d) \qquad (6).$$

In practice, to measure D_i one covers the set with boxes of linear size d keeping track of the mass m_i in each box, and calculates the information entropy I(d) from the summation in (2). If the set is fractal, a plot of I(d) versus the logarithm of d will follow a straight line with a negative slope equal to $-D_i$.

At the beginning of this section, we noted that the information dimension differs from the box dimension in that it weighs more heavily boxes containing more points. To see this, let us write the number of occupied boxes N(d) and the information entropy I(d), in terms of the masses m_i contained in each box:

$$N(d) = \sum_i m_i^0 \qquad (7)$$

$$N(d) = -\sum_i m_i \log(m_i)$$

The first expression in (7) is a somewhat elaborate way to write N(d), but it shows that each box counts for one, if $m_i > 0$. The second expression is taken directly from the definition of the information entropy (1). The number of occupied boxes, N(d), and the information entropy I(d) enter on different ways into the calculation of the respective dimensions, it is clear from (7) that $D_b \leq D_i$. The condition of equality between the dimensions is realized only if the data set is uniformly distributed on a plane.

Mass Dimension

Draw a circle of radius r on a data set of points distributed in a two-dimensional plane, and count the number of points in the set that are inside the circle as M(r). If there are M points in the whole set, one can define the 'mass' m(r) in the circle of radius r as:

$$m(r) = \frac{M(r)}{M} \qquad (8).$$

Consider a set of points lying on a smooth line, or uniformly distributed on a plane. In these two cases, the mass within the circle of radius r will be proportional to r and r^2 respectively. One can then define the mass dimension D_M as the exponent in the following relationship:

$$m(r) \approx r^{D_M} \qquad (9).$$

In practice, one can measure the mass m(r) in circles of increasing radius starting from the center of the set and plot the logarithm of m(r) versus the logarithm of r. If the set is fractal, the plot will follow a straight line with a positive slope equal to D_M. As the radius increases beyond the point in the set farthest from the center of the circle, m(r) will remain constant and the dimension will trivially be zero. This approach is best suited to objects that follow some radial symmetry, such as diffusion-limited aggregates. In the case of points in the plane, it may be best to calculate m(r) as the average mass in a number of circles of radius r.

It can be shown that the mass dimension of a set equals the box dimension. This is true globally, i.e., for the whole set; locally, i.e., in portions of the set, the two dimensions may differ. Let us cover the set with N(d) boxes of size d, and let us define the mass, or probability, in the i-th box m_i as:

$$m_i = \frac{M_i}{M} \qquad (10)$$

where M_i is the number of points in the i-th box and M is the total number of points in the set. We can now write the average mass, or probability, in boxes of size d as m(d), the average m_i in the N(d) boxes:

$$m(d) = \frac{1}{N(d)} \sum_{i=1}^{n(d)} m_i = \frac{1}{N(d)} \tag{11}$$

(the sum of all the masses m_i is obviously one). As the operation of calculating the mass contained in a box of size d is the same as calculating the mass in a circle of radius r, we can write our definition of mass dimension (9) in terms of d rather than r:

$$m(d) \approx d^{D_M} \tag{12}$$

By using (4) and re-arranging terms, we obtain:

$$N(d) = \frac{1}{d^{D_M}} \tag{13}$$

which is the definition of the box dimension; thus, the mass dimension equals the box dimension.

Results and Discussion

In all the 90 cases a straight line was evidenced, so the three different approaches to estimate the fractal dimension works well. As an example, we show the plot to estimate the information dimension for 'Coatlicue', the Aztec god of life and death (shown in Figure 2).

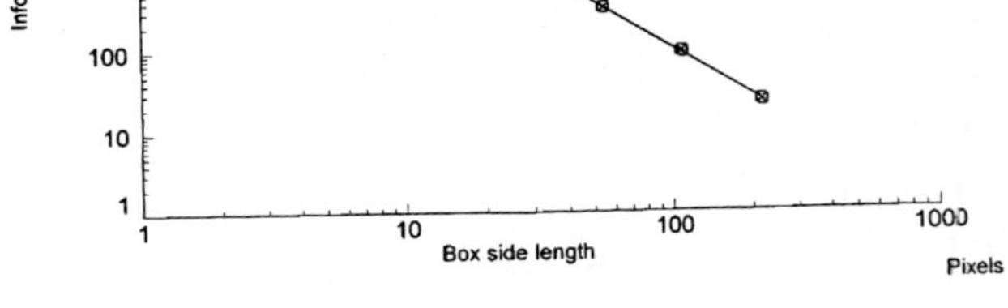

Figure 2. Log-log plot for 'Coatlicue'. It can be appreciated a straight line with a negative slope $-D_i = 1.906 \pm 0.006$.

The calculated fractal dimensions are reported in Table 1. For all the 90 cases the fractal dimension values were high from a $D_b = 1.803 \pm 0.023$ for the left and superior side of the 'Vase of seven gods' (MC, Group X), to a $D_M = 2.492 \pm 0.195$ for the left side of the 'Door to underworld of the Temple 11, platform' at Copán (MC, Group I). This late case could be related to the Mayan vases, which are integrated in the Group X in Table 1, are less complex than the other figures and groups because they contain wider empty but painted rectangular or squared spaces. Certainly, there is unknown the sequence in which the lines were traced in those works having high D_M values as the left and the superior side of the 'Door to underworld of the Temple 11, platform' at Copán (MC, Group I), which contains a lot of human like figures representing gods and ancestors but they are not concentrically distributed in a trapezoidal plane explaining its high D_M value surpassing the dimension of the plane. In this figure the traces are in fact irregularly distributed which makes really a complex composition able to fill the trapezoidal plane, and this characteristic is common to other works from the same civilization an Aztec culture (Table 1) such as the whole and parts of the 'Temple of foliated cross tablet' (MC, Group I); the whole center east of the 'Ball Game Tablet' at Chichen-Itzá (MC, Group I); 'Mural of the 4 Ages' at Toniná (MC, Group I); the whole 'Tablet of 96 Hieroglyphs' at Palenque (MC, Group III); 'Temple of the Cross, Door panel, Glyphs 2 and 14' (MC, Group III); 'Temple of the Sun, superior view' (AC, Group IV); 'Temple of the Sun' at Palenque (MC, Group IV); 'Pyramid of the Wizard' at Uxmal (MC, Group IV); 'Pyramid Temple' at Tulún (MC, Group IV); 'Palace of Hochob' at Tabasqueña (MC, Group IV); 'Dresden Codex, page 13b' (MC, Group VI); 'Borgia Codex, ritual 2, page 34' (AC, Group VI); 'Aztlán Annals', page 3 (AC, Group VI); 'Stela F' at Quirigua (MC, Group VII); 'Stela A' at Copán (MC, Group VII); 'Humboldt Disc' (AC, Group VIII); 'Huaquechula Disc' at Puebla (AC, Group 8); 'Jaguar, portico 10, jaguars joint, zone 2' at Teotihuacan (AC, Group IX); 'The Inferior Face of West Side of Chamber 1 of Murals' at Bonampak (MC, Group IX); 'Mural of the battle' at Chichen-Itzá (MC, Group IX); 'mayan vase with drawing of moon god with snake roll up' (MC, Group X); 'mayan vase' of Naranjo (MC, Group X); and 'disc of the Cenote sagrado' at Chichen-Itzá (MC, Group X).

Table 1. Box (D_b), information (D_i), and mass (D_m) dimension, and their standard deviations (SD) for different Mesoamerican artistic and architectural work types.

Work Type	n	D_b±SD	D_i±SD	D_m±SD
Group I. Tablets from Palenque and other sites	15	1.918±0.010	1.932±0.002	2.018±0.111
Group II. Maya and other stelae	9	1.923±0.007	1.940±0.001	1.887±0.060
Group III. Maya hieroglyphs	15	1.910±0.008	1.903±0.003	2.036±0.088
Group IV. Frontal view of Maya pyramids, temples and other buildings	8	1.919±0.007	1.923±0.002	1.998±0.138
Group V. Calendar pages (tonalamatl) from codex	7	1.921±0.008	1.926±0.002	1.937±0.051
Group VI. Dresden and other codex pages		1.918±0.009	1.924±0.003	2.038±0269
Group VII. Frontal view of great stone monuments	8	1.917±0.009	1.914±0.003	1.954±0.053
Group VIII. Circular astronomic and calendar great stones	7	1.900±0.006	1.877±0.003	1.975±0.047
Group IX. Murals of Mesoamerica	9	1.919±0.006	1.929±0.002	1.964±0.058
Group X. Maya vases (roll out) and other	12	1.883±0.013	1.888±0.003	1.966±0.214
Overall average	***90***	***1.912±0.009***	***1.916±0.002***	***1.98[illegible]±0.117***

Curiously, a few of the circular astronomic and calendar great stones from Aztec culture (Group VIII), which really contain a lot of information radially distributed are well characterized by D_M values, that is, these values are similar to D_b and D_i values. Clearly, this occurs for 'Aztec Calendar' or 'Sun Stone' (D_b = 1.92±0.005, D_i = 1.9±0.005, D_M 1.901±0.008); 'Tizoc Disc' (D_b = 1.906±0.008, D_i = 1.882±0.004, D_M 1.866±0.008); and 'Chalco Disc' (D_b = 1.885±0.006, D_i = 1.858±0.002, D_M 1.842±0.01). What deserve be mentioned is that this approach, to estimate fractal dimension, works well in a few artistic or architectural works from the Groups I, IX and X. It is remarkable that Martínez del Sobral [5] has been described all these astronomic and calendar works by taking into account golden rectangles. Thus our result suggests the usefulness of D_M when artistic and architectural works contain information radially distributed, so we prefer to use it on that type of works.

Martínez del Sobral [5] pointed out that many pages from codices such as 'Mendocino Codex' 'Borbonic Codex', 'Borgia Codex' and 'Dresden Codex' are geometrically described by golden rectangles, and we find that codex pages (Groups V and VI) are well characterized by D_b and D_i. Examples are '1-wind 13th' from 'Borbonic Codex' (D_b = 1.932±0.004, D_i = 1.931±0.001), 'Page 1' from 'Mendocino Codex' (D_b = 1.938±0.004, D_i = 1.926±0.003), 'Page 13b' from 'Dresden Codex' (D_b = 1.909±0.004, D_i = 1.908±0.0008), 'Page 55' from 'Borgia Codex' (D_b = 1.949±0.01, D_i = 1.940±0.008).

From Group IV, Pyramids and Temples, Martínez del Sobral (2000) also described the following works through golden rectangles: 'Temple of the Sun' at Teotihuacan (D_b = 1.913±0.003, D_i = 1.93±0.0009), superior view of the 'Temple of the Sun' at Teotihuacan' (D_b = 1.923±0.004, D_i = 1.913±0.003), superior view of the 'Temple of Inscriptions' (D_b = 1.959±0.008, D_i = 1.954±0.005), 'Pyramid Temple I' at Tikal (D_b = 1.910±0.012, D_i = 1.905±0.002), 'Pyramid of 365 Niches' at Tajín (D_b = 1.914±0.004, D_i = 1.935±0.001), superior view of the 'Pyramid of 365 Niches' at Tajín (D_b = 1.926±0.007, D_i = 1.91±0.002).

Figure 3. 'Coatlicue' the Aztec god of life and death as drew by León y Gama (from Martínez del Sobral [5]) (left) and architectural design of the 'Pyramid of the Sun' at Teotihuacan (right).

From the Group VII, we characterize the following works. 'Olmec Colossal Head, Monument 1' at San Lorenzo (D_b = 1.905±0.009, D_i = 1.914±0.002), described by de la Fuente [1] and Martínez del Sobral [5] through golden rectangles. Also, Martínez del Sobral has characterized the following works by using golden rectangles: 'Coatlicue' (D_b = 1.922±0.002, D_i = 1.906±0.006), 'Pacal Sarchophagus' cover at Palenque (D_b = 1.924±0.011, D_i = 1.953±0.0003), 'Stela A' at Copán (D_b = 1.937±0.006, D_i = 1.934±0.003).

In general, our results could be suggesting that Mesoamerican artists and architects used specific patterns and they preferred works with higher box (1.912±0.009) and information (1.916±0.002) fractal dimensions as appreciated in Table 1. In figure 3, we show two of the analyzed works for a readers' best appreciation.

Conclusions

Fractal geometry and Complexity are present in different cultures and in different centuries (Bovil, 1996; Briggs, 1992; Sala and Cappellato, 2003). Many of the Mesoamerican art and architectural works have an high fractal dimension.

Meaningfully, Mesoasoamerican artistic and architectural works are characterized by a box fractal dimension D_b = 1.912±0.009, and/or by an information fractal dimension D_i = 1.916±0.002.

There is a lack of studies to elucidate with a best precision the range for each type of fractal dimension to characterize the Mesoamerican artistic and architectural works once it has been discovered most of them are included in a series of golden rectangles that is connected to an aesthetic sense.

References

[1] Bovil, C. Fractal Geometry in Architecture and Design. (Birkhauser, Boston, 1996).

[2] Briggs, J., Fractals - The Patterns of Chaos: a New Aesthetic of Art, Science, and Nature. (Touchstone Books, 1992).

[3] de la Fuente, B., Los Hombres de Piedra. Escultura Olmeca. (2nd Edition, Universidad Nacional Autónoma de México, Dirección General de Publicaciones. México, D.F. 1984). p.390

[4] Hilgemeier, M., *One metaphor fits all: a fractal voyage with Conway's audioactive decay*. In C. A. Pickover (ed.), *Fractal Horizons: The Future Use of Fractals*. (St. Martin's Press. New York, USA. 1996). pp. 137-161.

[5] Martínez del Sobral, M., Geometría Mesoaméricana. (1st Edition, Fondo de Cultura Económica, México, D.F. 2000). p.287

[6] Miller, M. E., *The Maya Art and Architecture*. (Thames and Hudson, London, 1999).

[7] Sala, N. and Cappellato, G., *Viaggio matematico nell'arte e nell'architettura*. (Franco Angeli, Milano, 2003).

[8] Stierlin, H., *The Maya: Palaces and pyramids of the rainforest* (Taschen, Köln, 2001).

[9] TruSoft Int'l Inc. Benoit, version 1.3: Fractal Analysis System. (20437th Ave. No. 133, St. Petersburg, FL 33704, USA).

[10] www.geocities.com/CapeCanaveral/Station/8228/arch.htm.

In: Chaos and Complexity Research Compendium
Editors: F. Orsucci and N. Sala, pp. 289-293
ISBN: 978-1-60456-787-8

Chapter 22

NEW PARADIGM ARCHITECTURE[1]

Nikos A. Salingaros*
Department of Applied Mathematics, University of Texas at San Antonio,
San Antonio, Texas 78249, USA

Charles Jencks wishes to promote the architecture of Peter Eisenman, Frank Gehry, and Daniel Libeskind by proclaiming it "The New Paradigm in Architecture". Supposedly, their buildings are based on the New Sciences such as complexity, fractals, emergence, self-organization, and self-similarity. Jencks's claim, however, is founded on elementary misunderstandings. There is a New Paradigm architecture, and it is indeed based on the New Sciences, but it does not include deconstructivist buildings. Instead, it encompasses the innovative, humane architecture of Christopher Alexander, the traditional humane architecture of Léon Krier, and much, much more.

According to Jencks, the new paradigm consists of deconstructivist buildings, typified by the Guggenheim Museum for Modern Art in Bilbao, Spain, by Frank Gehry, and including other work and unbuilt projects by Peter Eisenman, Daniel Libeskind, and Zaha Hadid. Jencks has just revised his popular book "*The Language of Post-Modern Architecture*", and has ambitiously re-titled it *"The New Paradigm in Architecture"* (Yale University Press, New Haven, 2002).

Jencks bases his proposed new paradigm on what he thinks are the theoretical foundations of those buildings he champions. He claims that they arise from, and can be understood with reference to applications of the new science; namely, complexity theory, self-organizing systems, fractals, nonlinear dynamics, emergence, and self-similarity. In my own work, I have used results from science and mathematics to show that vernacular and classical architectures satisfy structural rules that coincide with the new science.

Jencks claims a new paradigm with the opposite characteristics of living structure. That's not what one expects from the new science, which helps to explain biological form. Trying to

---

1 This essay is a shortened version of "Cherles Jencks and the New Paradigm in Architecture", a chapter in the author's book "Anti-architecture and Deconstruction" (Umbau-Verlag, Solingen, 2004). Dr. Salingaros is considered as a leading theorist of architecture and urbanism, and an authority in applying science and mathematics to understand architectural and urban form.

* E-mail address: salingar@sphere.math.utsa.edu, Homepage: http://www.math.utsa.edu/~salingar

get a perspective on this contradiction leads one to a witches' brew of confused concepts and statements. Jencks does not provide a theoretical basis to support his claim of a new paradigm. An architecture that arises from the new science represents the antithesis of the deconstructivist buildings that are praised by Jencks. Clearly, we cannot have totally opposite and contradictory styles arising from the same theoretical basis.

As a scientist who has taken an interest in architecture, I have worked with Christopher Alexander, and with coauthors who are scientists and mathematicians, some of them very eminent. Alexander's new work "*The Nature of Order*" (Center For Environmental Structure, Berkeley, 2003) is an important and integral part of the new science. Our contributions to architecture are an extension of science into the field of architecture, beyond mere scientific analogies. The deconstructivists belong outside science altogether, and, despite their claims, do not come anywhere near to establishing a link with the new science.

Instead, the deconstructivist architects draw their support from the French deconstructivist philosophers. Here we have two monumental problems: (*i*) deconstruction is rabidly anti-science, as its stated intention is to replace and ultimately erase the scientific way of thinking; and (*ii*) the spurious logic of French deconstructivist philosophers was exposed with devastating effect by the two physicists Alan Sokal and Jean Bricmont ("*Fashionable Nonsense*", Picador, New York, 1998). How can we therefore accept claims for a new paradigm in architecture, based on science, if it is supported by charlatans who moreover are anti-science? A critical investigation into the pervasive and destructive influence of anti-scientific thought in contemporary culture is now underway.

It turns out that there is a basic confusion in contemporary architectural discourse between *processes*, and *final appearances*. Scientists study how complex forms arise from processes that are guided by fractal growth, emergence, adaptation, and self-organization. All of these act for a reason. Jencks and the deconstructivist architects, on the other hand, see only the end result of such processes and impose those images onto buildings. But this is frivolous and without reason. They could equally well take images from another discipline, for this superficial application has nothing to do with science.

To add further confusion, Jencks insists on talking about cosmogenesis as a process of continual unfolding, an emergence that is always reaching new levels of self-organization. These are absolutely correct descriptors of how form arises in the universe, and precisely what Christopher Alexander has spent his life getting a handle on. Any hope that Jencks understands these processes is dampened, however, when he then presents the work of Eisenman and Libeskind as exemplars of the application of these ideas of emergence to buildings. None of those buildings appears as a result of unfolding, representing instead the exception, forms so disjointed that no generative process could ever give rise to them.

It appears that perhaps the deconstructivist buildings Jencks likes so much are the intentional products of *interrupting* the process of continual unfolding. They inhabit the outer limits of architectural design space, which cannot be reached by a natural evolution. We have here an interesting example of genetic modification. Just like in the analogous cases where embryonic unfolding is sabotaged either by damage to the DNA, or by teratogenic chemicals in the environment, the result is a fluke and most often dysfunctional. Should we consider those buildings to be the freaks, monsters, and mutants of the architectural universe? Hasn't the public been fascinated with monsters and the unnatural throughout recorded history as ephemeral entertainment?

The key here is *adaptation.* I have looked into how Darwinian processes act in architecture on many distinct levels. A process of design that generates something like a deconstructivist building must have a very special set of selection criteria. No-one has yet spelled out those criteria. What is obvious, however, is that they are not adaptive to human needs, being governed instead by strictly formal concerns. Some factors responsible for the high degree of disorganized complexity in such buildings are: (*i*) a willful break with traditional architecture of all kinds; (*ii*) an expression of geometrical randomness and disequilibrium; and (*iii*) ironic statements or "jokes". Trying to avoid the region of design space inhabited by traditional solutions, which are adaptive, pushes one out towards novel but non-adapted forms.

By employing scientific terms in an extremely loose manner Jencks erodes his scientific credibility. As an example, he talks of "twenty-six self-similar flower shapes" used by Gehry in the Bilbao Guggenheim. As far as I can see, there are no self-similar shapes used in that building. As to resembling flowers, they don't, because flowers adapt to specific functions by developing color, texture, and form, all within an overall coherence which is absent here. There is a tremendous difference between a mere visual and a functional appreciation of fractals. The Guggenheim Museum is disjoint and metallic, and as far removed from any flower as I can imagine. Jencks then refers to these non-self-similar shapes as "fluid fractals". I have no idea what this term means, as it is not used in mathematics. A third term he uses for the same figures is "fractal curves". Again, those perfectly smooth curves are not fractal.

I was puzzled to read an entire chapter in Jencks's book entitled "Fractal Architecture" without hardly seeing a fractal (the possible exceptions being decorative tiles). I can only conclude that Jencks is misusing the word "fractal" to mean "broken, or jagged" — even though he refers to the work of Benoît Mandelbrot, he has apparently missed the central idea of fractals, which is their recursiveness generating a nested hierarchy of internal connections. A fractal line is an exceedingly fine-grained structure. It's not just zigzagged; it is broken everywhere and on every scale (i.e. at every magnification), and is nowhere smooth.

Jencks himself admits that: "*The intention is not so much to create fractals* per se *as to respond to these forces, and give them dynamic expression*". What does this mean? He refers to a building that has a superficial pattern based on Penrose tiles, and calls it an "exuberant fractal". Nevertheless, the Penrose aperiodic pattern exists precisely on a single scale, and is therefore not fractal.

Jencks discusses with admiration unrealized projects by Peter Eisenman, which both claim are based on fractals. But then, Jencks adds revealingly: "*Eisenman appears to take his borrowings from science only half-seriously*". Science, however, cannot be taken only half-seriously; one can only surmise that we are dealing with a superficial understanding of scientific concepts that allows someone to treat fundamental truths so cavalierly. Jencks cites Eisenman's Architecture Building for the University of Cincinnati as an example of what he proposes as new paradigm architecture. However, from a mathematician's perspective, there is no evident structure there that shows any of the essential concepts of self-similarity, self-organization, fractal structure, or emergence. All I find is intentional disarray.

As is admitted by its practitioners, de(con)struction aims to take form apart — to degrade connections, symmetries, and coherence. This is exactly the opposite of self-organization in complex systems, a process which builds internal networks via connectivity. For this reason, deconstructivist buildings resemble the severe structural damage such as dislocation, internal tearing and melting suffered after a hurricane, earthquake, internal explosion, fire, or (in an eerie toying with fate) nuclear war.

Architecture and urbanism are prime examples of fields with emergent phenomena. Cities and buildings with life have this property of incredible interconnectedness, which cannot be reduced to building or design components. Every component, from the large-scale structural members, to the smallest ornament, unites into an overall coherence that creates a vastly greater whole. Deconstructivist buildings, however, show the opposite characteristics where each component degrades the whole instead of intensifying the whole. This is easy to see. Does a structural piece intensify the other pieces around it? Is the total coherence diminished if it were removed? The answer is *yes* in a great Cathedral, but *no* in a deconstructivist building. I think that everyone will agree with me that each portion of today's fashionable deconstructivist buildings detracts from and conflicts with every other portion, which is the opposite of emergence.

Traditional architects such as Léon Krier and others have been using timeless methods for organizing complexity, and attribute their results to knowledge derived in the past. It is only very recently that we have managed to join two disparate traditions: (*i*) strands of various architectures evolved over millennia, and (*ii*) theoretical rules for architecture derived from a drastically improved understanding of nature. The new paradigm is a revolutionary understanding of form, whereas the forms themselves tend to look familiar precisely because they adapt to human sensibilities. Most architects, on the other hand, wrongly expected a new paradigm to generate strange and unexpected forms, which is the reason they were fooled by the deconstructivists.

The buildings that Jencks prefers all have a high degree of *disorganized* complexity. This quality is arrived at via design methods mentioned previously. One can also include the use of high-tech materials for a certain effect, which is carefully manipulated to achieve a negative psychological impact on the user. This last feature is best expressed by Jencks himself in describing a paradigmatic building: "*It is a threatening frenzy meant, as in some of Eisenman's work, to destabilize the viewer ...*". I don't think that anyone is going to consider the common theme of disorganized complexity as constituting sufficient grounds for claiming a new paradigm.

Jencks suggests that we are supposed to get excited because a computer program that is used to design French fighter jets is then applied to model the Bilbao Guggenheim. We are also expected to value blobs (which mimic 19C spiritualists' ectoplasm) as relevant architectural forms simply because they are computer-generated. This fascination with technology is inherited from the modernists (who misused it terribly). When the technology is powerful enough, one may be misled into thinking that the underlying science can be ignored altogether. Most informed people know that one can model any desired shape on a computer; it is no different than sketching with pencil on paper. Just because something is created on a computer screen does not validate it, regardless of the complexity of the program used to produce it. One has to ask: what are the generative processes that produced this form, and are they relevant to architecture?

We stand at the threshold of a design revolution, when generative rules can be programmed to evolve in an electronic form, then cut materials directly. There exists an extraordinary potential of computerized design and building production. Architects such as Frank Gehry do that with existing software, but so far, no fashionable architect knows the fundamental rules that generate living structure. A few of us, following the lead of Alexander, are discovering those rules, and we eventually hope to program them. Others working within traditional architecture have always known rules for generating living structure; now they are ready to generalize them beyond a specific style. When the scientific rules of architecture are universally adopted, the products will surprise everyone by their innovation combined with an intense degree of life not seen for at least one hundred years.

Much of what I have said has already been voiced by critics of deconstructivism. And yet, like some mythical monsters, deconstructivist buildings are sprouting up around the world. Their clients, consisting of powerful individuals, corporations, foundations, and governments, absolutely want one of them as a status symbol. The media publicity surrounding deconstruction reinforces an attractive commercial image. I admit that the confused attempts at a theoretical justification, misusing scientific terms and concepts haphazardly, succeed after all in validating this style in the public's eye. It appears that something is clearly working to market deconstructivism, and Jencks's efforts help towards this promotion.

Architects today are told that the new science supports and provides a theoretical foundation for deconstructivist architecture. Nothing appears to justify this claim. On the contrary, I believe the evidence shows that there does exist a new paradigm in architecture, and it is supported by the new science. Charles Jencks is in part correct (though strictly by coincidence, since his own proposal for a new paradigm is based on misunderstandings). Nevertheless, this new paradigm architecture does not include deconstructivist buildings. The new paradigm encompasses the innovative, humane architecture of Christopher Alexander, the traditional, humane architecture of Léon Krier, and much, much more.

In: Chaos and Complexity Research Compendium
Editors: F.F. Orsucci and N. Sala, pp. 295-305
ISBN 978-1-60456-787-8

Chapter 23

SELF-ORGANIZED CRITICALITY IN URBAN SPATIAL DEVELOPMENT

Ferdinando Semboloni
Department of Town and Regional Planning, and Center for the Study of Complex Dynamic - University of Florence, Italy

Abstract

The micro-dynamic models of urban development, usually conceive the evolution as a continuous process of diffusion. Nevertheless, in many cases the changes of the urban fabric depend on the chains of causation which give rise to a great number of little projects and to a very few number of great urban projects. In this paper I present a model simulating the urban development which highlights these phenomena. In fact, in this model the dynamic depends on the accumulation of a potential energy which is suddenly released. In addition, a reaction chain is stimulated by a diffusion process in the neighborhood such in the sandpile model. The model is developed in a 3-D spatial patter, composed of cubic cells which take a limited number of states: un-built, housing, retail and industry. The changing of state happens when the potential energy accumulated overcomes an established threshold, and depends on local and global causes. The global causes are responsible for the accumulation of energy. In turn local causes stimulate the reactions chain resulting in the urban avalanche. The model is experimented in a growth period, and in a stability period. The power law distribution of urban avalanches is analyzed. A parameter is further applied to the effects of the chains of causation, and the results obtained with the variation of the parameter are evaluated in relation to the the sensitivity to the initial conditions.

Introduction

The growth of an urban cluster is usually conceived as an addition of elements to the existent cluster in relation to the state of the elements in the spatial neighborhood in the previous step. Nevertheless in many cases changes happen simply by imitation of previous changes. In other words, the elements change their state in relation to the state of the surrounding elements, as well as in relation to the variation of it. This functional relation generates the domino effect: the falling down of one element is able to originate a chain of variations which can continue ad infinitum (figure 1) [1].

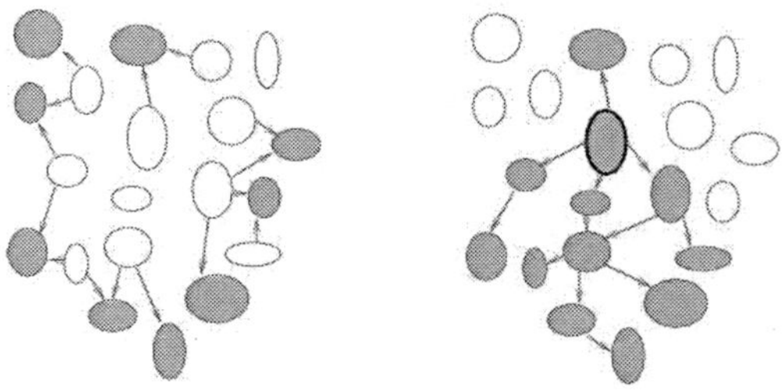

Figure 1. The addition of cells versus the chain of changes.

In the urban dynamic these phenomena may happen in a planned or unplanned way: a gentrification process is a typical unplanned transformation of an entire urban area. In other cases huge transformations are planned and the building or the renovation of an urban area can be completely designed, even if it is usually supposed a start up of the project, for instance the investment of a public company, and a following process of imitation by other investors. The dynamic of a city is characterized by these chains of imitations which give rise to a set of changes involving urban areas of different size, and the size distribution of these urban areas is similar to a power law distribution. In other words the stable state of a city is a critical state in which projects without a typical size can be produced [6]. Even if Alexander [1] has anticipated such vision of the urban dynamic at least for designing purposes, the theoretic background of the present approach refers to the theory of self-organized criticality which was formulated by Bak et coauthors ([2]). The sandpile model, utilized in order to study the properties of similar systems, is resumed hereafter. This model is usually experimented in a 2-D space, organized in squared cells which can take two states, say 0-1. In this model a cell at random receives a grain. This action is considered as a perturbation of the system. When the number of grains in a cell overcomes an established threshold, the cell changes state and the grains located in the cell are distributed in the surrounding four cells. Normally the threshold is equal to the number of cells in which grains are redistributed. For this reason in each surrounding cell, only one grain is received from the cell which has changed its state. The falling down of the grains is suspended if the number of grains in one cell attains the threshold, thus generating a chain of changes, generally called "avalanche". In other words, the number of perturbations is minimized and each avalanche is not connected with the following. In fact after a site has changed state it comes back to the previous state and is ready to be eventually invested by the following avalanche.

In relation to the urban dynamic, the grains can be conceived as the opportunities to invest due to the increase of land rent. When these opportunities overcome a threshold the cell is built, thus influencing the surrounding cells. The potentiality to invest is zero in the central cell after the investment has been performed while the opportunities of the surrounding cells increase because the risks decrease. Anyway it is not possible to completely transfer

[1]The coloured figures can be downloaded from: http://fs.urba.arch.unifi.it/cclpap/index.html

the sandpile model in the explanation of the urban dynamic. In fact in the urban dynamic a site is frozen after a building process, at least for an established period which, in essence, depends on the period needed for the amortization of the investment. For this reason not all the areas of a city are in a critical state. In addition in the sandpile the avalanches are sequentially distributed, in turn, in the urban development they may happen contemporaneously. These differences are considered in-depth in the following section where the model is explained.

The Model

The model is organized in a 3-d squared grid of cubic cells, as in [4] (figure 2). Each

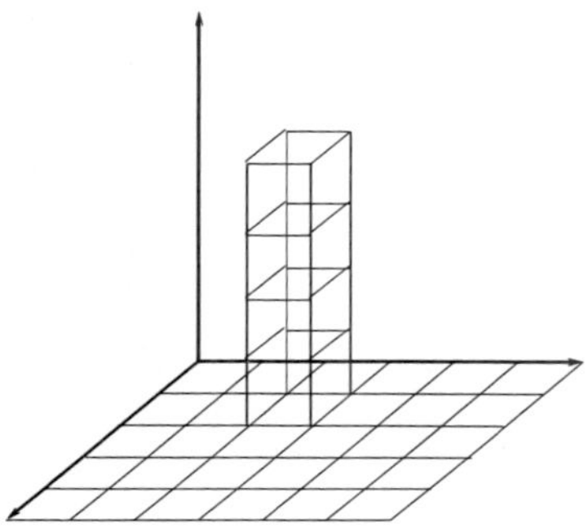

Figure 2. The spatial patter in 3-D. The distances are calculated in 2-D, as row flies. The building in the third dimension is submitted to the constraint that the underlying cell was already built.

cell can take a state, otherwise stated, it can be occupied by an use (figure 3) and the model dynamic is based on the transition of each cell from one state, or use, to another.

The global dynamic is constrained to total values for each use established exogenously [5]. If the current number of cell for a specified use is lower than the established, then some cells are stimulated to change state. In fact, the global values are constrained, but the spatial distribution of it is totally managed by the model.

In essence the functioning of the model is the following. The grains are specialized in relation to the relevant uses, i.e.: housing A and B, retail, and industry. In other words each cell has a number of containers of grains equal to the number of possible relevant uses (figure 5). The grains in each container represent the potentiality for a cell to be utilized for the corresponding use. At each step a grain is added to some cell in dependence to the difference: global desired quantity minus existent quantity of each use. These grains are distributed in relation to the suitability of this cell for the use in question. When the number of grains in a container of a cell related to an use reaches the threshold, set equal to 5, the cell is assigned to the use and the potentiality of the containers of the cell is decreased by a quantity equal to the threshold. In fact, after the change of state, the potentiality of further variation in the cell is null or negative. Further, the potentiality related to the assigned use is distributed into the surrounding cells. These surrounding cells are the four bordering plus

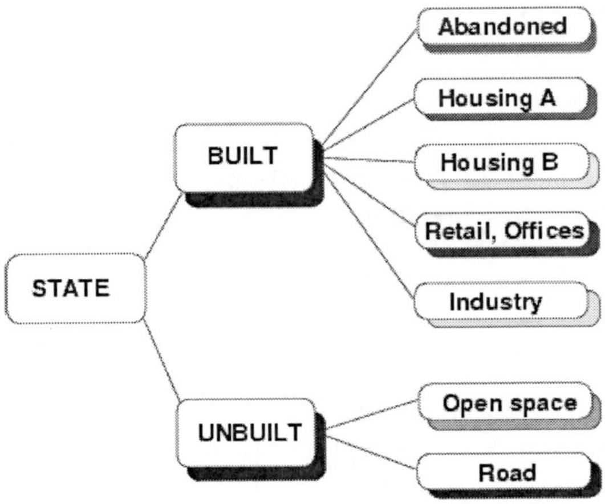

Figure 3. The allowed states for each cell. States are divided in build and un-built. The built states include two types of housing, A, and B, retail and offices, and industry. The income of housing A is considered higher than the income of housing B. In addition a cell is abandoned when it is built but empty as use. The unbuilt states include roads and open spaces.

the upper cell, because the pattern is in 3-D (see figure 4), and the threshold is equal to the number of cells in which the grains are distributed.

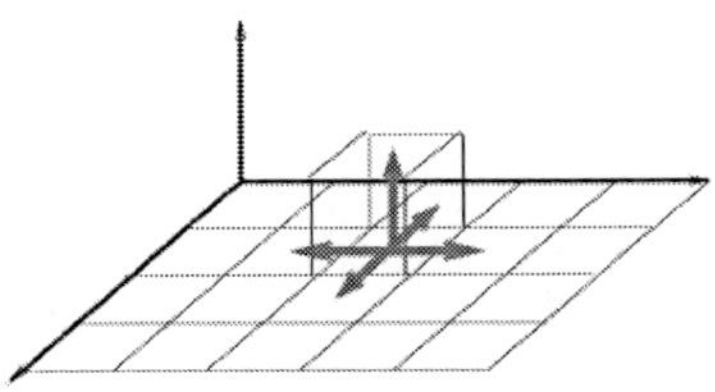

Figure 4. The distribution of the grains into the four surrounding cells and in the upper cell.

Let us consider the method for the distribution of grains. In the sandpile model the grains are assigned randomly, while in the present model they are assigned in relation to the suitability of a cell for the specified use (figure 5).

The suitability is calculated in relation to the surrounding uses in a radius of 6 cells, and depending on the distance from the central cell. The slope and the nearby to the roads are considered ([5]), as well as the cost of building in relation to the floor [4]. The assignment of a grain is based on the comparison of the suitability for different uses. For this reason a weight is included. It accounts for the importance of the use, or in other words for the ability of the use to compete in the land market. Finally the result is multiplied by a random factor which simulates the uncertainty in the evaluation of suitability. In conclusion the suitability for a cell c_{ijk} to be in state p is calculated by using the following equation:

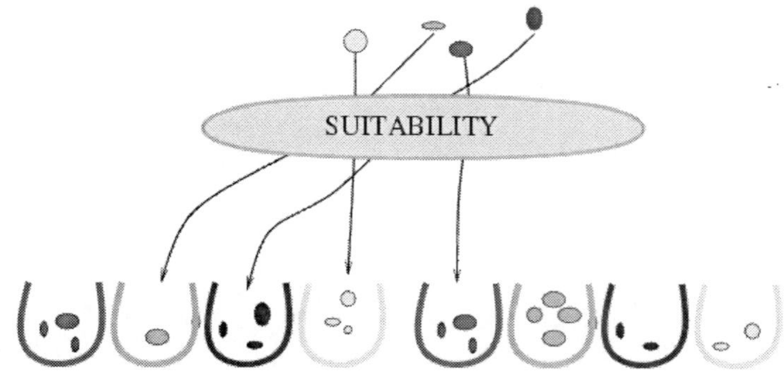

Figure 5. The grains fall down in the containers of cell i or j depending of the suitability of each cell for the specified use.

$$S_p = \frac{\sum_{q,d} m_{p,q,d} I_d - s_{ij}}{C_k} W_p r \tag{1}$$

where:

S_p is the suitability for the state p in cell c_{ijk};

$m_{p,q,d}$ is the weight connected to the cells in state q at distance d from c_{ijk} in relation to state p;

$I_d = 1$ if he state of the cell distant d from c_{ijk} is equal to q, $I_d = 0$ otherwise;

C_k is the building cost for a cell at floor k;

s_{ij} represents the difficulty to build in relation to the slope of ground;

W_p is the weight related to the state p;

r is a random factor: $r = 1 + [-\ln(rand)]^\alpha$, $\alpha = 3$.

In order to avoid an huge computation time, the suitability is calculated for a set of cells in which are included abandoned cells plus some cells chosen at random among unbuilt cells as well as built and assigned to an use from an established period which is about 200 steps. This set, which in the following experiments represents about 10% of the total quantity of cells, does not include cells in critical state i.e. cells having a potentiality equal to the threshold. In other words the addition of grains does not disturb the acting avalanches. After having calculated the suitability, the grains are distributed with the following method. Grains are assigned beginning by the maximum suitability till the global quantity of each grains, exogenously established, in relation to the desired quantity of each use, is reached. Finally cells are abandoned after an established period, (set equal 400 steps) and an abandoned cells is demolished if it is not occupied after an established period (set equal to 400 steps). The entire process is represented in figure 6.

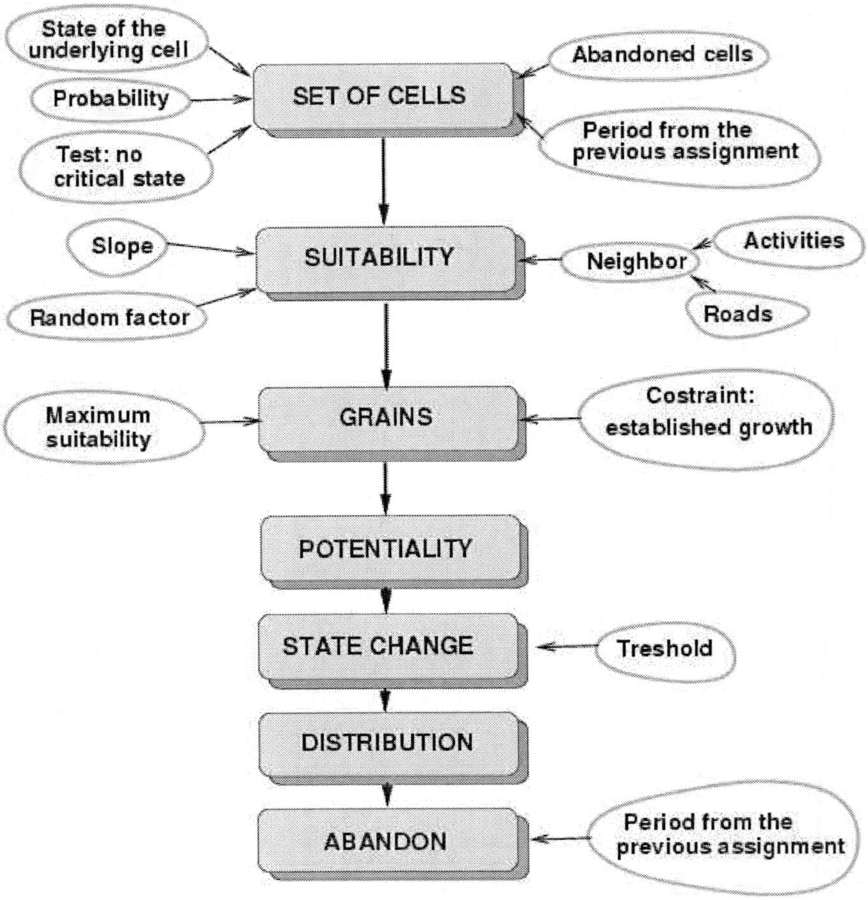

Figure 6. The process for the change of the state of a cell.

Results

The experiments have been performed on a squared grid of 50x50x10 cells of 200 meters sized. In order to shorten the number of grains in relation to the stimulated avalanches, each period, which is supposed to correspond to one year, is divided in 10 sub-periods. In each experiment, iterations have been 4000, corresponding to 400 years. In the first 200 years (i.e. 2000 steps) the number of built cells grows, while in the second period the number of built cells is stable, and only the abandoned or demolished cells are replaced by the model dynamic.

The maximum quantity of each use, as well as the values of the weights W_p are established as in table 1.

The quantity of each use is supposed equal to 1 at the beginning and increases linearly till the maximum after 200 years. In fact a seed is established almost in the center of the area. At each period the expected quantity of each use is calculated. This quantity is utilized in order to establish the number of grain to distribute. The values of the parameters $m_{p,q,d}$

Table 1. The quantity of cells per use, and the corresponding quantities of inhabitants and employees. In the last column are included the values utilized for the weights W_p

.

Use	Cells	Inhabitants or employees per cell	Total in-habitants or employees	W_p
Housing A	200	500	100 000	10
Housing B	200	500	100 000	5
Retail, offices	60	600	36 000	50
Industry	10	300	42 000	1

are shown in figure 7, and the resulting spatial pattern is shown in figure 8.

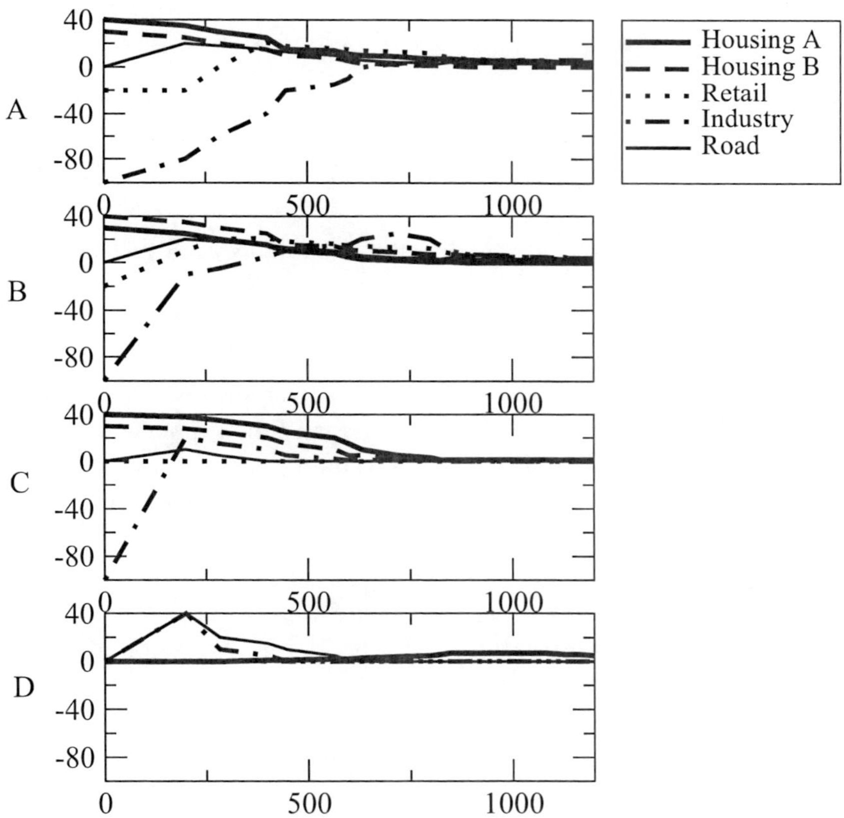

Figure 7. Variation of weights $m_{p,q,d}$. Graph A: X axis, distance (d), Y axis weight of cell in state q (states are listed in the legend) in connection with Housing A use. Graph B: Y axis weight of cell in state q in connection with Housing B use. Graph C: Y axis weight of cell in state q in connection with retail use. Graph D: Y axis weight of cell in state q in connection with industrial use.

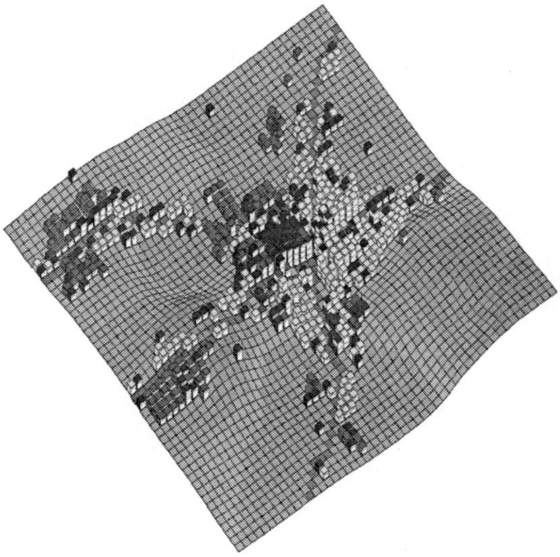

Figure 8. The spatial pattern after 400 years. Black: retail, gray: housing A, light gray: housing B or industry.

Because the dynamic is based on the transfer of grain of potentiality from one cell to the surrounding cells, the size of each avalanche is calculated by recording the chain of causation from the start up cell to the other cells. In order to evaluate the distribution of the size of avalanches, these have been ranked by size. In other words we have estimated the cumulative distribution function (CDF). In case of a power law distribution, the probability distribution function (PDF) can be obtained by increasing the exponent of the CDF by one. The size has been plotted in relation to the rank, and the result is shown in figure 9. The estimated function is $s \propto r^{-0.28}$, where s is the size of the avalanche, calculated by using the number of cells included in the avalanche, and r the rank (being equal 1, the rank of the greatest avalanche). The CDF is $P(s' > s) \propto s^{-1/0.28}$ and the PDF is $P(s) \propto s^{-(1/0.28+1)}$. This result means that avalanches of great size are very limited in relation to the small avalanches. In addition from figure 10 it results that during the period of stability the size of avalanches increases.

Discussion

In order to evaluate the impact of the sandpile method, a probability to distribute the grains of potentiality in the 5 surrounding cells, has been included, as parameter, in the model. The application of this parameter results in a change of the urban cluster. This change is evaluated by using the centroid of the urban cluster during the simulation. Twenty simulations have been performed by varying the seed of the random number generator. The first ten by using a probability to distribute the grains equal to one, i.e. by using the normal model, and the second ten by using a probability equal to 0.1. The resulting spatial pattern of one of the second set of experiments is shown in figure 11, while in figure 12 are shown

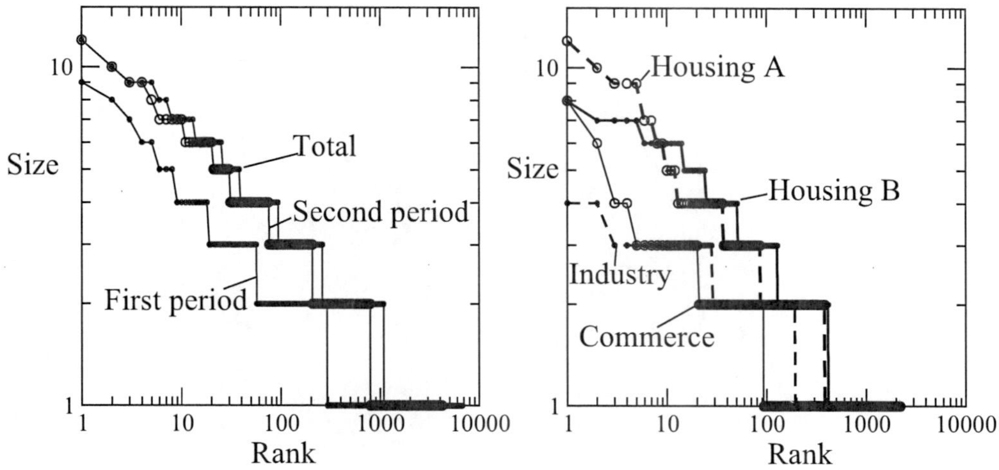

Figure 9. The rank-size distribution of avalanches. X axis: rank, Y axis: size. Left side: the rank-size distributions obtained considering all the avalanches, the avalanches which happen during the first period of growth, and during the second period of stability. Right side, the rank-size distributions of the avalanches per use.

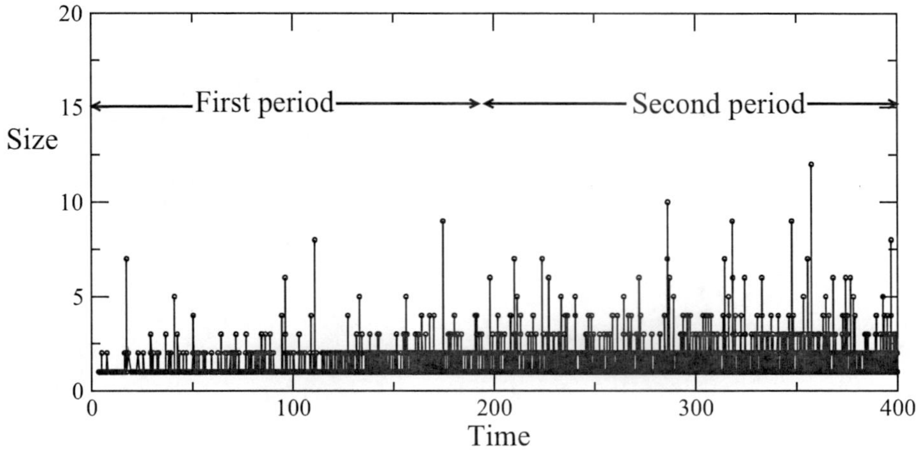

Figure 10. The temporal series of avalanches. X axis: time, Y axis: size of avalanches.

the paths of the centroids in the first and second set of experiments during the steps of the simulations.

As figure 12 shows, the paths of the second set of experiments are less scattered. In essence the more the process of distribution of grains is activated the more the final result is dependent on initial conditions. In other words the chaotic behavior of the dynamic system depends on the chains of causation established by the sandpile method.

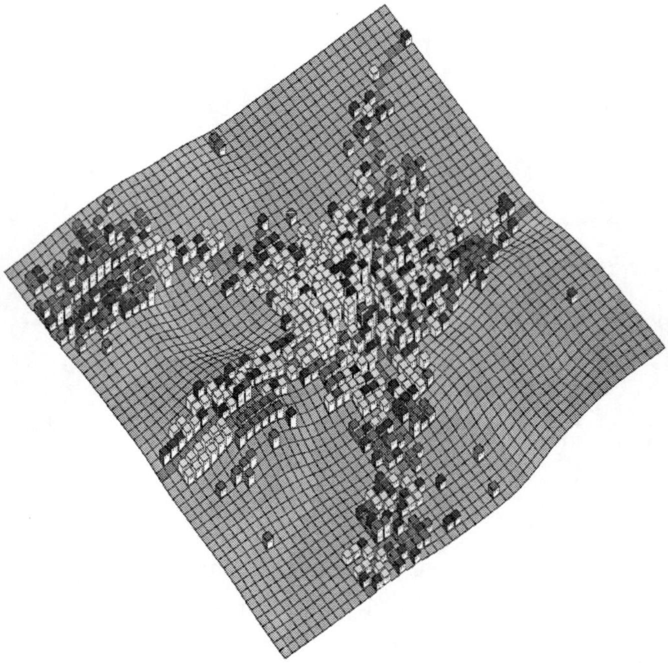

Figure 11. The spatial pattern after 400 years. Black: retail, gray: housing A, light gray: housing B or industry.

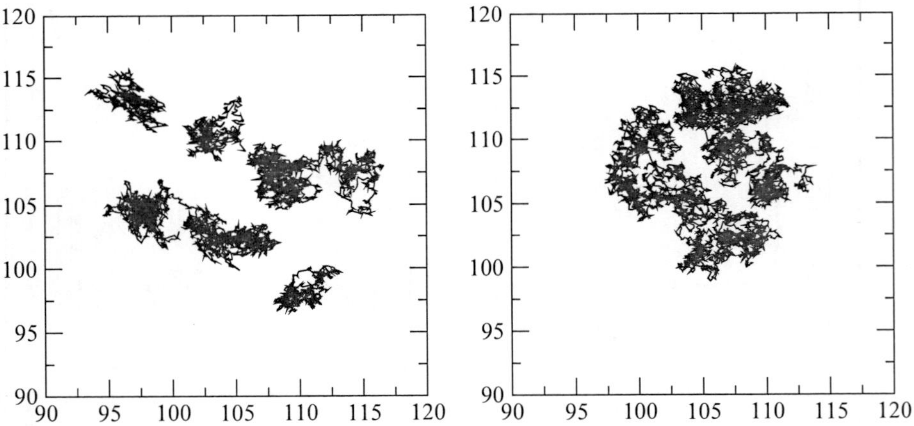

Figure 12. The path of the centroid of the urban cluster, in the final period of the simulation. Ten simulations obtained by varying the seed of the random number generator. Left side probability equal 1, high variability. Right side probability equal 0.1, low variability.

Conclusion

The sandpile method has been applied to the simulation of the urban development. The urban dynamic can be well simulated as a system in which the steady state is characterized by a self-organized criticality, and it has been shown that the chaotic behavior of the simulated urban dynamic depends on the chain of causations generated by the sandpile model.

Acknowledgments

I thank Prof. Franco Bagnoli, Faculty of Engineering, University of Florence for interesting discussions. Disclaimers apply as usual.

References

[1] C. Alexander. *A New Theory of Urban Design.* Oxford University Press, New York, 1987.

[2] P. Bak, C. Tang, and K. Wiesenfeld. Self-organized criticality. *Phisical Review A*, **38**:364–374, 1988.

[3] M. Batty and Y. Xie. Self-organized criticality and urban development. *Discrete Dynamics in Nature and Society*, **3**:109–124, 1999.

[4] F. Semboloni. The dynamic of an urban cellular automata model in a 3-d spatial pattern. In *XXI National Conference Aisre: Regional and Urban Growth in a Global Market*, Palermo, 2000.

[5] R. White, G. Engelen, and I. Uljee. The use of constrained cellular automata for high-resolution modelling of urban land-use dynamics. *Environment and Planning B: Planning and Design*, **24**:323–343, 1997.

[6] F. Wu. A simulation approach to urban changes: Experiments and observations on fluctuations in cellular automata. In P. Rizzi, editor, *Computers in Urban Planning and in Urban Management on the Edge of the Millennium. Cupum99.* , F.Angeli, Milano, 1999.

In: Chaos and Complexity Research Compendium
Editors: F. Orsucci and N. Sala, pp. 307-320
ISBN: 978-1-60456-787-8

Chapter 24

GENERATION OF TEXTURES AND GEOMETRIC PSEUDO-URBAN MODELS WITH THE AID OF IFS

Xavier Marsault*

UMR CNRS MAP, "Modèles et simulations pour l'Architecture, l'urbanisme et le Paysage"
Laboratoire ARIA, Ecole d'Architecture de Lyon
3, rue Maurice Audin, 69512 Vaulx en Velin

Abstract

Geometric and functional modelling of cities has become a growing field of interest, raised by the development and democratisation of computers being able to support high-demanding graphics in real time. Actually, more and more applications concentrate on creating virtual environments. ARIA has been working for two years, within the DEREVE project (DER, 2000), on pseudo-urban textures and geometric models generation, by means of fractal or parametric methods.

This paper explains our attempt to capture inner coherence of urban shapes and morphologies, by fractal analysis of 2D½ textures (top view + height) of real and synthetic city maps. The basic ideas lean on autosimilarity detection, fractal coding of regions, and processing with Iterated Function Systems (IFS). We introduce a genetic-like approach, allowing interpolation, alteration and fusion of different urban models, and leading to global or local synthesis of new shapes. Finally, a 3D reconstruction tool has been developed for converting textures to volumes in VRML, simplified enough for real time wanderings, and enhanced by some automatically generated garbage dump and decorated elements. Programs and graphic interface are developed with C++ and QT libraries.

Keywords: Fractal city, Urban pattern, IFS. Image, 2D½ and 3D model, Genetics, Fusion, Level of detail, Shape filtering, VRML

* E-mail address: Xavier.Marsault@lyon.archi.fr, Homepage: http://www.aria.archi.fr

1. Introduction

1.1. Fractal Cities?

Usually, geometric models of town patterns or whole cities can be generated with the aid of spatial growth simulators, or temporal simulators based upon a scenario (ex: Sim City), or by means of static shapes (Parish, Muller, 2001). Many related works deal with fractality: some of them use cellular automata (Torrens, 2000), other ones use DLA (diffusion limited by aggregation) (Bailly, 1998) or organic models inspired by physical laws (Makse, 1996).

Indeed, some recent studies reveal the fractal nature of many urban structures at large scales and some architectural objects (Sala, 2002), (Batty, Longley, 1994), (Frankhauser, 1994). Focusing on the near scale of buildings and built patterns, we have shown that some urban shapes exhibit a local property of autosimilarity, while they lose it in a larger analysis. In this context, one way of research was to attempt to use IFS (that share this property) to analyse and generate new urban morphologies. Two cities belonging to the suburbs of Lyon (St Genis and Venissieux, fig 7a) and two synthetic maps (fig. 6) have been used during all developments and tests.

1.2. Iterated Function Systems (IFS)

IFS theory is totally based upon the “scale change invariance” property (SCE), and thus allow the generation of fractal objects with a set of contractive functions showing this property, called Iterated Function System (IFS). It has been studied by Hutchinson within the mathematical frame of autosimilarity (a mathematical object is said to be autosimilar if it can be split into smaller parts calculated from the whole by a “similar transformation”) (Hutchinson, 1981), and by Barnsley within the frame of fractal geometry (Barnsley, 1988), leading to image compression applications (Barnsley, 1992-1993), (Jacquin, 1992).

Image Compression

Since usually a given image is not a fractal object, it is unlikely to find a whole fractal generator of it. But there is an interesting application of IFS to image compression allowing fractal coding, where small regions are coded from contractive SCE transformations (called *lifs*, for “local ifs”) of other larger regions. The most common algorithm was developed by Jacquin (Jacquin, 1992). Given a square uniform pavement of N range blocks r_i of size B and a pool of domain block d_i for matching research (Fig. 1), it tries to find a function $i \to \alpha(i)$ and N *lifs* ω_i such as $r_i \to \hat{r}_i = \omega_i(d_{\alpha(i)})$, so that $\sum_i \left\|\omega_i(d_{\alpha(i)}) - r_i\right\|^2$ is minimum for each i (local collage). The famous “collage theorem” ensures that the decoded image is an approximation of the original image, and gives a maximum measure of the error. Each ω_i is set with a transformation projecting a domain block di of size D at the place and scale of the range block ri (decimation of pixels + isometry), followed by an affine transformation on grey levels of pixels : $\hat{r}_i = iso_i(\sigma_i . r_i + \beta_i)$.

The isometric transformation iso_i is chosen among the 8 possible transformations of a square block (identity, –90, 90 or 180 degrees rotations, x, y, or diagonal axis symmetries).

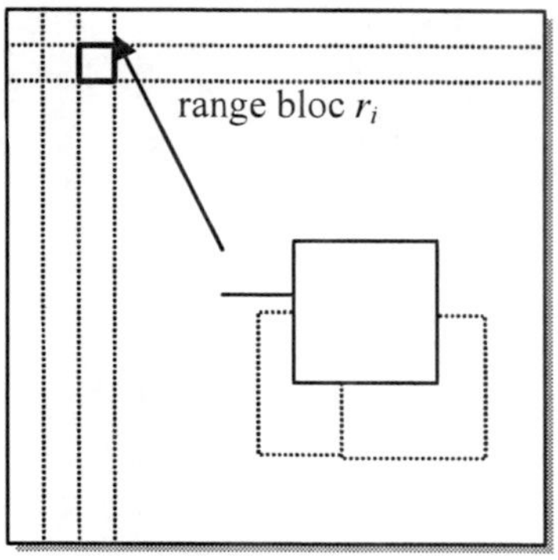

Figure 1. Transformation from a domain block to a range block (lifs).

The image I is decoded by calculating the attractor of *lifs* w_i from any random image I_0.

We note : $W(I)=\bigcup_{i=1}^{N}\hat{r}_i=\bigcup_{i=1}^{N}\omega_i\left(d_{\alpha(i)}\right)$ and $W^{k+1}(I)=W\left(W^k(I)\right)$.

The attractor is then defined by : $A(I)=\lim_{k\to+\infty}W^k(I_0)$.

The more local collages are better, the more the result of the attractor is a good approximation of the original image (collage theorem). A little number of iterations is needed for the decoding process to converge. In order to ensure the uniform convergence, we limit $|\sigma_i|<1$ for each block, even if it has been shown this limit can be exceeded (Hutgen, Hain, 1994).

1.3. Are Urban Patterns Autosimilar?

This question appear to be essential, since one can rarely observe such a global property for objects in nature or in the real world, especially for urban structures. Nevertheless, the use of IFS for generating town-like patterns has been described by (Woloszyn P, 1998), who illustrates how the iteration of a simple substitution rule from an initial and basic pattern leads to an image that looks like an urban structure (Fig. 2). Fragmented patterns are recursively generated from only one transformation applied to all shapes ; they approximate the streets, the secondary networks, and the buildings, and are totally fractal. But this technique can't reproduce real city patterns, more irregular and more complex in their structure, and not totally autosimilar.

On the other hand, a local property of autosimilarity can be more or less put in evidence with real urban patterns, and we can envision the application of local IFS to the generation of urban fabric. We've demonstrated this result by computing autosimilarity measures with Jacquin's IFS on 256 x 256 grey level images of urban patterns.

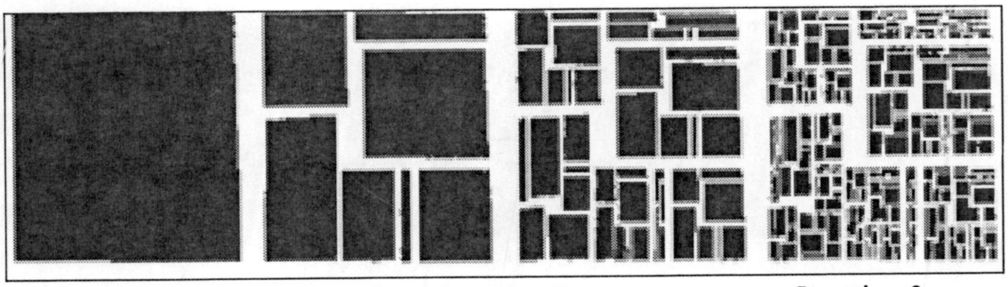

Original pattern Iteration 1 Iteration 2 Iteration 3

Figure 2. Example of pseudo-urban fractals (Woloszyn, 1998).In that purpose, we developed mathematical functions allowing some local or global measures of autosimlaity within an image, at the expense of computing time, because the algorithm still examines all square blocks configurations. The entire range and domain block pools {B} and {D} are parsed (not only those concerning Jacquin's pavements). For every pixel p∈B, all possible appariement configurations (B,D) are tested, avoiding the contraction limit $\forall i, |\sigma_i| < 1$, which speeds up the calculus. The appariement of D and R blocks is based upon the minimum of the local error measure defined by $\mu(R,D) = \sum_{p \in B} (\omega_p(D) - R_p)^2$, normalized between 0 and 1. The more μ stretches towards 0, the better the appariement ; but it seems difficult to fix a threshold that decides there is autosimilarity or not. Therefore, we propose to define a local measure of this approximation, by searching the best couples (B,D) for appariement. Both cases with parameter B fixed or variable have been experimented.

When B is fixed, let's set up the D parameter, and an upper limit for D starting from B+1. In the case where D parameter is fluctuant, we keep a D block so that $\{\mu(R,D) \min imum, D > B\}$. Then, locally for each pixel p, we can define an average autosimilarity measure : $\tilde{\mu}(p) = \frac{1}{card\{R, R \supset p\}} \sum_{R \supset p} \min_{D>B} \mu(R,D)$. But, this average can potentially mask an existing D block for which the appariement is exact, or almost exact. So, we also calculate (Fig. 3) the minimal measure : $\mu_{\min}(p) = \min_{R \supset p}\left(\min_{D>B} \mu(R,D) \right)$. When B is fluctuant, we could locally consider the higher value $B_{max}(p)$ of B for which the $\mu_{\min}(p)$ measure is minimum, and propose another measure (1- $\mu_{\min}(p)$).$B_{max}(p)$, which grows both with B and the appariement quality. But such a task should require a tremendous computing time, even for a 256 x 256 image. One can also wander which information could be gathered from the study of the function D→ $\mu(R,D)$. For example, the decrease of this function could help characterizing a typical behaviour.

Our goal being the generation of urban shapes and structures that look like real ones, we decide to lean on real city plans, and use the IFS as an analysis and synthesis tool. Because IFS operate on a continuous space of shapes, allowing interpolation, alteration and fusion, and integrate as a whole approach analysis and generation of global or local new shapes, we expect them to produce good results.

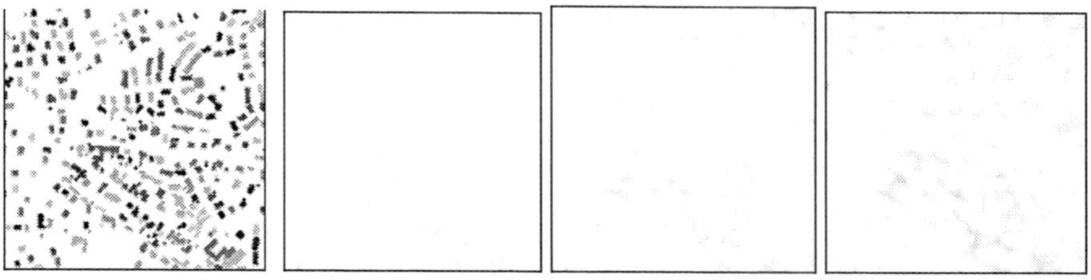

part of Saint Genis μ for B=5, D=20μ for B=7, D=20 μ for B=8, D=20

Figure 3.

2. From IFS to Urban Textures and Geometric Models

2.1. A Simplified Coding Method

Urban scale concerns the spatial distribution of buildings within a certain piece of landscape. It can be described with a restrictive approach by a set of more or less simplified volumes, especially for fast rendering. The image compression technique described in 1.2 allows an approximated coding of an image from local transformations of parts of itself. It can be used to encode urban pattern with IFS if we decide to convert geometrical 3D volumes to images. In this 2dD½ approach, the grey levels represent the heights of buildings. Then, we use Jacquin's fractal compression technique for coding the ground shapes and heights of buildings which populate a city map. We get an autosimilar approximation of the map, whose accuracy depends on the nature and the choice of the initial pavement of the map, and on the number of local *lifs* given to approximate the local diversities of the shapes.

We have proposed some adaptations for urban pattern analysis: initial and static regular square pavement for the range blocks of size B, exhaustive research within the domain block pool (with varying size of D for each block B), pre-calculus of range blocks similarities, accelerated appariement by classification of range and domain blocks (uniform, outline), elimination of ground blocks and a « topological collage » for matching (see below).

Towards a Spatial-Coded Model

When several domain blocks D are candidates to the best appariement for a given range block B, the question of choice appears, whereas it is not significant in the frame of image compression. In that case, the algorithm should select the D block whose neighbourhood if the closest to the one of the B block. It expands successively the outline of each block by on pixel until it finds the minimum among all the proposed range blocks. This option that we call « *topological collage* » slowers the processing time, but is the only one that can really take account of the topological links between B and D blocks. It has been successfully applied to our appariement algorithm used by the "asymmetric fusion" operator (see 2.3).

2.2. General Processing Scheme

Some pseudo-convex interpolation, mutation, fusion and filtering operators have been designed to generate new urban models leaning on existing ones (real or synthetic), or to add some modifications to them. For this purpose, we use the genetic analogy introduced by (Vences, Rudomin, 1997) (see 2.3).

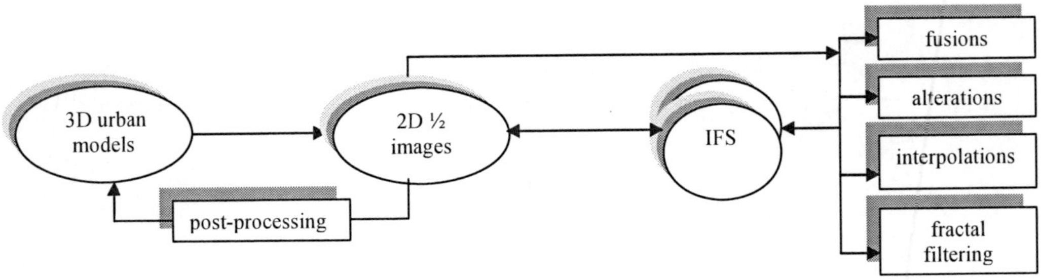

Figure 4. General processing scheme.

Pseudo-Convex Interpolation of IFS

Given two distinct images encoded by IFS_1 and IFS_2, our first idea was to define a λ-parametric convex IFS leaning on IFS_1 and IFS_2. Indeed, if the iteration semi-group is convex, its attractor is λ-continuous (Gentil, 1992). As this is not the case with the semi-group composed by the 8 isometries of the square, we should rather speak of pseudo-convex interpolation. And the awaited results are disappointing, since we get in fact the same as basic image interpolation. Nevertheless, depending on other suitable choices for the type of pavement used, IFS interpolation could become possible.

2.3. A Genetic-Like Formalization

The genetic analogy proposed by Vences and Rudomin first in the frame of image compression, is very powerful for exploring new ways of creation, and let us envision applications to the generation and the alteration of urban geometric models. Assuming the notation $IFS = (\omega_1, \omega_2, ..., \omega_N)$, where ω_i are the *lifs*, we consider the IFS as a chromosome (genotype), the *lifs* as genes, and the attractor (image or 3D model) as a phenotype. This analogy can be justified in several ways. First, the information for decoding an image fragment is distributed among many *lifs*. Some lifs alterations can have consequences on numerous zones, or not. Moreover, the whole body of *lifs* represents a highly non-linear and complex system.

Following this scheme, we apply the general fusion mechanism (Renders, 1995) which consists in generating a large population of IFS models sharing the same genes (inherited from two parents), while the mutation allows the alteration of genes during the crossing process or the exchange of genes along the same chromosome. The following paragraphs describe some ways we used to implement the fusion process.

Direct or Asymmetric Fusion

We follow the genetics analogy, where the fusion process, even highly combinatorial, does not take any genes at random. Since *lifs* are coding zones whose content may be very different, the process of fusion must be guided by an appariement step between IFS parents.

Indeed, without any control, the direct fusion *lifs* to *lifs* or their copy from one zone to another (a kind of mutation), give very bad results. So, we decided to keep the original distribution of *lifs* and to attempt to group them, before fusion occurs between both IFS.

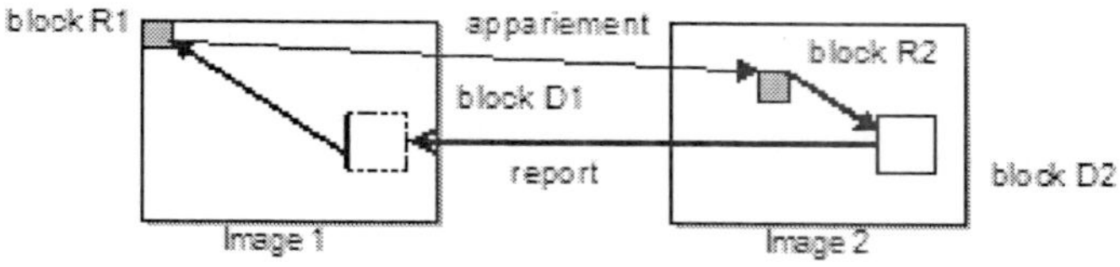

Figure 5. Principle of asymmetric fusion process.

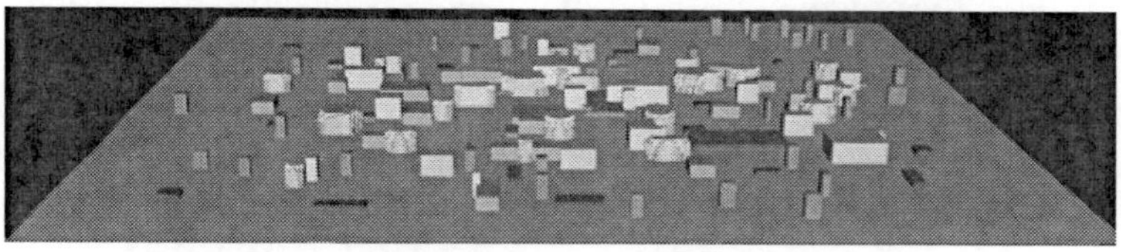

synthetic town model A

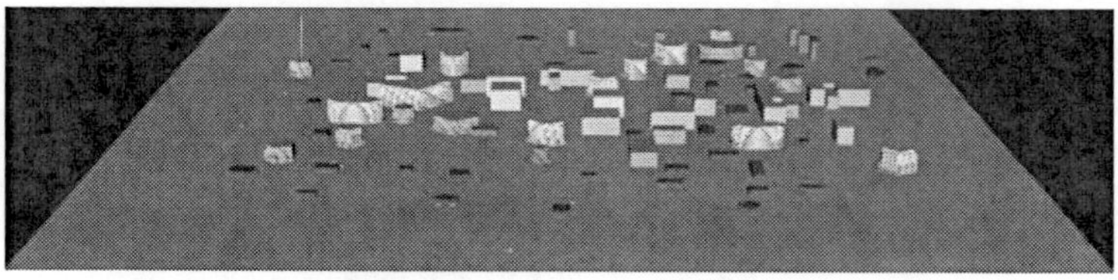

synthetic town model B

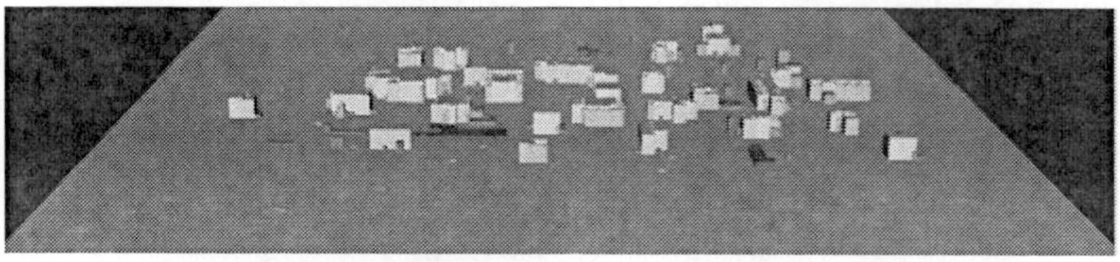

asymmetric fusion AmodB for B=4 and D=8

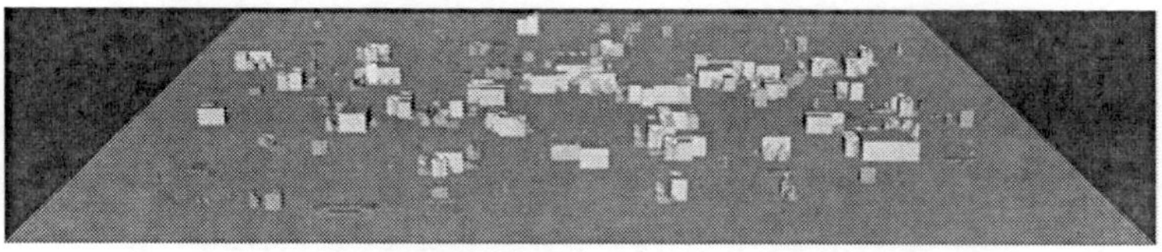

asymmetric fusion BmodA for B=4 and D=8

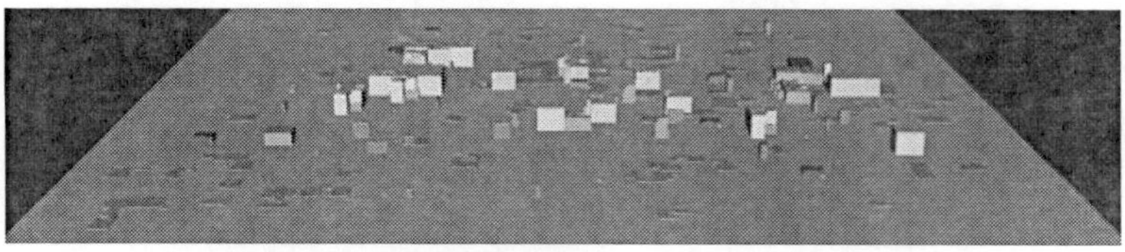

asymmetric fusion AmodB for B=8 and D=16

Figure 6. Continued on next page.

asymmetric fusion BmodA for B=8 and D=16

Figure 6. 3D models of asymmetric fusion with synthetic towns A and B.

A first step we proposed to take this mechanism into account is to calculate appariements of range blocks between both images, based on the images content. The process is asymmetric, since we associate to each image a list of range blocks and their *lifs* counterparts in the other one (fig. 5). A range block R_1 from image 1 is linked (by appariement) to a range block R_2 of image 2, which is encoded by an *lifs* based on domain block D_2. The process of fusion consists in replacing *lifs1* by *lifs2*, and leads to two fusion images : 1mod2 et 2mod1. One can observe on figure 6 the effect of asymmetric fusion : the generated distribution of shapes look like their two parents, modulated each one by the other.

Pavement-Based Fusion

Given a unique Jacquin square pavement for both IFS, and a size B for range blocks, we define square macro-blocks (or pavements) of size multiple of B, in order to group several connex range blocks. The pavement-based fusion consists in crossing spatially grouped sequences of *lifs* (rather than isolated ones) between both IFS, in order to preserve topology. The process alternately keeps some *lifs* from the first IFS and the second one. Possible discontinuities only appears at the borders of the macro-blocks. Our technique lets the algorithm first inject some macro-blocks of important size, and finishes with smaller ones, like a town planner who first deals with higher scales of the city before looking at the content of the neighbourhoods. Moreover, the fact of varying the size of injected macro-blocks allows the modulation of the crossing scale. While varying the minimum and maximum limits of the macro-blocks size, we modify the model topology by authorizing more or less discontinuities.

The macro-block locations and the fill rates of each IFS are provided by the user or generated by a pseudo-random generator. The user also enters upper and lower limits for the macro-blocks size. The algorithms first fills up the entire image with one IFS. Then, it takes the other one, and will alternate until a break-test is verified. A margin is entered by the user to let the algorithm have a tolerance while matching the fill rate criterion. This margin progressively diminishes each time the filling IFS is changed, while the size of the macro-blocks is lowered of one-pixel. This is a heuristic allowing to create on the fly a new genotype from both parents' ones, guided by constraints depending on their phenotypes. For convenience, we added as fill rate criterion a fusion parameter λ, ranging from 0 to 1, allowing a sort of IFS interpolation between both models.

We define a non-intersection criterion allowing to label as "admissible" each macro-block whose outline does not intersect buildings more than a tolerance threshold *S*, given by the user. This criterion is computed on the grey level differences around the outline. Moreover, it also takes account for the previous crossing steps of the algorithm, leading to a better continuity in the phenotype shapes. Nevertheless, this precaution does not guarantee that all domain blocks will belong to preserved zones, but this is a first and serious limitation to this problem. The *S* parameter plays an essential role in the appearance of new and original shapes.

Pavement-controlled fusion gives some very good graphic, and provides new local or global shapes and distribution of shapes, whose details are the consequence of crossing models. The pavement choice is controlled by preservative criterions, and the number of map inputs is not limited for this process.

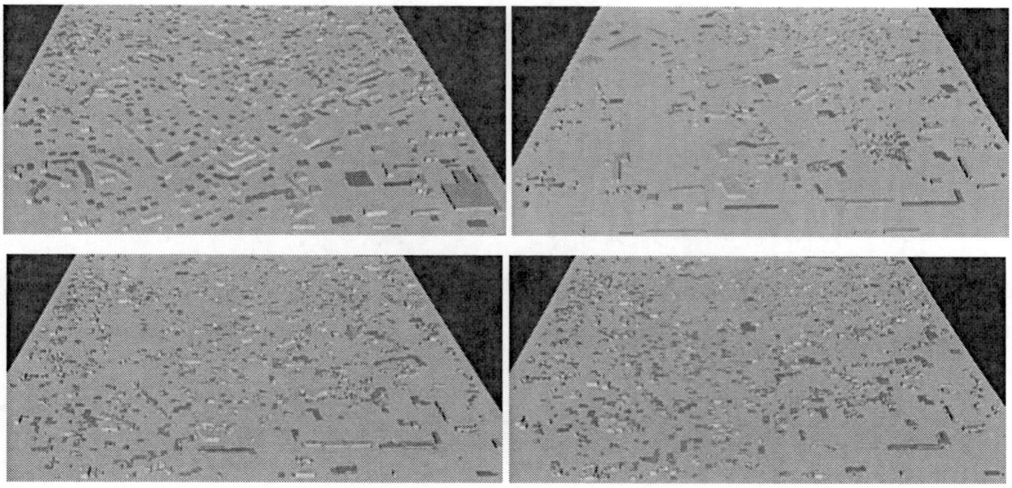

Figure 7. Example of pavement-controlled fusion (down) on real cities (Saint Genis and Vénissieux,up).

2.4. Shape and Detail Filtering

Adjusting IFS Scale for Detail Filtering

Since IFS coding does not take account for dimensions, it is easy to mathematically rebuild its attractor at any scale. This property leads to what has been called "fractal zoom", that can be used with values greater than one (creation of fractal detail), or less than one (shape simplification). So we denote a correspondence between the fractal zoom and the generation of continuous "levels of detail" for objects, that can be used within a real-time wandering (fig. 8).

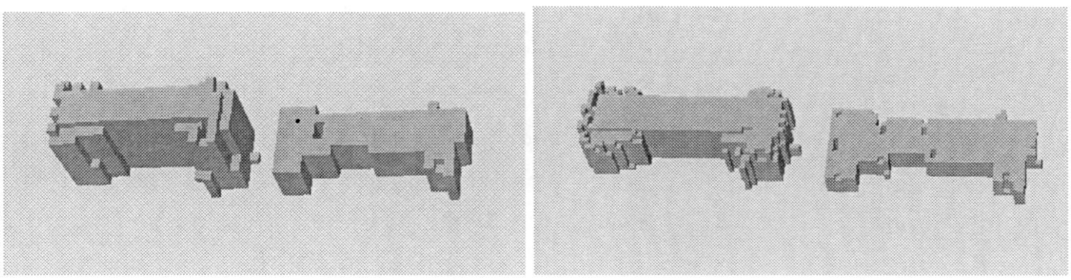

Figure 8. Two versions of the same buildings (before and after a 2x fractal zoom).

Shape Filtering within the Domain Blocks Pool

On the other hand, the encoding of range blocks from domain blocks being surjective, some domain blocks may be used more than other ones, and our experiments confirm this property, that gives an indication on the quantity of generative information used to approximate an urban fabric. Therefore, our idea was to implement a low-pass filtering on the IFS domain block calling frequencies, estimated for each range block containing the analysed pixel of the image, and then to recalculate the attractor of the IFS. This can lead to drastic geometric simplifications, depending on the value of the cut frequency *fc* (fig. 9).

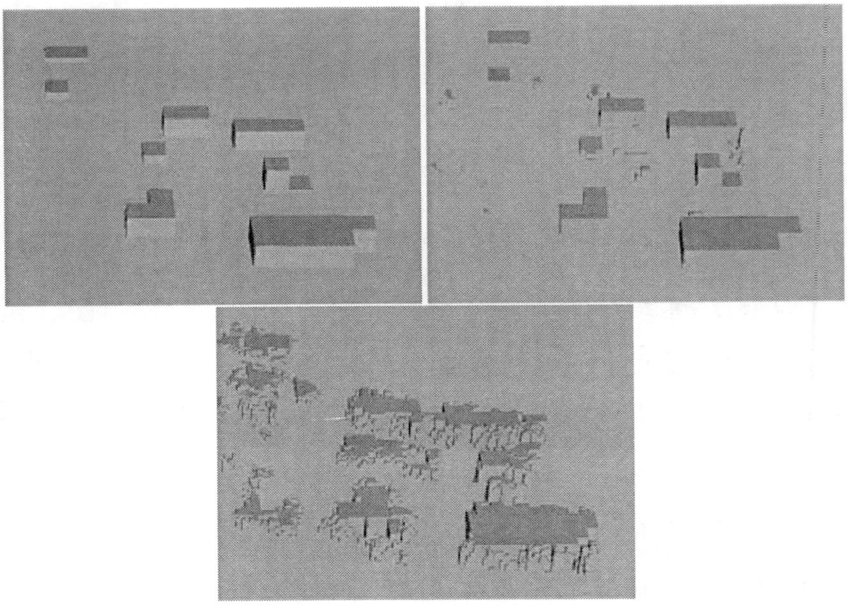

Figure 9. Low-pass filtering with IFS (fc=1 ; fc=10 ; fc=100).

3. From Generated Scenes to Real World

3.1. Towards Urban Shapes Interpretation and Classification

Because IFS do not take account for dimensions, it is first necessary to provide the correspondence between the pixel and grey level units and the size of the objects in the real world. Then, a general method of correspondence between virtual objects and real world ones has been proposed, based on the mixed criterion (surface, height), and allows a primary classification of generated urban objects, completed by some quick shape analysis to help identifying the type of a building, for example. This typology contains 7 types of objects : buildings and houses (blue grey), urban furniture (light orange), trees and vertical vegetation (light green), ground-levelled objects (swimming pools, parkings, lawns, ponds ; dark green), fountains (dark blue), shelters of garden (brown), electrical posts and public lamps.

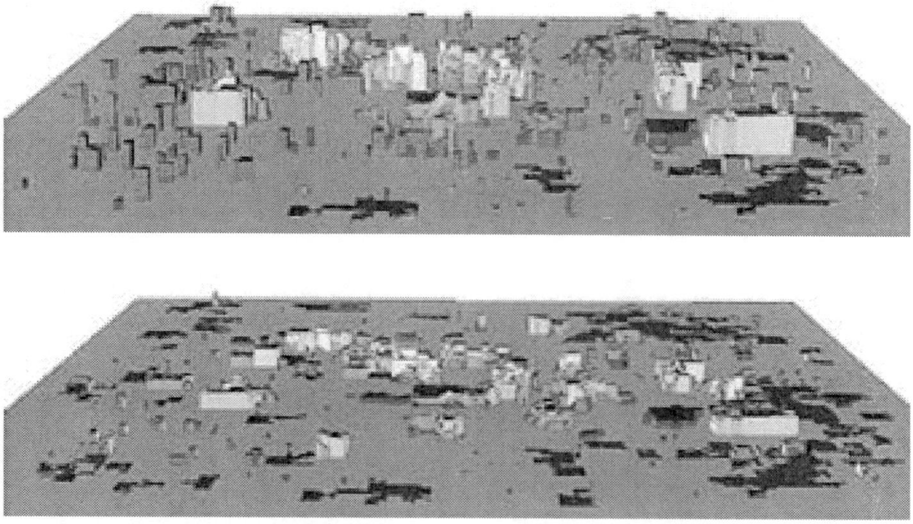

Figure 10. Two examples of classification of urban objects with a colour table.

Instead of using raw objects as they are generated, we could operate some substitutions with other ones in typical libraries representative of certain urban atmospheres, for example. But this solution hasn't been already implemented.

3.2. Simplifying and Smoothing Generated Shapes

Some algorithm developments have been required for smoothing irregular distribution of pixels, due to the jaggy appearance of vectorized pixels and the fractal nature of generation, and for obtaining simplified geometric shapes. Our work involves many existing simplification algorithms (Douglas-Peucker, characteristic vertex extraction), combined in a robust approach, introducing the notion of "significant geometric detail", with a scale tolerance factor (Fig. 6).

Another promising way of research concerns local adjustment of pre-defined configurations of "common angles" in architecture, up to global adjustment with constraints (for example, for placing roof shapes).

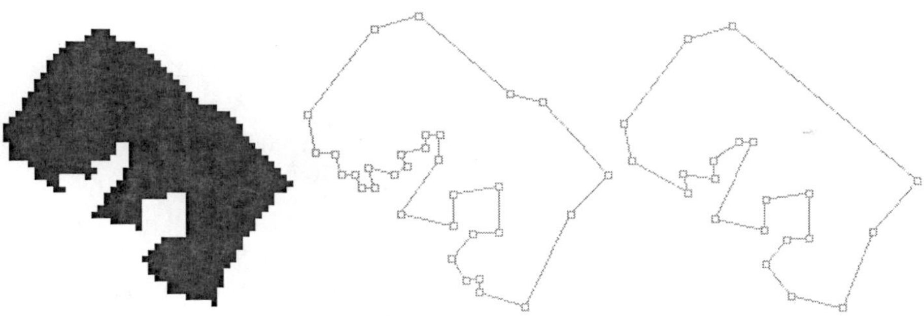

Figure 11. A noisy building – outline with details – simplified outline (tolerance = 1/20).

3.3. Adding Automatic Garbage Dump to the Scenes

Automated generation of streets and places graphs from urban imprints maps is an interesting research topic. This "raster approach" of the problem is rather new comparing to the one dealing with vector objects, and avoids the difficult task of shape vectorisation in noisy environments. Our still progressing work is done in three stages :

- extraction of geometric structuring characteristics from the maps : graphs of streets and places, combined within a technique for spatially grouping houses and buildings in neighbourhoods ;
- geometric generation of corresponding smooth 3D objects in VRML ;
- search for heuristic methods to qualitatively identify plausible elements of garbage dump networks (ex : boulevards, avenues, alleys, water streams).

We're still working on opened and closed connex graphs of street network. We apply some « mathematical morphology » basic tools for extracting homotopic skeletons of the ground zones and streets width (fig. 12). We obtain two types of graph, depending on the possibility to connect the city to its environment (opened city), or not (secluded city). To improve the quality of morphological processing, we work on super scanned images (a factor of 4 seems to be sufficient for 256 x 256 or 512 x 512 images).

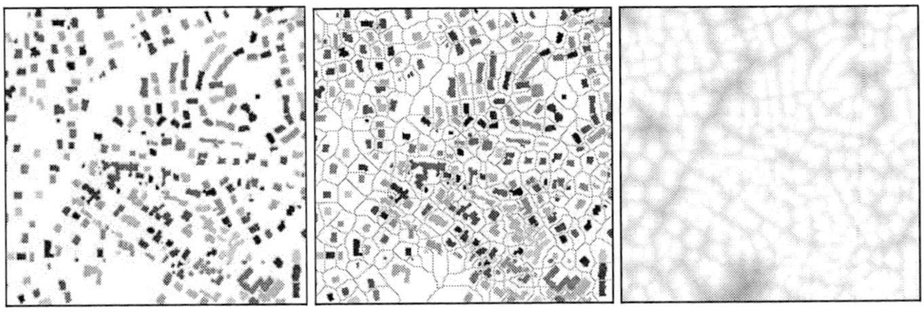

Figure 12. Homotopic opened skeleton (b)of part of Saint Genis city (a) and its street-width map (c).

4. Conclusions and Future Works

4.1. Discussion

We have shown in this paper how it is possible to encode simplified 2D½ city models using an IFS compression technique derived from Jacquin's algorithm. Cutting urban maps with the aid of a square pavement allows local control on the content of the split zones. In this purpose, we initiated a genetic-like approach to share information between IFS coding two (or more) city models, in order to compose new urban and architectural shapes by fusion and mutation. The possibility to deal with real or synthetic urban fabrics opens a wide field of creation, and many ideas have been suggested for that. We mostly obtain orthomorphic geometric models, because of the approach of converting blocks of connex pixels as cubes with their borders. Recent references on city modelling use such objects (Parish, Muller,

2001), where buildings are designed with the aid of L-systems. One can also observe the similitude between the geometric aspect of our models and the famous « architectones » of Kasimir Malevich (Figure 13).

But, up to now, we've just used a uniform and general square pavement. The fact we did not consider other pavements more suitable for the process of isolated buildings or groups of buildings results in discontinuities and loss of topological identity, even if we strove to minimize them in our pavement-based fusion algorithm. Indeed, a related difficulty is the adjustment of the range size parameter B : if B is too small, only the outline of the objects are coded by *lifs*, and if B is too high, it can be very hard to find some blocks similarities. On the other hand, the IFS approximation quality requires a sufficient resolution for images. These two constraints result in higher computation time for *lifs*, even with the uniform square pavement.

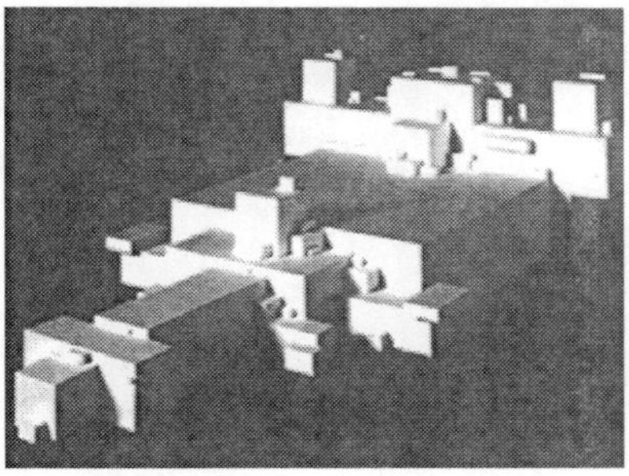

Figure 13. A famous cubic architecton of Kasimir Malevich (1926).

4.2. Remaining Investigations

From a scientific point of view, several ways of research remain :

Our experiments still suffer from a lack of theorical developments on IFS coding and partitioning for the use of fusion between several city models. It is important to search for better pavements of range and domain blocks, well fitted to match models and process properties.

A semi-synthetic approach for IFS will be explored, in order to reproduce given models of spatial distribution of shapes, using "condensation IFS" that allow the import of extern objects in non coding blocks.

Some experiments with genetic algorithms have to be done to optimise the fusion results, given some shape or statistical distribution criterions extracted from real cities, or by applying the famous "universal distribution law" (Salingaros, 1999).

We also envision the study of location and recombinant mutations in order to increase the size of the IFS original extent, by distributing some *lifs* or groups of *lifs* to other places. From a simulation point of view, this would become a first step towards infinite generation of non-repetitive urban fabric.

References

Bailly, E. (1998) *Fractal geometry and simulation of urban growth, UMR CNRS Espace* **6012**, Nice.

Barnsley, M. (1993) *Fractal image compression*, AK Peters, Ltd, Wellesley.

Barnsley, M. (1992) Image coding based on a fractal theory of iterated contractive image transformation, *IEEE transactions on image processing*, **1**:18-30.

Batty, M., Longley, P.A. (1994), *Fractal Cities: A Geometry of Form and Function*, Academic Press, London and San Diego.

DER (2000), Développement d'un Environnement logiciel de REalité Virtuelle Elaboré, *Projet de recherche DEREVE de la région Rhône-Alpes,* LIGIM, Université Lyon I.

Frankhauser, P. (1994) *La Fractalité des Structures Urbaines*, *Collection Villes*, Anthropos, Paris.

Frankhauser, P. (1997) *L'approche fractale : un nouvel outil de réflexion dans l'analyse spatiale des agglomérations urbaines*, Université de Franche-Comté, Besançon.

Gentil, C. (1992) *Les fractales en synthèse d'images : le modèle IFS*, Thèse, LIGIM, Université Lyon I, Lyon.

Hutgen, B., Hain, T. (1994) On the convergence of fractal transforms, *Proceedings of ICASSP*, **561**-564.

Hutchinson, J. (1981) Fractals and self-similarity, *Indianna Universiry Journal of Mathematics*, **30**:713-747.

Jacqui,n A.E. (1992) Image coding based on a fractal theory of iterated contractive image transformations, *IEEE transactions on image processing,* **1**(1):18-30.

Makse, H.A. (1996) Modelling fractal cities using the correlated percolation model, *Fractal and granular media conference.*, session C18

Marsault, X. (2002) Application des Iterated Function Systems (IFS) à la composition de tissus urbains tridimensionnels virtuels, *Autosimilarités et applications*, Cemagref, Clermont Ferrand.

Parish, Y., Muller, P. (2001) Procedural modelling of cities, *SIGGRAPH.*

Renders, J.M. (1995) *Algorithmes génétiques et réseaux de neurones*, Editions Hermès.

Sala, N. (2002) The presence of the self-similarity in architecture : some examples, in M.M.Novak (ed), *Emergent Nature, World Scientific*, **273**-283.

Salingaros, N. (1999) A universal rule for the distribution of sizes, *Environment and Planning B : planning and Design*, **26**:909-923, Pion Publications.

Torrens, P., (2000) *How cellular models of urban systems work*, CASA, Angleterre.

Vences, L., Rudomin L. (1997) *Genetic algorithms for fractal image and image sequence compression*, Instituto Tecnologico de Estudias Superiores de Monterrey, Camus Estado de Mexico, Computation Visual.

Woloszyn, P., (1998) Caractérisation dimensionnelle de la diffusivité des formes architecturales et urbaines, Thèse, Laboratoire CERMA, NANTES.

In: Chaos and Complexity Research Compendium
Editors: F. Orsucci and N. Sala, pp. 321-330
ISBN: 978-1-60456-787-8

Chapter 25

PSEUDO-URBAN AUTOMATIC PATTERN GENERATION

Renato Saleri Lunazzi*
Architecte DPLG, DEA informatique et productique, master en industrial design.
Laboratoire MAP aria UMR 694 CNRS – Ministère de la culture
et de la Communication

Abstract

This research task aims to experiment automatic generative methods able to produce architectural and urban 3D-models. At this time, some interesting applicative results, rising from pseudo-random and l-system formalisms, came to generate complex and rather realistic immersive environments. Next step could be achieved by mixing those techniques to emerging calculus, dealing whith topographic or environmental constraints. As a matter of fact, future developments will aim to contribute to archeological or historical restitution, quickly providing credible 3D environments in a given historical context.

1. Introduction

Since the end of the 70's, the "fractality" of our environment raised as an evidence, pointing some peculiar aspects of everyday phenomena. Some micro and macro-scopic internal-arrangement principles appear to be similar or even auto-similar, leading the reasoning through general explanatory theories. Physicians and biologists regularly discover fractal processes through natural morphogeneses such as cristalline structures or stellar distribution. Human creations also seem to be ruled by fractal fundamentals and since 15 years, the "fractality measure" of some human artefacts can be somehow achieved.

Fractal investigation through urban patterns mainly focused on two subsequential aspects : the direct analysis of spatial organisation, and thus the formalization of self-generating geometrical structures. The growth of urban models is at this time fulfilled either

* E-mail address: renato.saleri@lyon.archi.fr

by time-based spatial simulators or by simple static generators. Spatial simulators are usually based on simple "life-game" (cellular automata) devices or even by "diffusion limited aggregation" formalisms (DLA). In this paper, we will mainly focus on some generative techniques involved in 2D and 3D automatic builders.

2. Research Task Context

The mainframe of this research task consists in real-time rendering of huge 3D databases. Different aspects of this goal have already been explored, considering from the top that rendering techniques should be optimal for a given applicative context. Therefore, the main aspect of MAP-aria participation in this project consists in building plausible urban structures related to some given historical or archeological context.

Early stages of our investigation pointed the discontinuous properties of growth phenomena. In other words we barely believed in the existence of a possible continous morphological development model, according to the evidence of micro and macro-scopic observable morphological differences on one hand and through bidimensional and threedimensional topological discontinuities on the other. In other words, we focused some "scale-based formalisms", related to specific urban scale-types, as listed in the following section.

3. Applications

The description of the following formalisms is broadly summarized. Further refinement on geometrical models, architectural primitives and morphological break-down are under development...

3.1. Random 2D – 3D Generators

Random or pseudo-random simple pattern generators applied to facades, according to buildings height or local floors indentations. Please note that the "hull filler" generator, mentioned in this section, is shortly described in section 3.3

This very first applicative experiment was only acquired to test some early combinational conjectures. Some 3D "hull-filled" objects are textured whith simple combinational patterns ensuring somehow an intrinsic global coherence in order to avoid 2D and 3D possible mismatch. This could be achieved by establishing for instance a common spatial framework, arbitrarly bounded here by 2,5 meters-sided cubes. As shown in the picture below, the intrinsic coherence of the texture itself depends on the pertinence of single texture patches positioning, known as inner, top, left, right and bottom occurrences : on the illustration, the gray-filled board zone invoke specific ledge-type instances as the inner white zones use generic tiles. Right underneath, some texture patches that come whith the 2D library and below, two facade variants.

These examples are here intended as “ironic standalone designs“ : the (im)pertinence of these random objects is obvious. Meanwhile, if coupled whith accurately-sized 3D objects, the visual impression could be effective, as shown on Figure 3.

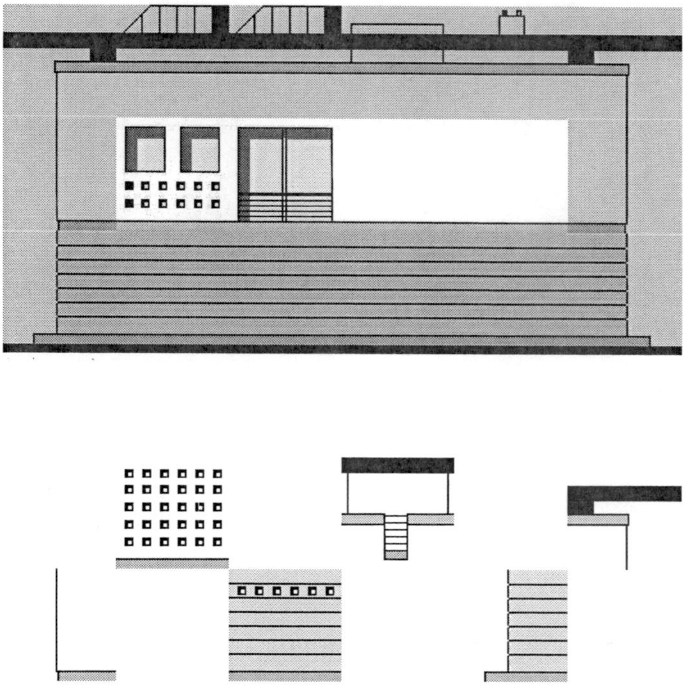

Figure 1. The automatic facade builder and some architecural tiles.

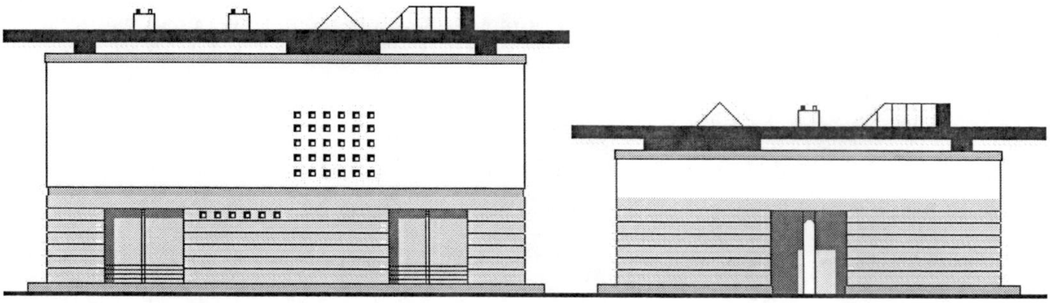

Figure 2. Some “automatic“ facades.

We recently improved this application capabilities through some Maya© Embedded Language developments. The synchronous object-texture pattern generator produces “on-click“ 3D architectural-like objects and plots them over a simple 2D grid, The main controls provide some expansion parameters such as linear spread-out and rotation constraints. This very first MEL application deals whith a single-input façade library ; a very next step will consider a wider variety of morpho-textural relevant matchings.

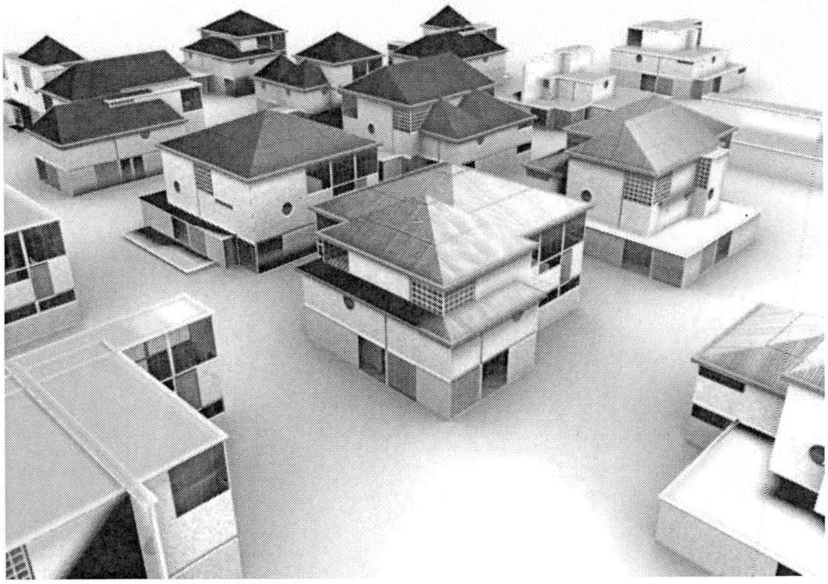

Figure 3. Applying and rendering colored tiles on random-generated rule-based 3D objects.

3.2. Graphtal or L-System Generator

Graphtal or L-System, Applied to Local Building and Block Propagation

The L-System, or Graphtal, starts from a simple recursive substitution mechanism. This rules-based generator, described in the late 60's by A. Lindenmayer (Lindenmayer 1968) can quickly provide complex geometric developments. It's charachteristic deal whith simple substitution rules, recursively applyed to a sprout, as shown below :

All we need to start is an alphabet, listed hereby : 0,1,[,]

In this example, 0 and 1 occurrences will "produce geometry" while [and] will provide a simple affine transformation (rotation and/or translation). We can now describe simple substitution rules, applied to alphabetic elements :

0 : 1[0]1[0]0 **1** : 11 **[** : [**]** :]

If we recursively apply those substitution rules to an initial sprout (applied from the top to the rule of letter "0") we obtain:

11 [1[0]1[0]0] 11 [1[0]1[0]0]1[0]1[0]0

Two "generations" or recursive steps later we obtain:

11 11 11 11 [11 11 [11 [1[0]1[0]0] 11 [1[0]1[0]0] 1[0]1[0]0] 1111 [11 [1[0]1[0]0] 11 [1[0]1[0]0]1[0]1[0]0] 11[1[0]1[0]0] 11 [1[0]1[0]0] 1[0]1[0]0] 11 11 11 11 [11 11 [11 [1[0]1[0]0] 11 [1[0]1[0]0] 1[0]1[0]0] 1111 [11 [1[0]1[0]0] 11 [1[0]1[0]0]1[0]1[0]0] 11

[1[0]1[0]0] 11 [1[0]1[0]0] 1[0]1[0]0] 11 11 [11 [1[0]1[0]0] 11 [1[0]1[0]0] 1[0]1[0]0] 11 11 [11 [1[0]1[0]0] 11 [1[0]1[0]0] 1[0]1[0]0] 11[1[0]1[0]0] 11 [1[0]1[0]0] 1[0]1[0]0

The "trick" consists here in replacing the brackets by specific 3D operations – typically affine transformations, such as rotations or translations - and the "0" and "1" occurrences by 3D pre-defined objects. We notice how the transformations and object creations are invoked in the following source code (obviously part of the main program, implemented within a "switch" JAVA object) The resulting output sourcecode is based on VRML 97, mimed with a CosmoPlayer© plug-in.

Depending on initial rules, such a model can quickly "run out of control" and generate huge 3D databases. Its specific initial generative inputs are the only condition for the whole evolution process – which is meanwhile eminently determinist; nevertheless, geometry partial overlaps are frequent and due to concatenated affine transformations previously described. Hereby we show a four-steps generated VRML model, made of solely 2 architectural primitives. Some extra visual artefact is provided by the height change of the objects, depending on their distance to the first geometric settlement.

Figure 4. A L-System-based growth engine.

Most of these generative models are developed whithin a web browser interface: a javascript code which dinamically generates a VRML source displayed by a CosmoPlayer plugin. We are studying by now other geometrical algorithms, in order to constrain these L-system, such as Voronoï diagrams or Delaunay triangulations.

3.3. Random or Pseudo-Random "Hull-Filler"

Random or Pseudo-Random "Hull-Filling" Generators for Single-Building Construction

The "hull-filling" model offers by itself rather interesting investigative perspectives: in this model the specific positioning of architectural types or sub-types could be guided by a prior analysis that tends to break down or disassemble some historically-contexted architectural types by a morphological factorization.

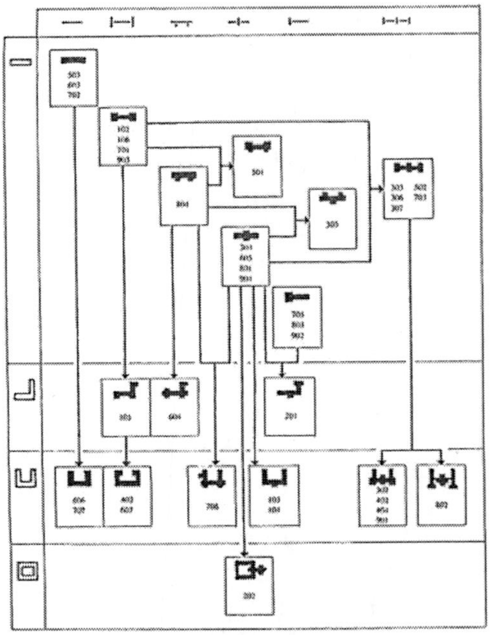

Figure 5. A graph-based morphological parser. Courtesy of “Laboratoire d’Analyse des Formes“

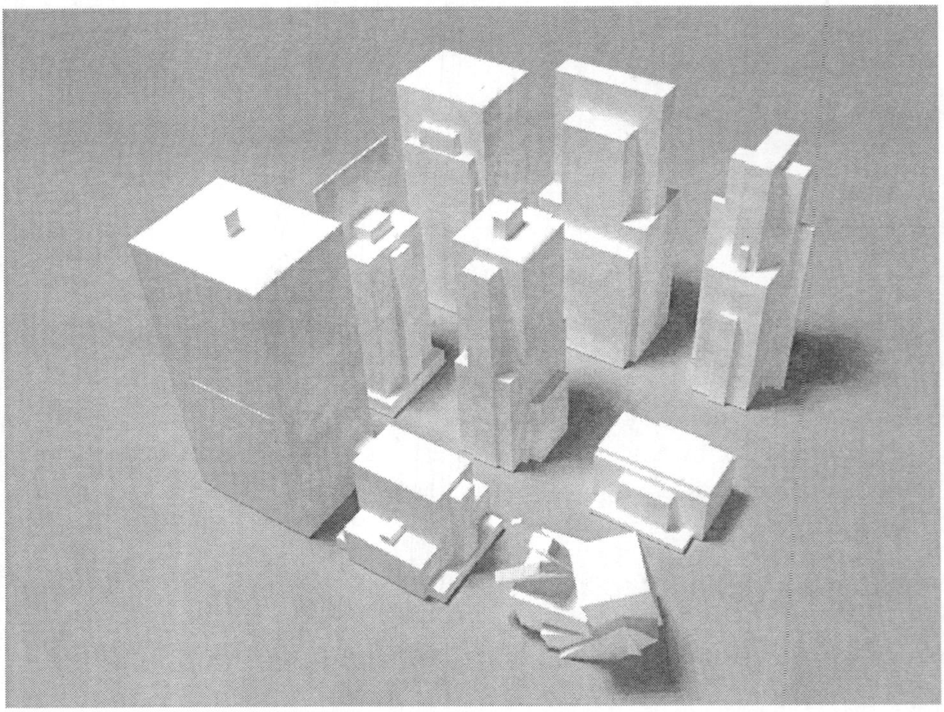

Figure 6. Some “hull-filled“ objects.

The process is obvously reversible and could be achieved by a rules-based grammar. The amazing Palladio 1.0 Macintosh© Hypercard Stack (Freedman 1990) is a noteworthy example of such a morphological synthesis. We also must here quote the scientific goal of the research team "Laboratoire d'Analyse des Formes" from the architecture school of Lyon that leads somehow this specific aspect of this research task (Paulin – Duprat 1991) Their aim is to identify major stylistics guidelines from distinct architectural families, dispatching them through pre-identified morphologic, functional, architectonic and compositional occurrences (Ben Saci 2000). A similar search will soon commence, leaning on Claude-Nicolas Ledoux (1736 – 1806) architectural production, whose factorizable characteristics appear as an evidence.

At the moment, this complex formalism is barely drafted; it is therefore interesting to point out the relevant difference of the "ugly duckling" bottom right object, that descends from the same construction formalism but differs from 1 single input attribute.

3.4. Multi-Scale Pattern Generator

A "top of the heap" wide range concentric propagator, whose aim is to distribute, filter and drop geometric locators above a given terrain mesh.

The deal is here to develop a "general land-scaled model", mostly a variant of the L-system model depicted above. The initial distribution of locators basically follows a concentric distribution. Their final positioning can be meanwhile modified by some disruptive factor, mostly depending on simple angular non-overlapping constraints. The graph below shows three different steps of the computation: locators displacement, neighbourhood tracking and plot drawing.

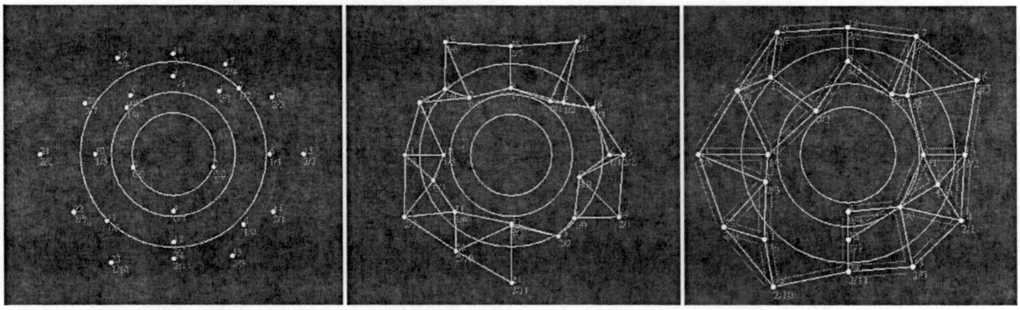

Figure 7. Deployment of a 2D geometric model.

A local geometric transformation transforms the initial structure to a position-related "constructible zone", starting from two initial input variables, named here d' and d'' At the moment, inevitable angular occlusions occur whith sharp and wide angles. This drawback should meanwhile be solved in a very next release of the applet.

Extracting the *n* closest neighbours and drawing the respective bijective connexions leads the entire process, and we can finally hybrid this bidimensional mesh to allocation rules and topopgraphic constraints, to produce the models shown on the figures below : the skeleton and the final rendering.

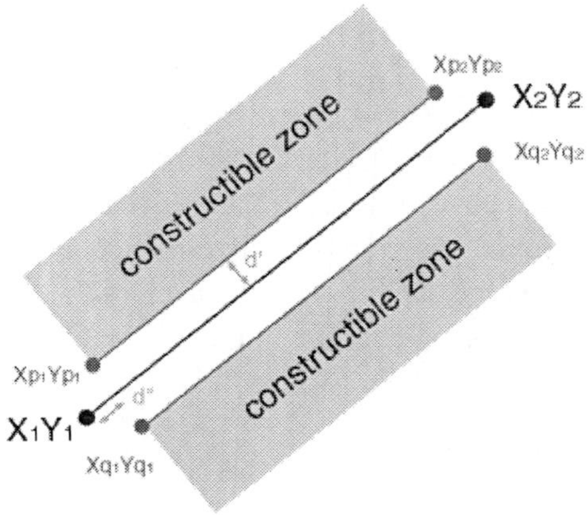

Figure 8. Geometric deduction of “constructible zones“.

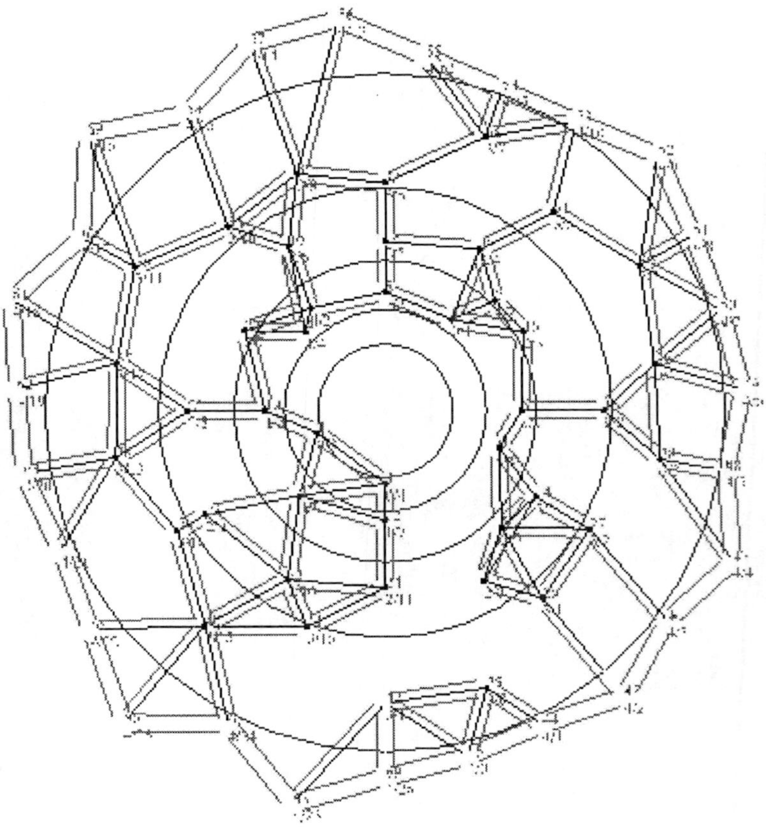

Figure 9. The geometric skeleton...

In this example, only four architectural primitives are distributed over the map ; a “hull – filling “ generator or som MEL-based architectural objects (both shortly depicted above) could be implemented to create a more realistic perceptive variety.

Figure 10. ...and it's 3D expression.

4. Conclusion

Virtual reality hardware and software costs and means are still relevant today. Trying to partially solve this peculiar aspect of leading 3D rendering techniques is part of the regional DEREVE project, whose aim is to build a convergent know-how, trying to extend hardware and software intrinsec performances through methodological and algorithmic applications, in terms of modeling and rendering. As a matter of fact, the specific involvement of the "MAP-aria" lab in this research task deals whith 3D scenes building, leaning on his specific architectonic culture and virtual reality previous experimentations.

References

Lindenmayer, A. (1968) "Mathematical models for cellular interactions in development", parts I-II. *Journal of Theoretical Biology* **18**: 280-315.

Freedman, R. (1990) "Palladio 1.0", Apple Macintosh© Hypercard Stack.

Paulin, M. and Duprat, B. (1991). "De la maison à l'école, élaboration d'une architecture scolaire à Lyon de 1875 à 1914", Ministère de la Culture, Direction du Patrimoine, CRML.

Ben Saci, A. (2000) "Théorie et modèles de la morphose", Thèse de la faculté de philosophie sous la direction de B. Deloche, université Jean Moulin.

Frankhauser, P. (1994) *La Fractalité des Structures Urbaines*, Collection Villes, Anthropos, Paris, France.

Frankhauser P. (1997) "L'approche fractale : un nouvel outil de réflexion dans l'analyse spatiale des agglomérations urbaines ", Université de Franche-Comté, Besançon.

Khamphang Bounsaythip C. (1998) "Algorithmes évolutionnistes" in "*Heuristic and Evolutionary Algorithms: Application to Irregular Shape Placement Problem*" Thèse - Public defense: October 9, (NO: 2336)

Heudin, J.C. (1998) "L'évolution au bord du chaos" Hermès Editions.

Horling, B. (1996) "Implementation of a context-sensitive Lindenmayer-System modeler" Department of Engineering and Computer Science and Department of Biology, Trinity College, Hartford, CT 06106-3100, USA.

Sikora S., Steinberg D., Lattaud C., Fournier C., Andrieu B. (1999) "Plant growth simulation in virtual worlds : towards online artificial ecosystems. Workshop on Artificial life integration in virtual environnements". European Conference on Artificial Life (ECAL'99), Lausanne (Switzerland), 13-17 september.

Barber, C.B., Dobkin, D.P., and Huhdanpaa, H.T., (1996) "The Quickhull algorithm for convex hulls," ACM Trans. on Mathematical Software.

Batty M., Longley (1994) P.A., "Fractal Cities: A Geometry of Form and Function", Academic Press, London and San Diego, CA.

Torrens, P. (2000) "How cellular models of urban systems work" , CASA.

In: Chaos and Complexity Research Compendium
Editors: F. Orsucci and N. Sala, pp. 331-337
ISBN: 978-1-60456-787-8

Chapter 26

TONAL STRUCTURE OF MUSIC AND CONTROLLING CHAOS IN THE BRAIN

Vladimir E. Bondarenko
Department of Physiology and Biophysics, School of Medicine and Biomedical Sciences, SUNY at Buffalo, 124 Sherman Hall, 3435 Main Street, Buffalo, NY 14214, USA
Igor Yevin*
Mechanical Engineering Institute, Russian Academy of Sciences, 4, Bardina, Moscow, 117324 Russia.

Abstract

Recent researches revealed that music tends to reduce the degree of chaos in brain waves. For some epilepsy patients music triggers their seizures. Loskutov, Hubler, and others carried out a series of studies concerning control of deterministic chaotic systems. It turned out, that carefully chosen tiny perturbation could stabilize any of unstable periodic orbits making up a strange attractor. Computer experiments have shown a possibility to control a chaotic behavior in neural network by external periodic pulsed force or sinusoidal force. We suggest that music acts on the brain near delta-,teta-, alpha-, and beta frequencies to suppress chaos. One may propose that the aim of this control is to establish coherent behavior in the brain, because many cognitive functions of the brain are related to a temporal coherence.

1. Introduction

Investigations of human and animal electro-encephalograms (EEGs) have shown that these signals represent deterministic chaotic processes with the number of degrees of freedom from about 2 to 10, depending on the functional state of the brain (awaking, sleep, epilepsy).

Recent investigations [1,2] revealed that music tends to reduce the degree of chaos in brain waves. For some epilepsy patients music triggers their seizures. Loskutov [3], Hubler and co-workers [4] and others studied control of deterministic chaotic systems. It was found

* E-mail address: yevin@online.ru, Phone: (095) 5760472

that carefully chosen tiny perturbation could stabilize any of unstable periodic orbits making up a strange attractor.

Computer experiments have shown the possibility to control chaotic behavior in neural networks by external periodic pulsed force or sinusoidal force [5,6]. We suggest that indeed the stable steps of music tonalities and appropriate chords are those tiny perturbations that control chaos in the brain. Any musical score might be considered as a program of controlling chaos in the brain. One may propose that the aim of this control is to establish coherent behavior in the brain, because many integrative cognitive functions of the brain are related to a temporal coherence [7].

2. Control of Chaos in the Brain by Sinusoidal or Periodic Pulsed Force

The neural network model is described by a set of differential equations [5,6]:

$$\dot{u}_i(t) = -u_i(t) + \sum_{j=1}^{M} a_{ij} f(u_j(t-\tau_j)) + e\sin\omega_e t,$$

$$i, j = 1,2,..., M, \tag{1}$$

where $u_i(t)$ is the input signal of the ith neuron, M is the number of neurons, a_{ij} are the coupling coefficients between the neurons, τ_j is the time delay of the jth neuron output, $f(x) = c\tanh(x)$, e and ω_e are the amplitude and frequency of the external force, respectively. We studied the case when the all τ_j are constant ($\tau_j = \tau$). The coupling coefficients are produced by random number generator in the interval from –2.048 to +2.048, the coefficient c is used to vary coefficients a_{ij} simultaneously.

The forth-order Runge-Kutta method, with the time step $h = 0.01$, is used for solution of equation (1). Small random values of $u_i(0)$ are chosen as the initial conditions. For the time t in the interval from $-\tau$ to 0, $u_i(t)$ are equal to zero. Time series of $N = 100000$ and $N = 8192$ points are analyzed after the steady state is reached. The frequency spectra are calculated using the ordinary digital Fourier transform. For the evaluation of the correlation dimension ν the Grassberger-Procaccia algorithm is used. According to this algorithm, the time series of single neuron's inputs are analyzed. The sampling frequency is chosen so that each significant spectral component should have at least 8-10 sample points on the time period.

For calculation of the largest Lyapunov exponent in M-dimensional phase space, two trajectories are computed from the equation (1): unperturbed $u_0(t)$ and perturbed $u_\varepsilon(t)$. For the calculation of perturbed trajectory after reaching the steady state, the small values εu_i are added to u_i. Here ε is in the range from 10^{-14} to 10^{-6}. The largest Lyapunov exponent is defined as

$$\lambda = \lim_{t\to\infty} \lim_{D(0)\to 0} t^{-1} \ln[D(t)/D(0)]$$

where

$$D(t) = \left[\sum_{i=1}^{M} (u_{i\varepsilon}(t) - u_{i0}(t))^2 \right]^{1/2}$$

$$D(0) = \left[\sum_{i=1}^{M} u_{i\varepsilon}(0) - u_{i0}(0))^2 \right]^{1/2}$$

are the distances between the perturbed and unperturbed trajectories at the current and the initial moments, respectively. The largest Lyapunov exponent λ is calculated from time series of $N = 100000$ points.

We start from the case when the amplitude of the external force $e = 0.0$. Under this condition, the neural network produces chaotic output with the correlation dimension $\nu = 5.2 - 7.1$ (depending on the ordinal number of the neuron) and the dimensionless largest exponent $\lambda = 0.017$. The peak frequencies in the cumulative spectra of 10 neurons are in the ratios of 0.12:0.28:0.46:1.04 (Fig. 1).

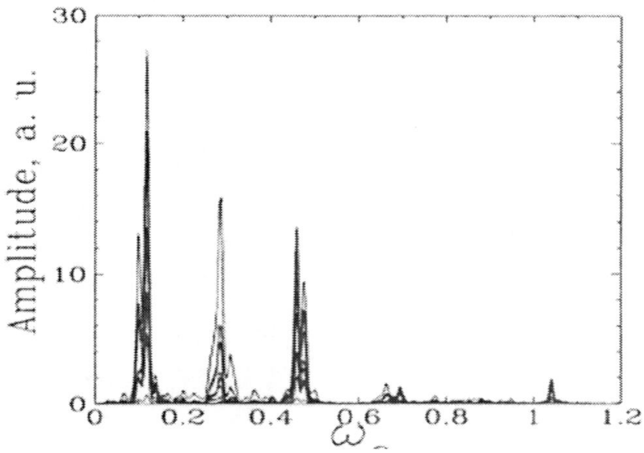

Figure 1. Spectra of the outputs for all ten neurons without an external action: $M = 10$, $c = 3.0$, $e = 0.0$, $\tau = 10.0$.

Similar ratios of main rhythms of the human EEG (delta-, theta-, alpha-, and beta rhythms) are observed in the experiments also: 2.3:5.5:10.5:21.5 [5].

Application of the external sinusoidal force to this neural network changes the output from relatively high-dimensional chaotic (ν ι $5 - 8$, $\lambda > 0$) to low-dimensional chaotic ($\nu \leq 3$, $\lambda > 0$), quasiperiodic ($\nu \leq 3$, $\lambda \approx 0$), or periodic ones ($\nu \approx 1$, $\lambda \approx 0$) [6].

As a rule, the low-dimensional outputs are observed when the frequency of the external force is close to the eigenfrequency of self-excited oscillations in the neural network without an external action (Fig. 2). One may expect, therefore, that music acts on the brain near these eigenfrequencies or its harmonics, because considerably smaller amplitudes of the external forces are necessary to suppress chaos in the case of resonance, than without resonance.

But our neural network has only four eigenfrequencies whereas piano has over 80 keys producing more than 80 different frequencies. In order to resolve this contradiction, the attractor network model of music tonality is proposed that is based on Hopfield's model of associative memory.

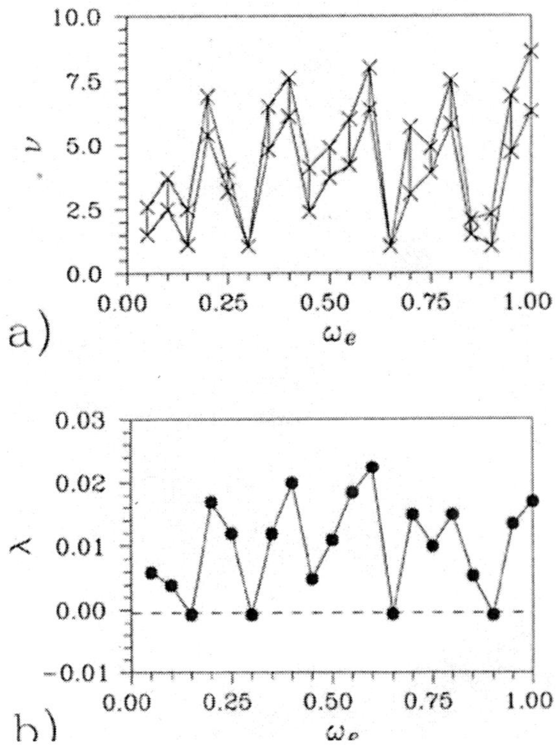

Figure 2. Correlation dimension ν (a) and the largest Lyapunov exponent λ (b) as functions of external force frequency ω_e: $M = 10$, $c = 3.0$, $\tau = 10.0$, $e = 7.0$.

3. Model of Music Tonality

Using Hopfield's model, we can consider pitch perception as a pattern recognition process. It gives us an ability to explain why notes with octave interval we hear as very similar. When we hear, for instance, note "C" in different octaves, we recognize very similar sound patterns, keeping in mind complex overtone structure of every musical note. In other words, sound patterns of notes divisible by octave are the most similar among all others notes and therefore belong to the same basin of attraction and precisely by this reason we hear notes divisible by octave as very similar.

Tonality is a hierarchy (ranking) of pitch-class. If the only pitch-class is stressed more than others in a piece of music, the music is said to be tonal. If all pitch-classes are treated as equally important, the music is said to be atonal.

Almost all familiar melodies are built around a central tone toward which the other tones gravitate and on which the melody usually ends. This central tone is the keynote, or tonic. Three stable steps of tonality: tonic, median, and dominant are prototype patterns or attractors of neural network model. Others steps of tonality: subdominant, submediant, ascending parenthesis sound, descending parenthesis sound play the role of recognizable patterns, gravitating to some or other prototype pattern [8,9].

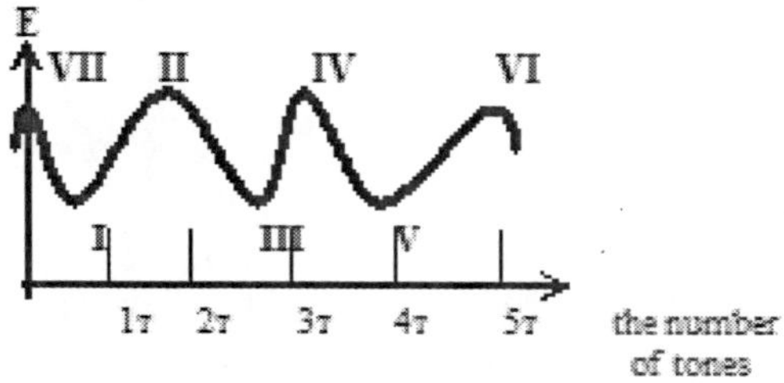

Figure 3. Hopfield's potential function *E* for major tonality in Western tonal music.

The degree of instability (the degree of gravitation to appropriate stable state) depends on distances between unstable and stable sounds. The strongest gravitation of VII step to I step and of IV step to II step are observed (Fig. 3).

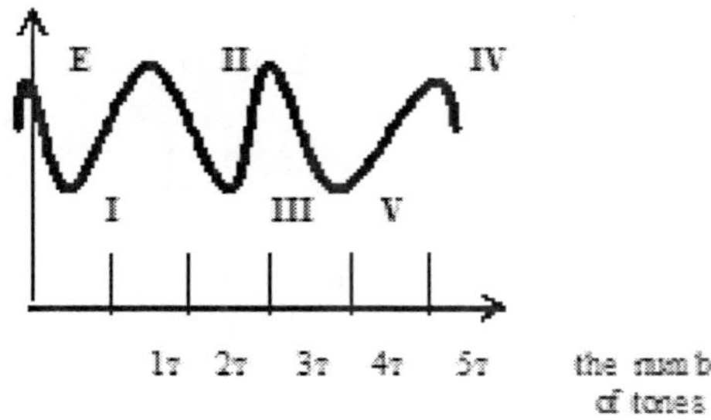

Figure 4. Potential function *E* for minor tonality in pentatonic scale.

There are no semitone (half step) intervals between notes in music of some Eastern countries (for instance, in China, Vietnam, Korea) (Fig. 4). Such pitch organization is called pentatonic. Though pentatonic is more ancient than modern Western tonality system (Fig. 3), we can formally obtain major and minor tonalities in pentatonic by removing IV and VII steps from diatonic major and minor tonalities. For the lack of minor seconds intervals in a pentatonic scale there are not such strong gravitation as in a natural scale [9,10]. Because western and pentatonic systems of tonalities recognition have the same potential function, we may suggest that this potential function is formed not by music, but is an inherent property of brain functioning.

It is reasonable to suggest that that all kinds of major tonalities gravitate to the one basin of attraction and all kinds of minor tonalities gravitate to the other basin of attraction.

4. Stable States of Tonalities and Resonance Action

Because music acts on the brain as external force we may depict the action of major tonalities through the auditory nerve on neural network in the following way (Fig. 5):

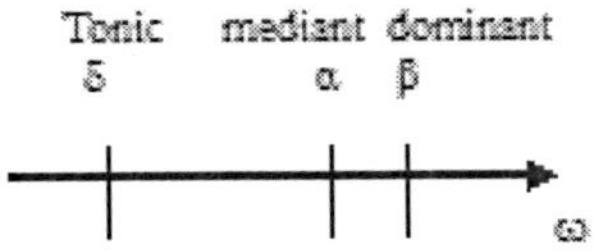

Figure 5. Resonance action of major tonalities, ω is the frequency of spike trains in auditory nerve.

It means that the frequencies of spike trains, corresponding to tonic, mediant, and dominant in major tonalities in auditory nerve coinside with the frequences of delta-, alpha-, and beta rhythms of the brain, respectively. We hope this is a plausible assumption.

The action of minor tonality on the brain we may depict as follows (Fig. 6):

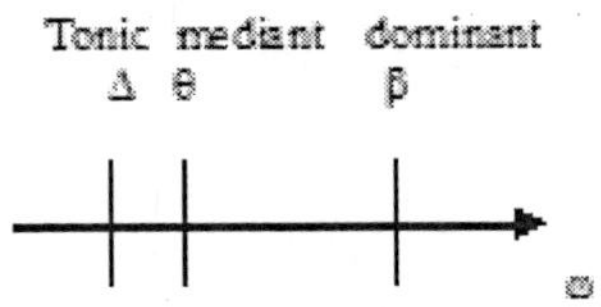

Figure 6. Resonance action of minor tonalities, ω is the frequency of spike trains in auditory nerve.

In this case the frequencies of tonic, mediant, and dominant of minor tonalities coincide with delta-, theta-, and beta rhythms in the brain.

The total action of music consisting of major and minor tonalities we may represent in the following way (Fig. 7):

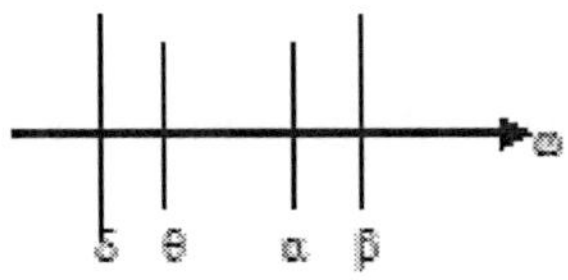

Figure 7. Resonance action of major and minor tonalities

Hence, we have four different music frequencies acting as external forces on four different eigenfrequencies of neural network.

As well known, interval structure of major and minor triads are the same as stable steps interval structure of corresponding tonalities. It means, that the action of these triads is reduced to simultaneous resonant action on delta-, teta-, alpha-, and beta frequencies.

References

[1] N. Birbaumer, W. Lutzenberger, H. Rau, G. Mayer-Kress, and C. Braun, "Perception of music and dimensional complexity of brain activity," *International Journal of Bifurcations and Chaos,* vol. 2, no. 6, pp. 267-278, 1996.

[2] A. Patel and E. Balaban, "Temporal patterns of human cortical activity reflect tone sequence structure," *Nature*, vol. 403, no. 6773, pp. 80-84, 2000.

[3] V.V. Alexeev and A.Yu. Loskutov, "The destochastization of a system with strange attractor by a parametric action" *Moscow University Phys. Bull.*, vol. 26, no. 3, pp. 40-44, 1985.

[4] A.W. Hubler and E. Lusher, "Resonant stimulation and control of nonlinear oscillation," *Naturwissenschaft,* no. 76, pp.67-74, 1989.

[5] V.E. Bondarenko, "Analog neural network model produces chaos similar to the human EEG. *International Journal on Bifurcation and Chaos*, vol. 7, no. 5, pp.1133-1140, 1997.

[6] V.E. Bondarenko, "High-dimensional chaotic neural network under external sinusoidal force," *Physics Letters A,* vol. 236, no. 5-6, pp. 513-519, 1997.

[7] W. Singer, "Neuronal representations, assemblies and temporal coherence," *Progress in Brain Research*, vol. 95, pp. 461-474, 1993.

[8] I. Yevin and S. Apjonova, "Attractor network model and structure of musical tonality," *Abstracts of the 9th Conference Society Chaos Theory in Psychology and Life Sciences,* Berkeley, CA, USA, July, 1999.

[9] I. Yevin, *What is Art from Physics Standpoint?* Moscow: Voentechizdat, 2000 (in Russian).

[10] I. Yevin, *Synergetics of the Brain and Synergetics of Art*, Moscow: GEOS, 2001 (in Russian).

In: Chaos and Complexity Research Compendium
Editors: F. Orsucci and N. Sala, pp. 339-348
ISBN: 978-1-60456-787-8

Chapter 27

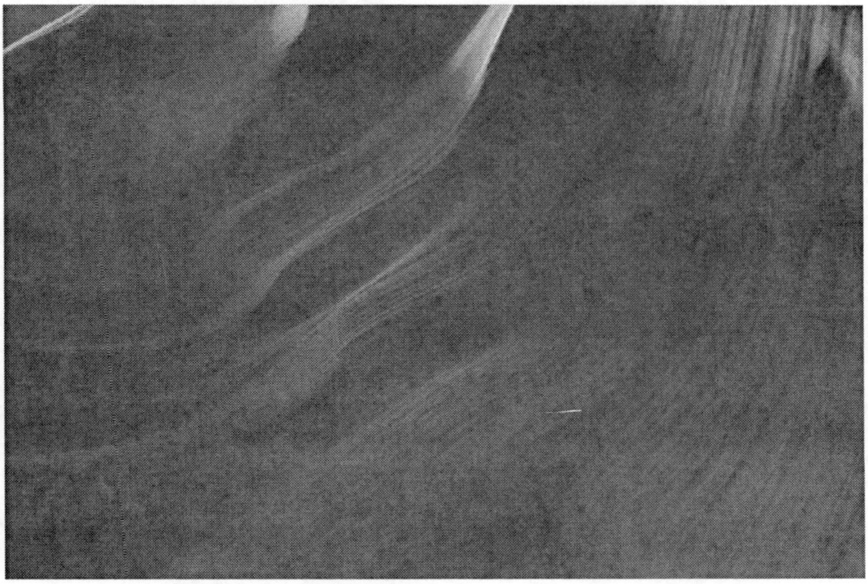

"Wavy Texture 2" Antelope Canyon USA photographed by Jin Akino

COLLECTING PATTERNS THAT WORK FOR EVERYTHING

Deborah L. MacPherson [*]
Independent Curator, 118 Dogwood Street, Vienna VA 22180-6394

Abstract

Would we even want a meta-methodology or collection such as "patterns that work for everything"? One simple evolving system of explanation and conceptual illustration? Where

[*] E-mail address: dmacp.mail@verizon.net 703 242 9411 and 703 585 8924, www.accuracyandaesthetics.com, www.contextdriventopologies.org

would these patterns reside? Who would interpret them? There are concepts being developed in the study of chaos and complexity that may help make arrangements for this collection. In particular, a glimpse at what the patterns might look like and act like. Maybe they also act like music, maybe we can discuss, present and interpret abstract information patterns the meticulous way we discuss, present and interpret abstract art. If you stuck a pin in today and drew back to the time when physics, chemistry and biology were one - what are we truly capturing about chaos and complexity for the corresponding point in the future? Are today's algorithm writers yesterday's alchemists and what is the best, least constrained and highest quality way to preserve the fundamental and esoteric qualities of this work for future studies? Can we imagine and develop an inherited collective memory for our machines, like language and culture are for us, to pass stories from one generation to the next? Even if they speak different languages and live in different places as we do, something we can all measure may be generated by providing an unsupervised opportunity for our machines to create or illustrate patterns we have not thought about yet, noticed or engineered. There is a story in the study of chaos and complexity that may be able to tell itself.

What Do We See and How Are We Telling This Story?

Below is Robert May's early glimpse presented in *American Naturalist* (1976). What if - even though so much high quality, rich and diverse information has been generated, presented and represented since the generation of this diagram - what if this is still an accurate portrayal of what we can see even with all of the new information? As the source of this diagram, we can assume that May cared deeply about this new science, that he was more convinced about his emerging ideas then anyone else could be, and he was committed to figuring these ideas out an accurate, arguable, mathematical way.

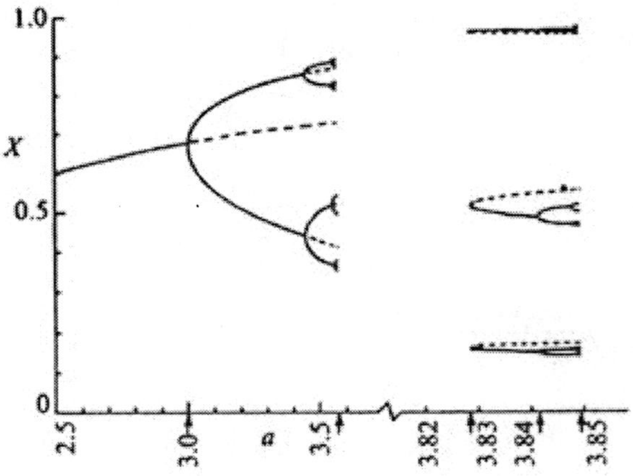

Figure 1. Bifurcations and Dynamic Complexity in Simple Ecological Models by Robert May in American Naturalist (1976).

Which elements of chaos and complexity studies are so fundamental and essential that together they sketch an overall? Which are the important intricacies? Each person will have a slightly different interpretation. These combinations and points of view about what is "important" are the never ending discussion and debate that signify progress in all domains. To capture legitimate progress and new ideas in the literal sense of preservation, we can also assume the most accurate record of chaos and complexity SCIENCE are the technical papers and any code we can still read. However, the literature alone does not completely describe this branch or attempt to explain why people have dedicated such passionate thought to it. One reason *Chaos, Making a New Science* (Gleick, 1987) was a best seller so long is that this is also such a compelling STORY. The concepts did seem new, and obvious, which is rare.

Stories are allowed to include pure commentary simply because these are interesting details, no other reason. A technical paper wants to eliminate unnecessary distractions. Methodology, related work and open questions are carefully annotated to form a context that justifies where the work belongs. Source code is now required with most journal submittals; words and .jpgs of algorithms and equations are no longer enough for a thorough review. The form of the continuous discussion and debate has changed as much as the topics being argued and we are not done yet.

Any scientific body of work has always been an evolving web of interconnections that is very complex, today we just have more efficient ways of looking. For example, there have been massive improvements in navigating related work. Scientific digital libraries such as Cite Seer and ScienceDirect are not only thick with searchable publications, but customized alerts for topics of interest are available, users can access techniques and contact the authors with questions. Dealing with specific, complex and abstract information has become a much more interactive and precise process. Extracting a research thread from a digital library is like running on a hamster wheel, one piece of evidence leads to three more. Fortunately, the convenient units that research threads can be now broken into, away from entire books and journals, makes the content much easier to sift through when pulling together and justifying a new whole.

Regardless of technical and communication improvements, the problem of deciding which work is related and why will never be "solved". If systems and machines are to help us contextualize reasoning, presumably like journal referees, they would also insist on more to analyze than text in/text out. It would be progressive to engineer and be able to manipulate algorithms in/algorithms out, imagery in/imagery out, transformations in/transformations out and of course mixing and matching different proportions and hierarchies of the essential components. Specific hierarchies and combinations could only be recognized in context, the most useful metric would be proportion because proportion often indicates design.

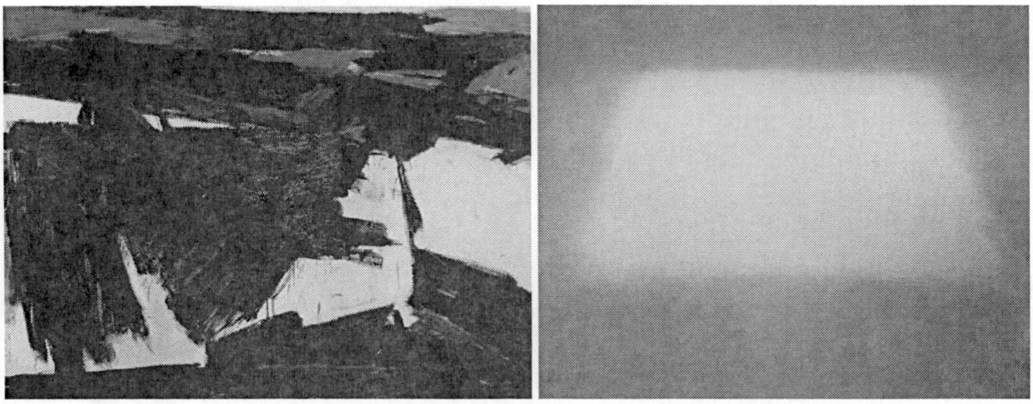

Figure 2. a) "Delaware Gap" by Franz Kline b) "Pollen from Hazelnut" by Wolfgang Laib.

These artworks are being compared because one has no color, one is all color. One is fixed, the other could blow away. One is on the wall, the other on the floor. One is exactly the same as an archive, the other changes form completely. Their proportions are a similar scale in relation to the viewer and they are both in the permanent collection of the Hirshhorn Museum and Sculpture Garden, Smithsonian Institution. How should they be digitized?

Sometimes, regardless of which systems, machines or measurements are available, the concepts themselves may be so abstract or complicated that it becomes an extraordinary challenge just bringing a sensible group together. It is nearly impossible to be objective relating new ideas to familiar ideas, this is part of what is making it a new idea. At a Roundtable Discussion held at the Kreeger Museum, Olga Viso (2003), Deputy Director and Curator of Contemporary Art at the Hirshhorn Museum and Sculpture Garden, described evaluating contemporary art:

> "Sometimes you are not sure what you are looking at, so you need re-look at it, then look at it again"

Olga Viso is not carefully examining these abstract complicated objects and ideas just to see or count them – her purpose is to make decisions and draw conclusions. Like theorists and detectives, a curator identifies or proposes new patterns, is engaged in a different kind of internal and external dialogue.

Our new ability to share deeply interpretive information also gives us new reasons to look again and again at these circular patterns and dialogues. We are on the verge of a new way to discuss which patterns and dialogues have value; which objects, information and ideas we should provide care for; try to stop time and conserve so they can be interpreted again later with a fresh perspective and historical comprehension.

A museum of any type has unlimited examples why critical selections and an interesting story are necessary with objects. Some objects museums are responsible for are quite fragile, it is safer to look at a copy, but there are already too many objects to look through let alone interpret, never enough resources to care for the originals not to mention the copies, therefore it serves very little purpose trying to "keep it all". Like scientific ideas going in and out of favor, eventually museums can only focus on high quality originals, try to cover as much as possible, fill in gaps and build bridges between different aspects of the collection. The science and story of chaos and complexity is like a collection with many interpretations that would be very difficult to keep in one place. Decisions about scientific relevance, or exactly what constitutes proof, are made by huge numbers of people over time. Only the media and machines are fragile, yet there is no reason to care for or conserve them, our digital culture demands they improve.

Machines

The Smithsonian Institution has consolidated a History of the Computer & Internet Resources. This "stuff" resides in many locations, is composed of a never ending diversity of encodings on various unstable materials and quaint artifacts that are not expected to perform. Many of the machine languages have been lost and most people do not miss them. The unwritten history also includes an astronomer's "after paper" 999 dimension data array sitting in a drawer collecting dust, going obsolete. It includes someone curious playing around with genetic algorithms just to see what happens. There is an enormous portion of potentially relevant, interesting, complex information that is only partially interpreted and therefore may not be upgraded to meet new standards. The number of inspiring occurrences that were never recorded is beyond measure. Does it mean this information, or potential information, is not valuable or possibly even important?

So many finely detailed histories, new sketches and views have been enabled by our fickle relationship with machines. They can really spark our imagination but never ask "What are you measuring? Why are you measuring it? What is your method? Justification? Reason? Do you have funding? Has anyone else measured this? What can you show me?" They do not wonder what the best, most accurate interpretative record of emergence, chaos and complexity is. They have no collective memory or inside influences, they just perform. Which components of this now well established science cannot be recorded, preserved or represented without machines? Possibly none, but where is their voice in this democracy? As they evolve, are abandoned and replaced, most of their imagery is still limited to a backlit screen, their languages are illegible, they never get enthusiastic or bored yet they are also readers, recording our information patterns, always there. People talk about feedback loops, self similarity, unlimited variables and the effect of initial conditions but the encoding and representation of information patterns of all types feel like we are always starting in the same place, the transactions are constrained to equal packets working on a clock. Certain ways of thinking cannot be captured this fractured, regimented way. Maybe the patterns themselves can show us how to characterize this kind of information to help us to see new ways it is related.

Presentation and Representation

People are always deliberately inventing new ways to express, figure out and present what we are thinking about. At "Look Up! "Chaos" Comes to New York" held at the CUNY Graduate Center December 2003, Jim Crutchfield and David Dunn described creating the Theater of Pattern Formation:

> "... a comprehensive strategy for the visual and auditory articulation of scientific and mathematical research in the fields of complex systems and nonlinear dynamics or "chaos." It explores naturally occurring patterns in nature and mathematics and how they can be seen within the aesthetic traditions of the arts."

To get this presentation to work, not only were there the technicalities of getting the audio and visual patterns to influence each other, but also issues related to "stitching" together views, removing or faking distortions for the dome, the originals had to be developed in a round space, not on a computer screen. The results presented inside the dome sound like they will be effective.

New kinds of presentations such as the Theater of Pattern Formation feel like they are getting more true to the form of certain patterns and are definitely more compelling both to people who understand the underlying mathematics and people it simply appeals to. These sounds and images are slowly entering our popular consciousness and how can that be turned into something useful? We didn't have the search engine Google before, now people cannot imagine being without this way of looking. We are quick to learn a new way when it is useful.

Figure 3. From the "Theater of Pattern Formation" by James P. Crutchfield and David Dunn (2003), a large-scale multichannel video-audio exploration of structure and emergence in the spatial and acoustic domains. The target venues are sensory-immersive all-digital dome theaters. Image used with permission from Dr. Crutchfield.

School groups go to into planetarium presentations and rip things apart with their enthusiasm and energy. Their adult counterparts do the same monitoring the literature. Our anthropocentric collective understanding is continually being clarified, explored, shredded, discarded, updated, and reflected through our modes of presentation and representation – these modes will not stay the same or ever be enough for developing and presenting new ideas.

Machine Aesthetics

How can machines help us ponder on and sort through patterns that might work for everything to help us establish standards and convenient units to interpret and preserve them in the future? We do not generate many tools to examine or establish overalls while we are still looking through little windows of order, generating and collecting pieces. How would a machine auto-measure context, conceptual relationships and overalls? To what extent are we comfortable with their style of brute force fussy dialogue going unsupervised? What might they notice and classify as interesting or relevant that is different than we would think of looking for?

If machines have some share in the responsibilities of cleaning our complex and chaotic information basement, deem something redundant and eliminate it, will it be that hideous sweater that truly, should never be worn again - or will it be a forgotten photograph? Can we trust them to consolidate what we are currently unable to perceive as either embarrassing or precious because we are in the middle of it and cannot see everything? What can they help us get rid of in a way we can accept?

Figure 4. "The Administration Building" by Michael Leyton.

Figure 5. Neolithic pottery from the Museum of Almería, Spain.

We cannot just "keep all of the patterns" nor does that serve any useful purpose. Even if we are not sure we are able to recognize fundamental or essential meaning in the data, information and patterns we have now, there is at least one time when one person and one machine evaluated something that looked interesting in the data. Maybe they were not even sure why, it just felt like it, maybe it was just easy for the machine to handle. We should protect these original combinations to look at again later with our new machines. A preservation effort of this type would not be to understand the past, but to participate in the future. The digitization and automated experiment craze presents a one-time opportunity to collect more now than will be proven to have value later when unfortunately, the traces we have left may be of such low quality that we accidentally infer the wrong things. We could put a broken piece of clay under sophisticated lighting pretending it is important only to discover later that more valuable works have been lost protecting this one.

One fragile video tape by Pavel Hlava might have been the only imagery of the first plane hitting the World Trade Center, now there are so many copies and we understand what happened that it will be preserved by perpetuation. Data may be able to auto-perpetuate as it is distributed but it cannot auto-contextualize without characterization. There is no reason to limit salient data features to unique identifiers. There are variations in texture, density and alignment that machines can register more precisely than we can identify.

Where can we establish boundless sets of endlessly intricate questions, experimental setups, data components, conclusions and patterns for curious creative people and our never ending parade of media and machines to fool around with just to see what might be sifted out? If patterns that work with and supposedly represent everything were to be collected, analyzed, compared or just reflected upon, where could they be assembled or kept together in groups without generating too many copies? We could save only the context since most virtual information is a copy already. An image, description and measurements of a painting will never be as good as the real painting by itself. Source code that compiles very nicely does not put the reader inside the scientist's head when the discovery was made. If information patterns that register this kind of thinking were anything like music, how can we use them to auto-eliminate noise aesthetically? Get machines to recognize the patterns we prefer, continue talking about and connecting with each other, get them to learn our aesthetics?

Redundant and Similar Information

If it is even possible to have a comprehensive body of chaotic and complex patterns to represent all fields of inquiry it would need to be limitless, open and not restricted to certain languages. The system would be more similar to the act of translation than any sets of natural and machine languages. Scientists, scholars and the curious are actively generating a limitless collection of obvious or elusive relationships just by thinking about, categorizing and engineering their data, turning it into information, trying to add meaning to it. Then everybody starts to discuss and debate it. Maybe we can devise a mechanism to let this change the way data is perceived. Throughout the process of discovery, acceptance, rejection and perpetuation of information, there are many components that are similar. If we can use these similarities and conflicts to streamline and train the information space to automatically defer to the denser, higher quality, more original information and auto-delete the copy; this will not only protect the combinations that actually work, but will also help us to decide about and preserve what is actually important.

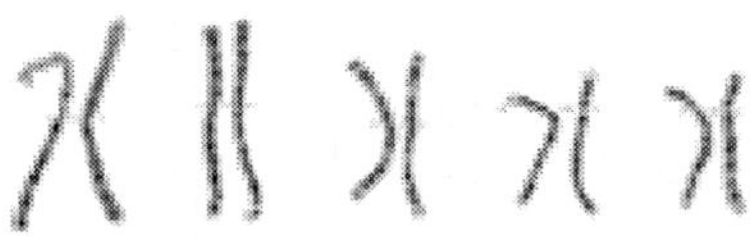

Figure 6. Human chromosomes from www.nature.com.

Conclusions

We have no current standards or shared systems to store and analyze unrelated chaotic or complex information in the partially interpreted state. Everyone is too busy, the patterns are confusing and there are too many. If we can get these patterns to play on their own, look at them again and again in a different relationship with our machines, maybe we can simplify them together. Patterns that work for everything are like intricate artifacts that will eventually become familiar. A collection of them might appear as mathematical patterns and meta-patterns to machines but could be transformed and presented to us any way we prefer. A systematic logic of hierarchy and flow to interpret these patterns on abstract levels is dependent on cycles, fading away and replacement. We should not keep these perplexing records locked in the chunks of granite that are the current style of metadata. Modern information patterns need to be more fluid and effect the other information around them. Like an artist working on a sculpture, as usual, there is too much there. Any system to collect patterns that work for everything would serve the explicit purpose of taking away, streamlining, making it elegant, beautiful, and not like something someone else already made. When complex or chaotic information qualifies for the last rounds of selection and we are left with only the context and essential components - each symbol, mark, word, arrangement, equation and level needs to count, be in their original state. There is no one "place" for context driven topologies, concept maps, or patterns that work for everything. They can only reside in our imagination, mathematical codes and communicative forms capable of binding these together. Techniques usually only improve, let us define a way for abstract information patterns to self-perpetuate, self-contextualize so we can keep only the highest resolution possible for the time when we are ready to see them.

Image Acknowledgments

"*Wavy Texture 2*" Antelope Canyon USA by Jin Akino May 2001, courtesy of the photographer

May RM and Oster G (1976) Bifurcations and Dynamic Complexity in Simple Ecological Models. *American Naturalist* 110, 573-599

"*Delaware Gap*" by Franz Kline (1958) and "*Pollen from Hazelnut*" by Wolfgang Laib (1998-2000)both from the permanent collection of the Hirshhorn Museum and Sculpture Garden, Smithsonian Institution

"*Sample Fractal*" from The Theatre of Pattern Formation by James P. Crutchfield and David Dunn (2003) at the Art & Science Laboratory. Image and text used with permission of Dr. Crutchfield.

"*The First Administration Building*" by Professor Michael Leyton, image provided by the artist

"*Argaric Neolithic Pottery*" on display at the Museum of Almería by Manuel Salas Barón

"*Human Chromasomes*" from www.nature.com

References

Gleick J (1987) Chaos, Making a New Science *Viking Penguin* ISBN 0 14 00.9250 1

May RM and Oster G (1976) Bifurcations and Dynamic Complexity in Simple Ecological Models. *American Naturalist* **110**, 573-599

Internet

Jin Akino http://www.internetacademy.co.jp/~yesaki/

CiteSeer http://citeseer.ist.psu.edu/cis

ScienceDirect http://www.sciencedirect.com/

Hirshhorn Museum and Sculpture Garden Smithsonian Institution http://hirshhorn.si.edu/

Smithsonian History of the Computer & Internet Resources http://www.sil.si.edu/subject-guide/nmah/histcomput.htm

Google www.google.com

Theater of Pattern Formation http://atc.unm.edu/research/asl/asl.html

Art & Science Laboratory http://www.artscilab.org/

Michael Leyton http://www.rci.rutgers.edu/~mleyton/homepage.htm

Museum of Almería by Manuel Salas Barón http://members.tripod.com/~indalopottery/history.htm

Human Chromasomes www.nature.com

INDEX

A

B

C

D

E

F

G

H

I

J

K

L

M

Q

R

S

T

U

V

W

Z